T0186860

PROGRESS IN DURABILITY ANALYSIS OF COMPOSITE SYSTEMS

PROCEEDINGS OF THE INTERNATIONAL CONFERENCE DURACOSYS 95
BRUSSELS/BELGIUM/16-21 JULY 1995

Progress in Durability Analysis of Composite Systems

Edited by

ALBERT H.CARDON
Free University Brussels, Belgium

HIROSHI FUKUDA
Science University of Tokyo, Japan

KEN REIFSNIDER
Virginia Tech Materials Response Group, USA

A.A.BALKEMA/ROTTERDAM/BROOKFIELD/1996

The texts of the various papers in this volume were set individually by typists under the supervision of each of the authors concerned.

Authorization to photocopy items for internal or personal use, or the internal or personal use of specific clients, is granted by A.A. Balkema, Rotterdam, provided that the base fee of US$1.50 per copy, plus US$0.10 per page is paid directly to Copyright Clearance Center, 222 Rosewood Drive, Danvers, MA 01923, USA. For those organizations that have been granted a photocopy license by CCC, a separate system of payment has been arranged. The fee code for users of the Transactional Reporting Service is: 90 5410 809 6/96 US$1.50 + US$0.10.

Published by
A.A. Balkema, P.O. Box 1675, 3000 BR Rotterdam, Netherlands(Fax: +31.10.413.5947)
A.A. Balkema Publishers, Old Post Road, Brookfield, VT 05036, USA (Fax: 802.276.3837)

ISBN 90 5410 809 6
© 1996 A.A. Balkema, Rotterdam
Printed in the Netherlands

Progress in Durability Analysis of Composite Systems, Cardon, Fukuda & Reifsnider (eds)
© 1996 Balkema, Rotterdam. ISBN 90 5410 809 6

Table of contents

3 *Damage*

4 *Accelerated testing*

Progress in Durability Analysis of Composite Systems, Cardon, Fukuda & Reifsnider (eds)
© 1996 Balkema, Rotterdam. ISBN 90 5410 809 6

Presentation

Composite Material Systems, the basic systems of nature, are the most promising advanced materials for future structural applications. The combination of different materials and interfaces in a composite system may result in better properties and performances, depending on the properties of the basic elements and the interaction level between those elements.

Many applications of composite systems are actually part of our daily technological environment. For more generalised structural applications of those systems, a crucial problem has still to be solved: the possibility to predict the long term behaviour in order to guarantee a safe residual structural behaviour after a complex, thermomechanical, loading history, interacting, for some applications, with environmental variations after an imposed life time.

This problem, the durability analysis of composite systems, has received increasing attention the last decade. It concerns not only aeronautical and space applications, known as advanced composite applications, but all load bearing applications in civil and mechanical engineering components.

Recent progress has been made and some interesting methods and techniques are now proposed. Some of them are limited to specific composite systems; others can probably be applied to more general composites.

Three principal axes for durability analysis can be identified:
– Those based on general or specific damage analysis methods;
– those based on structural analysis starting from the identification of the control elements of failure, the critical element method;
– those based on the prediction of the evolution of the thermomechanical properties of the composite system starting from short term results under the influence of some 'accelerating' factors.

In 1990 an International Colloquium on Durability Analysis of Polymer based Composite Systems (Durability 90; Aug. 27-31, 1990) was organized in Brussels.

In 1993, after the ICCM-9 in Madrid, an International Invited Informal Seminar on Durability Analysis of Composites took place in Porto (I³SODUR 93, July 17-20, 1993).

One of the conclusions of this seminar was the need for a new international meeting on Durability Analysis of Composite Systems in the Summer of 1995, and we have organized the International Conference on Progress in Durability Analysis of Composite Systems (DURACOSYS 95), July 16-21, 1995 – Brussels.

This conference brought together some eighty participants from 19 countries, to discuss 8 general invited lectures, 3 special lectures and 33 oral or poster presentations.

These proceedings contain 39 papers of the total of 44 presented at the conference. Those papers are divided in 6 sections:

- General aspects and methods (6);
- Fatigue behaviour (7);
- Damage analysis (6);
- Accelerated testing methods (6);
- Influence of water on properties (8);
- Specific systems and structures (6).

We wish to acknowledge the financial support for this conference provided by the Belgian National Science Fund (NFWO), the host university VUB, the Solvay Company, The United States Army Research, Development and Standardisation Group UK, The Virginia Tech Materials Group and the Science University of Tokyo.

Ken Reifsnider
Virginia Tech Materials
Response Group, USA

Albert H. Cardon
Composite Systems and
Adhesion Research Group
of the Free University
Brussels, Belgium

Hiroshi Fukuda
Science University of Tokyo,
Japan

Progress in Durability Analysis of Composite Systems, Cardon, Fukuda & Reifsnider (eds)
© 1996 Balkema, Rotterdam. ISBN 90 5410 809 6

Organization

INTERNATIONAL ORGANIZING COMMITTEE

Chairman
Albert H. Cardon, VUB, Brussels, Belgium

Co-chairmen
Ken Reifsnider, Virginia Tech, USA
Hiroshi Fukuda, Science University Tokyo, Japan
Jürgen Brandt, Daimler Benz AG, Germany
Ramesh Khana, Boeing, USA
J. Matsui, Toray Industries, Japan

INTERNATIONAL SCIENTIFIC ADVISORY COMMITTEE

R. Adams, UK
T. Blaszczynski, Poland
A. Bogdanovich, Latvia
H. Brinson, USA
O. Brüller, Germany
A. Torres-Marques, Portugal
A. Vautrin, France
G. Verchery, France
A. Vlot, Netherlands

J.A. Manson, Switzerland
L. Nicolais, Italy
G. Papanicolaou, Greece
Cl. Roy, Canada
K. Schulte, Germany
Z. Hashin, Israel
N. Himmel, Germany
P. Hogg, UK
S. Johnson, USA

BELGIAN HOST COMMITTEE

A.H. Cardon (VUBrussel)
R. Dechaene (U.Gent)
I. Verpoest (KULeuven)
J.M. Rigo (U. Liège)
R. Keunings (UCLouvain)
J. Charlier (ULBruxelles)
S. Boucher (Fac. Pol. Mons)

LOCAL ORGANIZING COMMITTEE

A.H. Cardon

M. Bourlau

F. Boulpaep

M. Bruggeman

P. Bouquet

J. De Visscher

L. Tirry

P. Bauweraerts

D. Debondt

A. Vrijdag

1 General aspects and methods

Progress in Durability Analysis of Composite Systems, Cardon, Fukuda & Reifsnider (eds)
© *1996 Balkema, Rotterdam. ISBN 90 5410 809 6*

A micro-kinetic approach to durability analysis: The critical element method

K. L. Reifsnider, S. Case & Y. L. Xu
Virginia Polytechnic Institute and State University, Blacksburg, Va., USA

ABSTRACT: This paper presents an outline of the "micro-kinetic approach," developed by the Materials Response Group at Virginia Tech to predict the remaining strength and life of composite material systems under long-term exposure to mechanical, chemical, and other environmental conditions. The approach consists of a philosophy and methodology for combining mechanistic models to predict the combined effects of time-dependent and cycle-dependent response such as fatigue, creep, and stress rupture. Validations of the approach have been completed for several material systems, and examples are given. Applications have been developed in association with several major industrial firms.

1 DURABILITY AND DAMAGE TOLERANCE

Durability and Damage tolerance is defined in terms of the remaining strength of a component after some period of service. Figure 1 illustrates the general concept.

Several important implications are associated with this concept. First, we are concerned with "long-term" behavior that involves extended exposure of the component to applied conditions that may include mechanical, thermal, and chemical environments. Second, since the "end of life" is defined by the reduction of strength to the level of applied loading, we can discuss both durability and damage tolerance in terms of strength, or more

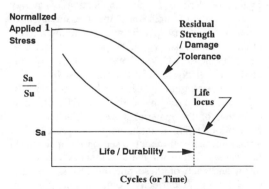

Figure 1   Conceptual illustration of the damage tolerance concept.

precisely, in terms of remaining strength after some period of service. Hence, we will be concerned with changes in strength caused by processes associated with extended service in various environments, processes such as damage development, oxidation, creep, and aging (in various forms).

A first element of our approach to this subject is to select remaining strength as a damage metric. Figure 2 shows the essence of this argument. We make the fundamental assumption that a service history that produces a given remaining strength results in damage that is "equivalent" to that produced by a different service history that results in the same remaining strength value (if both cases began with the same initial strength). The most important result of this assumption is the resulting damage accumulation rule implied by the definition.

As shown in Fig. 2, if n_1 cycles of loading are applied at cyclic amplitude Sa^1, then the "equivalent" number of cycles at some other amplitude, Sa^2, is found by locating the cycles at the second level that produces the same strength reduction, and the remaining cycles to failure is easily determined. It should be noted that this summation rule is <u>nonlinear</u> unless the remaining strength curves are straight lines. Hence, it is quite different from the familiar linear ("Miners") rule, and the result of "summing damage" by summing

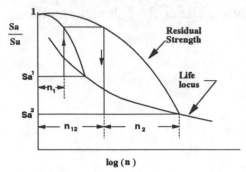

Figure 2 Illustration of remaining strength as a damage metric.

changes in remaining strength is affected by the *sequence* of the increments of applied conditions. The objective of our approach, then, is to calculate the remaining strength curves for arbitrary applied conditions, and to use those curves to "sum" the damage produced by changing applied conditions or changes in material condition during service.

We begin by asking what we mean by "strength," i.e., what strength are we talking about? We will construct an answer to that question from several parts. The first part is the "intrinsic" or "principle strength(s)" of the composite. As an example, we could discuss the tensile strength of a unidirectional lamina in the direction of the reinforcing fibers; we will call it X_t. This is a "principle material (tensile) strength in the fiber direction." We can measure this number in the laboratory. It is a well defined engineering quantity. But that is not sufficient for our model, since we would like to be able to discuss composite strength in terms of the constituent properties and performance in order to (later) discuss the effect of time- or cycle-dependent processes at the constituent level on global behavior. To accomplish this, we will use micro-mechanical representations of strength to combine the effect of many such processes in a physically correct manner, without the artifice of "partitions" or other phenomenological assumptions.

For tensile strength, the simplest example of such a representation is the rule of mixtures, i.e., we could claim that

$$X_t = v_f X_f + v_m X_m \qquad (1)$$

where X_f and X_m are the fiber and matrix strengths, and the coefficients are the volume fractions of those constituent phases. The utility of this type of formulation is obvious, especially when we wish to determine the strength of the composite when, say,

the fibers are being oxidized, or the matrix strength is decreasing because of matrix cracking, etc. Of course, equation (Razvan & Reifsnider 1991) is not a sufficiently rigorous representation of X_t; more complete models are available, and represent the tensile strength in terms of the geometry, arrangement, properties and performance of the constituents and the interphase regions between them (Razvan & Reifsnider 1991; Lesko 1994; Xu & Reifsnider 1993; Gao & Reifsnider 1993; Weitsman & Zhu 1992; Schwietert & Steif). For example, we have recently shown that the following representation includes many of the critical physical factors and affects that control tensile strength in fiber-dominated systems.

$$X_t = \sigma^{\frac{m}{m+1}} \left(\frac{2\tau_o L}{\hat{D}} \right)^{\frac{1}{m+1}} \left(\frac{2}{m+2} \right)^{\frac{1}{m+1}} \left(\frac{m+1}{m+2} \right)$$
$$\frac{(1+m)^{\frac{1}{m}}}{[C_n^m + C_n^{m-1} + \ldots + 1]^{\frac{1}{m}}}$$

$$(2)$$

In this representation, σ_o is the characteristic strength of the fibers, m is the Weibull shape parameter of the statistical distribution of fiber strengths, L is the characteristic length of the material, τ_o is the interfacial shear strength, $\hat{D}$ is a fiber diameter, and C_n are the local stress concentration numbers when n neighboring fibers are broken (as a precursor to the fracture event).

However, tensile strength models in the fiber direction are not a complete representation of composite strength. Most composites, especially fibrous ones, are anisotropic. There is a (tensor) array of principle material strengths that represent the collective "strength of the material." Figure 3 illustrates such an array. Each of the tension, compression, and shear strengths depicted in that figure are defined by the application of a very specific state of stress and the measurement of "strength" in a very specific way. In addition to the in-plane values shown, there are also values of strength measured "out of the plane" of the unidirectional ply or "plate" usually used to define ply-level strength. These values, collectively, represent the strength of the material, or, more correctly, of the material system. Each of these principle material strengths may be represented by a micromechanical model based on the mechanism that causes failure for the stress states that define them, and each of these micromechanical models can be used to correctly combine the effects of changes (with time, cycles, etc.) of the constituent properties, performance, or arrangement. But we

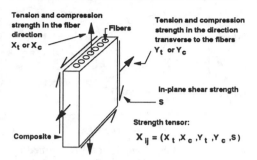

Five in-plane strength values for fiberous composites:

Tension and compression strength in the fiber direction
X_t or X_c

Fibers

Tension and compression strength in the direction transverse to the fibers
Y_t or Y_c

In-plane shear strength
S

Composite ►

Strength tensor:

$$X_{ij} = (X_t, X_c, Y_t, Y_c, S)$$

Figure 3 Illustration of the definition of "principle strengths" of a fibrous composite.

still need a method of discussing damage tolerance using these quantities together, especially if the applied stress state is multiaxial.

However, if we have a multiaxial stress state, and the tensor array of material principle strengths is available to us, we can construct a "failure function" that compares those two arrays. This is a familiar enterprise. Such functions are the basis of the classical formulations of "failure criteria." For composites, Tsai-Hill, Tsai-Wu, or maximum strain criteria are typically used. We will use the notation

$$Fa = Fa\left(\frac{\sigma_{ij}}{X_{ij}}\right) \qquad (3)$$

Such a failure function will allow us to discuss damage tolerance, since it will assemble the effect of applying a general state of stress compared to the array of principle material strengths. *Hence, we will formulate our damage tolerance problem in terms of changes in a failure function, Fa, during the life of a component, caused by degradation processes such as fatigue, creep, and stress rupture.*

2 THE CRITICAL ELEMENT METHOD

In our approach, so far, we have used a failure function, Fa, to define the "strength" of a composite system, and we have described how that function is defined in terms of principle material strengths (which define the "state of the material"). Now it is necessary to determine what to use for the stress components in the numerator of Fa. Are those global values, or local values? What mechanics

problem must we solve to determine them?

We claim that the boundary value problems we need to solve to find the controlling local stress states are, exactly, the boundary value problems we need to solve to represent the micromechanics of our principle material strengths, defined earlier. We further claim that the mechanics problem to be solved for that purpose is defined by the failure mode of the material system for a given loading condition, i.e., we set a boundary value problem based on a representative volume that defines that failure mode. For fibrous composites, the number of unique failure modes is generally small, especially for fiber dominated systems. Figure 4 shows several typical ones for tensile and compression loading. The most striking feature of those modes is their local nature, i.e., most often, we should use the local stresses that control the final local fracture event that triggers the global fracture of the composite in our failure function.

For example, in equation (2), the stress concentration factors, Cn, are determined from the solution for the stresses in fibers adjacent to a group of n broken fibers, at the micromechanical level. These are the stresses that control the tendency for the material to initiate fracture for this failure mode, defined by an unstable sequence of successive multiple fiber fractures.

Using careful laboratory tests, we can define a basic set of failure modes for a composite system by identifying the unique set of material strengths needed to define the "state of strength of the material system." Therefore, we know exactly what boundary value problems to solve, and we only need to determine which of the failure modes control the failure in a given application to define the problem we want to solve, uniquely. We will, however, require that a separate problem be solved (and a

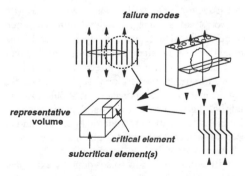

failure modes

representative volume

critical element

subcritical element(s)

Figure 4 Typical failure modes that define the representative volume and controlling boundary value problem for fibrous composites.

prediction of remaining strength) be made for each individual failure mode. The results of several such predictions may be compared to estimate the controlling failure mode, or, the controlling failure mode may be determined from experimental observations.

Now we know what stress state and material state to use as inputs to the failure function we use to define strength, and we have decided that the inputs should be local (if the failure event is triggered by a local event such as a critical number of fibers broken together, as is typically the case). The next essential question is, how is this failure function (defined at the local level for most failure modes) related to the global strength that we want to predict?

As we indicated in Fig. 4, the boundary value problem to be solved for a given failure mode is identified by the "critical element" in the representative volume. The critical element, in turn, is defined, physically, by the fact that global fracture occurs when the critical element fails. Hence, if we define the problem in the manner described, our local calculations will define the global behavior, i.e., the local-global transition is achieved by the critical element concept and the associated definition of the boundary value problem that is used to provide inputs to the failure function.

3 EVOLUTION

Now we are ready to address the question of how to use this formulation to predict remaining strength and life over a history of exposure of a component to mechanical, thermal, and other environmental applied conditions. The first observation is that the stress states and material states that define our failure function, Fa, will change with time (or cycles, etc.) since damage / degradation will alter those quantities. Hence, we can depict our problem, graphically, with an altered version of Fig. 2 (Fig. 5). In this diagram, it can be seen that the failure function, that replaces the global applied stress, Sa in Fig. 2, is no longer a constant. It should also be noted that the failure function, Fa, does account for multiaxial stress states and material strengths, and is chosen to properly represent a given failure mode. But how can we calculate the changes in Fa?

We will demonstrate how this is done by examples in subsequent sections. In general, changes in stress state are induced by damage (such as matrix cracking, delamination, etc.) in the "subcritical elements" (noted in Fig. 4). Those changes are

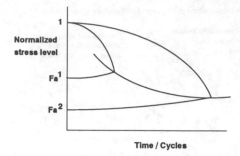

Figure 5 Remaining strength and life defined by the boundary value problem in the (local) critical element.

calculated using micromechanics to determine the changing stress state in the critical element as a function of the progress and rate of the contributing events. Changes in stress state may also be induced by time-dependent behavior (such as viscoelasticity) in both the critical and subcritical elements. Those changes can also be introduced into the local boundary value problem, but we are reminded that time will be introduced, explicitly, by this addition. Material (principle) strengths may be altered by processes such as oxidation or micro-damage within the critical element (such as micro-fracture or wear of the fiber-matrix interface during cyclic loading). These changes enter Fa directly, and are also dependent on the rates and progress of those processes. In fact, we can see from equation (2) that changes in such things as the fiber matrix interface (or interphase region), changes in the intrinsic fiber strength, or changes in the local stress concentration around a broken fiber (due to such things as viscoelastic relaxation) are "integrated" by the physics and mechanics represented in the micromechanical representations of principle material strengths (like equation (2)), without the artifice of "partitions," "damage rules," or other heuristic arguments. This is a major strength of the present philosophy.

For our general formulation, we are left only with the question of how to actually calculate the global strength using the local arguments constructed above. For this purpose, we appeal to the familiar field generally called "damage mechanics."

Following the general concepts advanced by Kachanov (1986), we will assume that the strength of our composite systems is defined by the (statistical) accumulation of micro-variations in the properties, geometry, and arrangement of the constituents and the interfaces / interphase regions between them. We construct our definition of a damage metric with the familiar scaler "continuity," ψ, defined as (1 - Fa) (or more generally, some function of Fa) where Fa is a mechanistically or phenomenologically based failure criterion defined in a "critical element" (which defines our domain) for which rupture defines rupture of the system (Reifsnider 1991a; Reifsnider & Stinchcomb 1986).

Then we construct a fundamental evolution equation for strength with the following rationale.

1. Helmholtz free energy $f = f(\psi, \epsilon_{ij})$, becomes $f = U\psi$ as in classical Kachanov theory [7]. Hence:

$$df = \sigma_{ij} d\epsilon_{ij} - Q d\psi \quad \text{where} \quad Q = -U \quad (4)$$

where Q is associated with the increment of entropy created by damage, and has the nature of energy released by degradation of the material state.

2. The central issue is the kinetic equation. We assume that the kinetics are defined by a specific (damage accumulation) process for a specific fracture mode, and define rates for all such processes of interest.

We start with the most general, common kinetic equation (a power law), such that:

$\frac{\delta \psi}{\delta \tau} = A \psi^n$, where τ is a normalized, generalized time variable (monotonic increasing), and n is a material constant.

Generally, $\tau = \frac{t}{\hat{t}}$ where $\hat{t}$ is the characteristic time constant for the process at hand. $\hat{t}$ could be a creep time constant, a creep rupture life, a fatigue life, etc., such that, for example, $\tau = \frac{n}{N}$.

Then:

$$\int_{\psi^o}^{\psi^i} d\psi = A \int_{0}^{t^i} (\psi(\tau))^n d\tau \quad (5)$$

The left hand side is

$$\psi^i - \psi^o = 1 - Fa^i - 1 + Fa^o \quad (6)$$

If we set A = 1, and approximate n = 1, then

$$Fa^i = Fa - \int_{0}^{\tau^i} (1 - Fa(\tau)) d\tau \quad (7)$$

which is the instantaneous value of the failure function for this process.

Then we define a "residual strength" Fr such that

$$Fr = 1 - \int_{0}^{\tau} (1 - Fa(\tau)) d\tau \quad (8)$$

where all quantities are defined in the critical element and for the process characterized by the characteristic time $\hat{t}$. A degenerate special case of equation 5 occurs for $\tau \to \frac{n}{N}$; $\hat{t} = N$ for which $Fa(\tau) \to \frac{S_a}{S_u}$, the ratio of unidirectional applied stress over unidirectional strength, whereupon $Fr = \frac{S_r}{S_u}$, and equation 5 integrates to

$$\frac{Sr}{Su} = 1 - (1 - \frac{Sa}{Su}) \frac{n}{N} \quad (9)$$

a linear degradation of strength from initial to final value. Equation 11 is also an identity in the sense that it satisfies the end points of the residual strength curve, i.e., it is correct at the limits. In general, however:

$$Fa = Fa \left(\frac{\sigma_{ij}(n)}{X_{ij}(n)} \right) ; \quad N = N(n) \quad (10)$$

or

$$Fa = Fa \left(\frac{\sigma_{ij}(t)}{X_{ij}(t)} \right) ; \quad N = N(t) \quad (11)$$

If we claim that the rate equation is explicit in generalized time and recast the basic kinetic law to read $\frac{\delta \psi}{\delta \tau} = \psi \tau^{j-1}$, we obtain the final kinetic equation in the form

$$Fr = 1 - \int_{0}^{\tau} (1 - Fa(\tau))(\tau)^{j-1} d\tau \quad (12)$$

which is the form we will use in the present program, and is essentially the form we first postulated in 1981 (Reifsnider & Stinchcomb 1983).

In many cases, it is appropriate to use the number of cycles to failure, N (under current local conditions) as the characteristic time of the controlling process.

Of course, it is possible that other processes control the life of the critical element, and that a "characteristic time constant" for those processes may be more appropriate to use in equation (12). Christensen discusses a kinetic theory of failure in which (using a generalization of Griffith crack instability concepts) the critical applied stress for instability is postulated as

$$\sigma^2 = \frac{2\Gamma}{hJ(\frac{\rho}{c})} \qquad (13)$$

in which ρ is a characteristic length, c is the flaw growth velocity, h is the initial flaw dimension, and J is the material creep function (Christensen 1981; Christensen 1979). If one assumes a creep function of the form

$$J(t) = \frac{J_o(1+\gamma_1 t^n)}{(1+\gamma_2 t^n)} \qquad (14)$$

and integrates equation (14) (taking only leading terms), one obtains the time to failure under constant (instantaneous) stress as shown in equation (15),

$$\hat{t} = \frac{1}{(\frac{1}{n}-1)\hat{\sigma}^{\frac{2}{n}}} \qquad (15)$$

where the applied stress is normalized by the instantaneous "fast fracture" (intrinsic) strength and n is a material constant. Christensen also discusses a rate equation and associated characteristic time for combined creep and chemical degradation.

Obviously, if the processes that control the failure of the critical element are clearly identified and carefully characterized in the laboratory, one can specify the form of $\hat{t}$. The important point in the present context is that the resulting expression depends on applied conditions and material constants; when applied to the critical element, those conditions and material parameters will, in general, be functions of time, cycles, or history of the component being modelled.

We remind ourselves that rate equations (as functions of applied stress, temperature, and environmental conditions) must also be determined for such things as matrix cracking and creep or other stress relaxation mechanisms (to modify the modulus at the local level) as well as rate equations that represent such things as (diffusion or chemical rate controlled) strength degradation of constituents or interphase regions (to modify the inputs to our X_{ij} correctly).

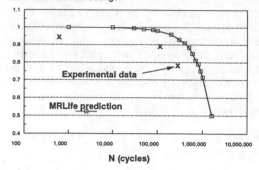

Normalized Residual Strength

N (cycles)

Figure 6 Remaining strength predictions and observations for a Quasi-isotropic center notched coupon fatigue loaded with R=-1.

4 EXAMPLES

Numerous predictions for polymer (and ceramic) composites have been successfully made (Reifsnider 1991b; Reifsnider 1992; Reifsnider & Gao 1991). As a first example, we consider an aromatic polymer composite (APC-2) with a polyetheretherketone (PEEK) matrix reinforced with AS-4 fibers. Unidirectional fatigue data were taken from Picasso and Priolo (Picasso & Priolo 1988). In order to use the a fatigue rate equation having the form

$$Fa = An - Bn*(\log N)^{Pn}$$

two regions of behavior were defined, such that

$$An = 1.3126 \, ; Bn = -0.1818 \, ; Pn = 1.0$$

$$An = 0.7865 \, ; Bn = -0.0425 \, ; Pn = 1.0$$

On the basis of these data and rate equations for micro-crack development, predictions of remaining strength were made for a quasi-isotropic laminate with fully reversed cyclic loading. Figure 6 shows an example of the predictions of remaining strength as a function of cycles for fully reversed loading of an AS-4/PEEK(APC-2) center-notched coupon, compared to experimental data. Stress relaxation around the hole was estimated on the basis of matrix cracking and a stress relaxation model developed earlier for a PMR-15 system. Hence, the code calculates the effect of the off-axis plies, applies a Whitney-Nuismer-type failure criterion to a "characteristic region" (with dimensions of 4.45 mm in this case), and estimates the remaining strength. The agreement is seen to be within about

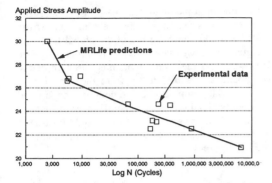

Figure 7 Comparison of predicted and observed fatigue lives of center notched quasi-isotropic coupons under fully reversed fatigue loading.

10 percent, which is typical of our experience for the prediction of remaining strength. The code also predicted that compression failure would dominate, and that was observed. An array of these predictions was made and used to construct a predicted fatigue life (S-N) curve, and compared to test data from Simonds and Stinchcomb (1989). The results are shown in Fig. 7.

Tests are also under way in our laboratory on the performance of high temperature polymer systems. An example of residual strength measurements as a function of cycles of loading for two amplitudes is shown in Fig. 8.

The predictions of the model follow the results well. In this case, the failure mode was greatly influenced by stress redistribution caused by delamination, which was, in turn, greatly affected by elevated temperature. In fact, at room temperature, the two remaining strength curves invert (as a function of amplitude), as do the predictions. For this material system and conditions, and this quasi-

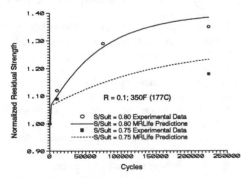

Figure 8 Observed and predicted remaining strength for a notched coupon of IM7/K3B subjected to fatigue loading.

isotropic fiber-dominated laminate, creep deformation did not play an important role.

Our last example illustrates the use of micromechanical strength representations to estimate the effect of micro-structural features on remaining strength and life. The composite material system is the Apollo 45-850 carbon fiber in an HC 9106-3 toughened matrix. Cross ply, center notched coupon specimens were prepared with two different fiber sizings, an "O" type derived from PVP thermoplastic, and an "A" type standard thermoset size. These coupons were subjected to fully reversed fatigue loading at 75 percent of the static ultimate strength in compression. The behavior of the two material systems, differing only by the fiber sizing, was remarkably different. The "A" sizing, which is thought to be less compliant and less tough, demonstrated a damage pattern that was sharply localized along a radial line perpendicular to the loading axis; many fibers were broken along that damage region. The material with the "O" sizing showed very little "transverse" damage development, but developed shear-damage regions tangent to the center hole, parallel to the load axis. Both material systems failed in compression, but the "O" material had lives over one million cycles, while the "A" material failed at an average life of 7,600 cycles.

This behavior was investigated with the help of the MRLife code, which combines the philosophy discussed in this paper. That code uses the micromechanical compression strength model of Xu and Reifsnider (1993). That model includes a representation of interfacial slipping, controlled by the parameter eta, which has the value 2 for complete fiber-matrix bonding, and decreases to a value of 1 for debonding in all regions of local tensile stress. Based on our observations of x-ray radiographs and examinations of SEM photographs, we estimated that the bonding changed in such a way during the life that eta varied between about 2 and 0.5 in the high-damage region over about 10,000 cycles for the "A" material, but changed by a much smaller amount over 100,000 cycles in the "O" system. Observed matrix cracking rates were also different in the two systems, resulting in different internal stress redistributions. The resulting predictions of remaining strength and life from the MRLife code are shown in Fig. 9.

Since we have not yet been able to measure the interface slip parameter directly, the predictions shown in Fig. 9 should be regarded as "best estimates," but the model predicts the same kind of (surprisingly large) change in remaining strength and life as the observations indicate.

The most important point made by this example is

9

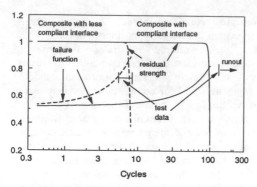

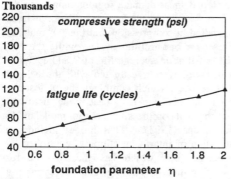

Figure 9 Predicted and observed effect of interface debonding on remaining strength and life of notched coupons under fatigue loading (R=-1).

that the micromechanics model provides the mechanism for interpreting the effect of microstructure and micro-defects on global behavior. Figure 10 shows how the bonding parameter (called

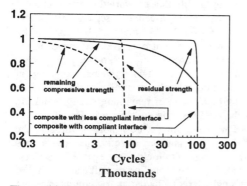

Figure 10 Comparison of global remaining strength with local compression strength variations, predicted by the MRLife code.

Figure 11 Variation of local compressive strength and global fatigue life with fiber-matrix bonding.

the "foundation parameter" in the figure) effects the compressive strength of the material, and (from MRLife predictions) the life at the global level.

Figure 11 shows that effect as a function of cycles. It is seen that the global remaining strength of the composite varies differently from the variation of the material compressive strength in the critical element (a small volume of material near the notch), as one might expect. (The same would be true if we were working with a crack growth model in "simple" metals.)

CLOSURE

We have described a unified micro-kinetic approach to the prediction of the remaining strength and life of composite material systems that uses mechanistic representations of principle material strengths and local stress redistribution to combine cycle dependent and time dependent effects. Distinctive features of the approach include:
● Sets a boundary value mechanics problem and representative volume on the state of stress and state of material defined by incipient fracture, for each failure mode
● Identifies a representative volume element, critical element, and subcritical element for each distinct failure mode
● Defines the damage-related state variable in Gibbs free energy of the system as a failure function written in terms of ratios of local stress components to corresponding principle material strength components
● Defines the damage metric in terms of remaining strength
● Uses mechanistic micromechanics representations for principle material strength to relate micro-details of properties, performance, geometry, and arrangement of constituents and interphase regions to global performance
● Predicts remaining strength and life in terms of the properties, performance, geometry, and arrangement of the constituents and the interphase regions between them.

Our approach is a framework that incorporates and combines representations of physical behavior. The predictions of the model will improve as those representations (and the understandings that support them) improve. There is a special need for increasing our understandings and representations of time dependent behavior such as creep, stress rupture, and aging. We also need to develop methods for predicting the reliability and damage tolerance of engineering structures based on

10

calculations at the materials level, such as those discussed above. Finally, there is a great need for characterizations and data, especially for material systems subjected to extreme conditions and environments. The present approach offers the hope of providing a systematic and mechanistic method of integrating and interpreting such information, to assist the designers of materials systems and structures.

ACKNOWLEDGEMENTS

The authors gratefully acknowledge the support of the National Science Foundation Science and Technology Center under grant no. DMR9120004, and NASA Langley Research Center under grant no. NAG-1-343.

REFERENCES

Christensen, R.M. 1981. Lifetime Predictions for Polymers and Composites Under Constant Load. *Journal of Rheology*. 25(5):517-528.

Christensen, R.M. 1979. *Mechanics of Composite Materials*. New York: John Wiley & Sons.

Gao, Z. & K.L. Reifsnider 1993. Micromechanics of Tensile Strength in Composite Systems. In W.W. Stinchcomb & N.E. Ashbaugh, (eds), *ASTM STP 1156*. Philadelphia: Am. Soc. for Testing and Materials. 4:453-470.

Kachanov, L.M. 1986. *Introduction to Continuum Damage Mechanics*. Boston:Martinus Nijhoff Publishers.

Lesko, J.J. 1991. *Interphase Properties and Their Effects on the Compression Mechanics of Polymeric Composites*. Dissertation. Blacksburg: Department of Engineering Science and Mechanics, Virginia Polytechnic Institute & State University.

Picasso, B. & P. Priolo 1988. Damage Assessment and Life Prediction for Graphite-PEEK Quasi-Isotropic Composites. ASME. *Pressure Vessels and Piping Division PVP*. New York. 146:183-188.

Razvan, A. & K.L. Reifsnider 1991. Fiber Fracture and Strength Degradation in Unidirectional Graphite/Epoxy Composite Materials. *Theor. and Applied Fracture Mechanics*. 16:81-89.

Reifsnider, K.L., Editor. 1991. *Fatigue of Composite Materials*. London: Elsevier Science Publishers.

Reifsnider K.L & W.W. Stinchcomb 1986. A Critical Element Model of the Residual Strength and Life of Fatigue-loaded Composite Coupons. In H.T. Hahn, (ed), *Composite Materials: Fatigue and Fracture*. ASTM STP 907. Philadelphia: Am. Soc. for Testing & Materials. 298-313.

Reifsnider, K.L. 1991. Performance Simulation of Polymer-based Composite Systems. In A.H. Cardon & G. Verchery, (eds), *Durability of Polymer-Based Composite Systems for Structural Applications*. New York: Elsevier Applied Science. 3-26.

Reifsnider, K.L. 1992 Use of Mechanistic Life Prediction Methods for the Design of Damage Tolerant Composite Material Systems. In M.R. Mitchell & O. Buck, (eds), *ASTM STP 157*. Philadelphia: American Society for Testing and Materials. 205-223.

Reifsnider K.L. & Z. Gao 1991. Micromechanical Concepts for the Estimation of Property Evolution and Remaining Life. In *Proc. Intl. Conf. on Spacecraft Structures and Mechanical Testing*, Noordwijk, the Netherlands, 24-26 April 1991: (ESA SP-321, Oct. 1991). 653-657.

Reifsnider, K.L. & W.W. Stinchcomb 1983. Cumulative Damage Model for Advanced Composite Materials. *Phase II Final Report to Air Force Materials Laboratory*. Wright Patterson AFB.

Schwietert, H.R. & P.S. Steif A Theory for the Ultimate Strength of a Brittle-Matrix Composite. *J. of the Mechanics and Physics of Solids*. 38(3): 325-343.

Simonds, R.A. & W.W. Stinchcomb 1989. Response of Notched AS4/PEEK Laminates to Tension/Compression Loading. In G.M. Newaz (ed), *Advances in Thermoplastic Matrix Composite Materials*. ASTM STP 1044, Am. Soc. for Testing and Materials. Philadelphia. 133-145.

Weitsman, Y.L. & H. Zhu 1992. Multifracture of Ceramic Composites. *Oak Ridge National Laboratory Contract Report*, ORNL-6703.

Xu, Y.L. & K.L. Reifsnider 1993. Micromechanical Modeling of Composite Compression Strength. Journal of Composite Materials. 27(6):572-588.

11

Progress in Durability Analysis of Composite Systems, Cardon, Fukuda & Reifsnider (eds)
© 1996 Balkema, Rotterdam. ISBN 90 5410 809 6

A micromechanical approach to time dependent failure of composite materials

T. Peijs, H. J. Schellens & L. E. Govaert
Centre for Polymers and Composites, Eindhoven University of Technology, Netherlands

ABSTRACT: The fracture behaviour of transversely loaded composites is investigated, assuming that fracture is matrix dominated. Since the stress and strain state of the matrix in composite structures is complex, the yield and fracture behaviour of a neat epoxy system is investigated under various multiaxial loading conditions. A good description of the multiaxial yielding behaviour of the matrix material is obtained with the 3-dimensional pressure modified Eyring equation. The parameters of this 3-dimensional yield expression are implemented into a constitutive model, which is able to describe the elasto-viscoplastic behaviour of polymers under complex loadings adequately. Subsequently, by means of a micromechanical approach, the matrix dominated transverse strength of a unidirectional composite material was investigated. Numerical simulations show that a failure criterion based on the maximum octahedral shear strain provides a good description for the rate dependent transverse strength of unidirectional glass/epoxy composites. Furthermore, such a strain criterion is also able to describe creep lifetime of transversely loaded unidirectional composites.

1 INTRODUCTION

Models for the prediction of strength of composite materials are generally directed to short-term failure using laminate analysis based on classical mechanics, coupled with a common failure theory such as maximum stress or maximum strain concepts (Tsai & Hahn 1980). The input data for these models is derived from standard tests on unidirectional laminates, normally in accordance with the American Society for Testing and Materials (ASTM) standards, and are typically low strain rate or quasi-static experiments. Subsequently, this data is often used by designers to analyze structures which are subjected to long-term static (creep), dynamic (fatigue) or shock (impact) loadings, which all differ strongly from the standard test conditions. Especially in off-axis loading situations, where the properties are strongly governed by the viscoelastic polymer matrix, a large influence of loading conditions can be expected.

Since almost all engineering components are subjected to load and environmental histories which differ strongly over the service time of the component, it is clear that there is a need for the development of new approaches for the prediction of failure of composites which enable us to include time-variable circumstances. In case of matrix dominated failure it seems therefore obvious to investigate the time and stress dependent failure of the polymer matrix. Failure of polymers is in most cases preceded by yielding. For brittle polymers, yielding leads almost instantaneously to fracture, whereas for ductile polymers, yielding initiates the formation of a neck which stabilization and propagation is the result of strain hardening (Ward & Hadley 1993). In polymer physics, relatively straightforward analyses have been developed which have proven their applicability to yielding of glassy polymers in multiaxial stress fields (Ward 1971, Duckett 1978, Govaert & Tervoort 1994). Previous studies recognised already the importance of a multiaxial stress state in the matrix for the initiation of transverse composite failure (de Kok et al. 1993, Berglund & Asp 1994). In order to describe yielding of the matrix under a multiaxial stress situation as encountered in composite materials, the yield behaviour of the epoxy system, which has been used as a matrix system for the glass fibre-reinforced composites, was characterized under various multiaxial loading conditions.

With recent developments in three-dimensional constitutive modelling of large strain plasticity in amorphous polymers (Tervoort et al. 1994), the numerical simulation of the behaviour of the polymer matrix under complex loading conditions is well within reach. In previous work a constitutive model, which describes yield in multiaxial stress and strain fields similar to the approach of Ward (1971) and Duckett (1978), was implemented into a finite element code (Smit 1994). In this research this technique was employed, in combination with micromechanics, to evaluate short-term and long-term transverse failure of unidirectional glass/epoxy laminates.

2 EXPERIMENTAL

2.1 *Materials*

A rather brittle epoxy system of Ciba Geigy (Araldite LY556/HY 917/DY070) was used as matrix system in this study. This epoxy system is based on diglycidyl ether of bisphenol-A with an anhydride curing agent. E-glass fibres (Silenka 0.84-M28) were used as reinforcement material.

For the characterization of the neat epoxy, plates were manufactured by casting the epoxy resin into a heated mould and curing for 4 hours at 80 °C and 8 hours at 140 °C. From these epoxy plates, samples were cut and milled into the various test specimens.

Unidirectional composites were manufactured by filament winding, where fibre strands were impregnated in a heated epoxy bath and wound uniformly on a framework. The impregnated fibres were subsequently placed in a mould and cured in a hot-press. The resulting composite plates had a thickness of 1.25 mm and a fibre volume fraction of approximately 50%. From these unidirectional laminates transverse three-point bending specimens were cut with a length of 50 mm and a width of 25 mm. The edges of all samples, both reinforced and unreinforced, were carefully polished.

2.2 *Testing*

The experimental part of this study consisted of mechanical tests on pure matrix material and on unidirectional composites. In order to characterize the yielding behaviour of the matrix material under complex stress and strain fields various tests are needed. In this research five different (multiaxial) tests were performed on the pure matrix material at strain rates varying over several decades, viz. uniaxial extension, uniaxial compression, planar extension, planar compression and simple shear tests. Uniaxial tensile tests were performed at different strain rates and were in accordance with ASTM D638-91. Uniaxial compression samples were end-loaded and measured 25x6x6 mm. Planar extension test samples had a thickness in the testing region of 2 mm over a length of 10 mm and a width of 50 mm. Planar compression tests were performed on dog-bone shaped specimens measuring 85x20x4 mm. In the test-section the specimens were supported by steel plates to create a plane strain condition. Simple shear tests, according to ASTM D4255-83, were performed on samples with a thickness of 2 mm and a length of 100 mm. The distance between the grips was 10 mm, resulting in an aspect ratio of 10.

The testing of the unidirectional composites consisted of transverse three-point bending experiments at different strain rates and a span-to-depth ratio of 32 and were in accordance with ASTM D790M-92. The composites were also tested under (constant load) creep conditions in three-point bending. Transverse three-point bending experiments were chosen instead of transverse tensile tests, because bending experiments are less sensitive for material flaws and therefore more accurate for the determination of the 'intrinsic' or 'true' composite strength.

3 CHARACTERIZATION OF MATRIX MATERIAL

In the case of matrix dominated failure modes the transverse strength of a composite is governed by the strength of the polymer matrix. In a composite the stress and strain fields in the matrix are complex. Consequently, in order to predict yielding of the matrix, expressions should be used which can describe yield under complicated stress situations. For the epoxy based composites studied in this research, a three-dimensional yield expression, usually referred to as pressure modified Eyring flow (Eyring 1936), is considered. This approach was initially proposed by Ward (1971) and Duckett (1978) and expresses the stress state and deformation rate at the yield point in terms of the octahedral shear stress τ_{oct} and the octahedral shear rate $\dot{\gamma}_{oct}$, respectively:

$$\dot{\gamma}_{oct} = \dot{\gamma}_0(T) \sinh\left[\frac{\tau_{oct} \upsilon - p \, \Omega}{k \, T}\right] \qquad (1)$$

where T is the absolute temperature, k is Boltzmann's constant, υ is the shear activation volume, P is the hydrostatic pressure, Ω is the pressure activation volume and $\dot{\gamma}_0(T)$ is a temperature dependent parameter. In this study the temperature dependence can be neglected since only isothermal loading conditions are considered. For large stresses this relation can be reduced to:

$$\tau_{oct} = \tau_{0,oct} + \frac{k \, T}{\upsilon} \ln \dot{\gamma}_{oct} + \mu \, P \qquad (2)$$

where $\tau_{0,oct}$ is a material constant and μ is the pressure dependence of the yield stress. Since, in general loading situations the hydrostatic pressure P will depend on the stress state at the yield point, it is convenient to express P as:

$$P = P_0 + \alpha \, \tau_{oct} \qquad (3)$$

where α is a coefficient which is characteristic for each loading geometry (see Table 1) and P_0 is the environmental pressure. In our case, P_0 equals 0.1 MPa and can therefore be neglected. Introduction of Eq. 3 into Eq. 2 leads to:

$$\tau_{oct} = \frac{\tau_{0,oct} + \frac{k \, T}{\upsilon} \ln \dot{\gamma}_{oct}}{(1 - \mu \, \alpha)} \qquad (4)$$

Table 1. The geometry factor α for each loading condition.

	α [-]
Uniaxial extension	$-\tfrac{1}{2}\sqrt{2}$
Uniaxial compression	$\tfrac{1}{2}\sqrt{2}$
Planar extension	$-\tfrac{1}{2}\sqrt{6}$
Planar compression	$\tfrac{1}{2}\sqrt{6}$
Simple shear	0

The results of the uniaxial extension, planar extension, uniaxial compression, planar compression and simple shear test are presented in Figure 1, where the octahedral shear stress is plotted versus the octahedral shear rate for all loading geometries. All experiments were performed at room temperature (295 Kelvin). The drawn lines are predicted by Eq. 4, using: $\tau_{0,oct}$ = 56.6 MPa, μ = 0.126 and υ = 3.5 nm³, showing clearly hat all experiments are in good agreement with this modified Eyring equation.

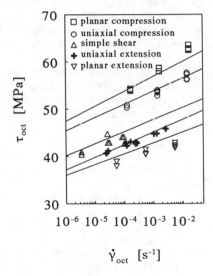

Figure 1. The octahedral shear stress versus octahedral strain rate for the various tests. The results are fitted with the pressure modified Eyring equation (Eq. 4).

15

4 MICROMECHANICAL ANALYSIS OF COMPOSITE MATERIALS

To account for the complex stress and strain situation in a composite, micromechanical simulations are performed with the Finite Element Method (FEM). The mechanical behaviour of the matrix system is modelled with the compressible Leonov model which has shown its applicability for the description of the mechanical behaviour of polymers (Tervoort et al. 1994). This model uses the same three-dimensional yield expression as presented in the previous Section.

4.1 Constitutive model

The characteristic material behaviour of polymers is presented in Figure 2a with the solid line. This behaviour exhibits an initial elastic deformation (I), which becomes fully plastic (II) with increasing strain. At large deformations, an increasing strain hardening effect results in a rise in stress (III), which finally leads to fracture. The mechanical analogy, shown in Figure 2b, was proposed by Haward and Thackray (1968) and consists of a Maxwell model with a linear spring and an Eyring dashpot to account for the strain rate dependence of the yield point. The strain hardening behaviour has been modelled by a Gaussian, rubber elastic, spring placed in parallel.

To facilitate finite element analysis a three-dimensional constitutive model is required. In this study the compressible Leonov model was used, which is described in detail elsewhere (Tervoort 1994). In this model the matrix material is regarded as a linear elastic, compressible solid up to the yield point, with a modulus of 3200 MPa and a Poisson's ratio of 0.37. The yield behaviour of the material is described as a generalised Newtonian fluid with a flow characteristic according to the data presented in Figure 1. The strain hardening behaviour was also modelled as a Gaussian, rubber elastic spring. The rubbermodulus was estimated at 31.5 MPa, as determined experimentally with dynamic mechanical thermal analysis.

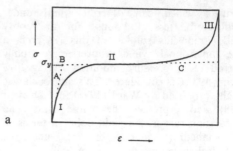

a

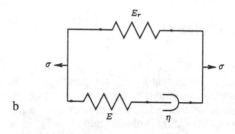

b

Figure 2. (a) Illustration of the characteristic behaviour of glassy polymer (solid line) with the fit of the compressible Leonov model (dashed line). True strain versus true stress is plotted. (b) Schematic representation of the compressible Leonov model.

4.2 Finite element model

In this study the calculations are based on a hexagonal fibre array, from which the repeating unit is shown in Figure 3a. Due to symmetry only a quarter of this repeating unit needs to be modelled. The finite element mesh, representing a fibre volume fraction of 50% is shown in Figure 3b. For the fibre 205 and for the epoxy matrix 675 generalized plane strain quadrilateral elements were used. Both mesh size and time increment were optimized by mesh refinement and time increment reduction.

4.3 Failure criterion

Since there is no limit to the deformation in the constitutive model an additional failure criterion is needed. In this study a limiting value of 15% for the octahedral shear strain was chosen. This value was estimated from a numerical simulation of a tensile experiment at a global strain rate of 10^{-4} s^{-1} by evaluating the local strain state in the composite at an externally applied stress equal to the experimentally observed tensile strength. For the prediction of failure at other strain rates or under the influence of a statically applied stress (creep), a micromechanical simulation of the desired test was performed up to a global state of deformation where the local maximum of the octahedral shear strain equals the critical value of 15%.

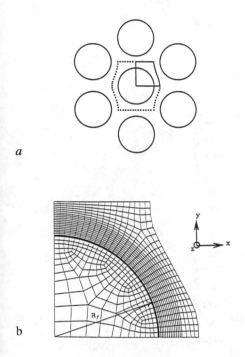

a

b

Figure 3. (a) Representation of the hexagonal stacking. (b) The used mesh. Due to symmetry only a quarter of the repeating unit has to be modeled.

5   MODEL VALIDATION

In this Section the results of the experiments on the unidirectional composites will be compared with the results of the numerical simulations. The experiments on the fibre-reinforced composites consisted of rate dependent strength measurements and creep experiments, both in three-point bending.

5.1 *Rate dependent transverse strength*

The unidirectional composites were transversely tested in three-point bending at constant strain rate. All tests were performed at ambient temperature and showed perfect linear elastic behaviour up to failure. The results of maximum stress versus strain rate of the composites in three-point bending are presented in Figure 4. The error bars are the standard deviation of the experimental outcome. The predicted strain rate dependence of the transverse strength is represented by the solid line and shows that the experimental results are well described by the strain criterion.

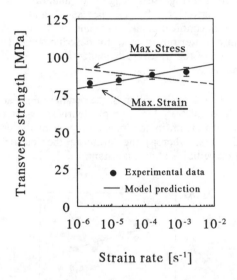

Figure 4. The strain rate versus the transverse strength. The results of three-point bending on transversely loaded composites (markers) are compared with numerical simulations using the octahedral shear strain as a failure criterion.

From these results it can be concluded that the octahedral shear strain criterion gives promising results for the description of failure of transversely loaded unidirectional composites. As expected, with increasing strain rate the globally applied strain increased using the strain criterion because at higher strain rates the polymer matrix has less time to localize.

5.2 Creep lifetime

Similar to the constant strain rate tests, the numerical predictions for creep failure were compared with experimental results. Again the local octahedral shear strain of the reference simulation (at a strain rate of 10^{-4} s^{-1}) is used to predict the creep time to failure for the different applied engineering stresses. The results of the experiments and the numerical simulations are shown in Figure 5. In all creep experiments the time needed to apply the constant engineering stress was negligible in comparison with the obtained creep time to failure. The applied stress was varied from 50 to 85 MPa. The solid line represents the predicted time to failure versus the engineering stress using the octahedral shear strain criterion. It can be seen that the micromechanical model in combination with this failure criterion is able to describe time to fracture of the transversely loaded E-glass/epoxy composites.

In Figure 6 the evolution of the octahedral shear strain is illustrated. The graph clearly shows that the local strain in the 'plastic' region increases with time, whereas the strain in the 'elastic' region remains almost constant.

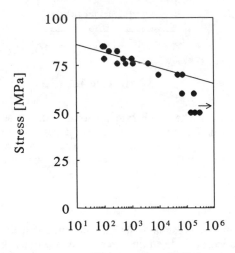

Figure 5. The applied engineering stress versus time to break, showing the results of creep in three-point bending on transversely loaded unidirectional composites (markers) in comparison with the model prediction (solid line).

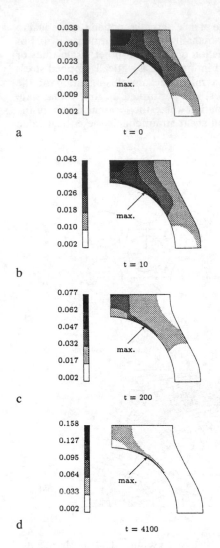

Figure 6. Contour plots of the evolution of the octahedral shear strain during creep at an engineering stress of 75 MPa. a-d, represent the various creep times.

6 CONCLUSIONS

In this study it is demonstrated that the modified Eyring equation can be satisfactory used for the description of the yield behaviour of the epoxy system under multiaxial loading conditions. The various multiaxial experiments also showed that the epoxy system clearly exhibits a pressure dependent yield behaviour. By introducing these parameters into the compressible Leonov model, numerical (FEM) simulations of the mechanical behaviour of the epoxy system could be performed. Micromechanical simulations of rate dependent transverse strength of unidirectional composites, showed the validity of a failure criterion based on a maximum local strain. Moreover, this failure criterion could be used for the prediction of creep time to failure.

REFERENCES

Berglund, L.A. and Asp, L., Fracture of epoxies subjected to uniaxial and triaxial stress fields in: *Proc. 9th Int. Conf. Deformation, yield and fracture of polymers*, Cambridge (The Institute of Materials, 1994) P55/1.

Duckett, R.A., The yielding and crazing behaviour of polycarbonate in torsion under superposed hydrostatic pressure, *The British Polymer J.,* 10 (March 1978) 11.

Eyring, H., Viscosity, plasticity, and diffusion as examples of absolute reaction rates, *J. Chem. Physics,* 4 (1936) 283.

Govaert, L.E. and Tervoort, T.A., Plastic instabilities in glassy polymers in: *Proc. 9th Int. Conf. Deformation, yield and fracture of polymers, Cambridge* (The Institute of Materials, 1994) P67/1.

Haward, R.N. and Tackray, G., *Proc. Roy. Soc. A.,* 302 (1968) 53.

Kok, de J.M.M., Meijer, H.E.H. and Peijs, T., The influence of matrix plasticity on the failure strain of transversely loaded composite materials, *Proc. ICCM/9*, ed. A. Miravete, (Woodhead Publ., 1993) 5, 242.

Smit, R.J.M., Numerical simulation of localization phenomena in polymer glasses. Eindhoven University of Technology, internal report WFW 94.046 (April 1994).

Tervoort, T.A., Brekelmans, M. and Govaert, L.E., A 3-D stress-strain relationship for glassy polymers in: *Proc. 9th Int. Conf. Deformation, yield and fracture of polymers*, Cambridge (The Institute of Materials, 1994) P66/1.

Tsai, S.W. and Hahn, H.T., *Introduction to composite materials*, Technomic Publ. Co. Inc. (1980).

Ward, I.M. and Hadley, D.W., *An introduction to the mechanical properties of solid polymers*, Chapter 11, John Wiley & Sons (1993) 206.

Ward, I.M., The yield behaviour of polymers, *J. Mater. Sci.*, 6 (1971) 1397.

Progress in Durability Analysis of Composite Systems, Cardon, Fukuda & Reifsnider (eds)
© *1996 Balkema, Rotterdam. ISBN 90 5410 809 6*

Characterization of nonlinear, time-dependent polymers and polymeric composites for durability analysis

R. A. Schapery
Department of Aerospace Engineering and Engineering Mechanics, The University of Texas at Austin, Tex., USA

ABSTRACT: An approach to modeling the mechanical behavior of fiber reinforced and unreinforced plastics with an evolving internal state is described. Intrinsic nonlinear viscoelastic and viscoplastic behavior of the resin matrix is taken into account along with growth of damage. The thermodynamic framework of the method is discussed first. The Gibbs free energy is expressed in terms of stresses, internal state variables (ISVs), temperature and moisture content. Simplifications are introduced based on physical models for evolution of the ISVs and on experimental observations of the dependence of strain state on stress state and its history. These simplifications include use of master creep functions that account for multiaxial stresses, environmental factors and aging in a reduced time and other scalars. An explicit representation of the strains follows, which is then specialized to provide three-dimensional homogenized constitutive equations for transversely isotropic, fiber composites. Experimental support for these equations is discussed. Finally, physical interpretation of some of the constitutive functions is given using results from a microcracking model as well as molecular rate process and free volume theories. It is shown that the present thermodynamic formulation leads to a generalized rate process theory that accounts for a broad distribution of thermally activated transformations in polymers.

1 INTRODUCTION

There is rapid growth in the use of structural plastics, with and without fiber reinforcement. Numerous current and potential applications involve their use in environments and at stresses for which creep and damage growth are life-limiting factors. The useful life may be limited by dimensional, stiffness or residual strength changes. Many applications involve relatively thick sections, thus requiring three-dimensional structural analysis for accurate predictions. Linear elastic or viscoelastic constitutive equations and structural analysis may be adequate for preliminary design purposes. However, in order to achieve structurally efficient designs that meet long-term performance requirements, nonlinear, time-dependent effects in constitutive equations frequently must be taken into account.

Here we shall be primarily concerned with only this aspect of the durability problem. Namely, nonlinear viscoelastic/viscoplastic constitutive equations will be discussed. Damage growth is included in the study, but prediction of structural lifetime is not directly considered. Nevertheless, the models described here should be useful in life predictions. Indeed, they provide some important relationships between short-term characterizations and long-term performance. This is

accomplished by taking advantage of a type of simplicity in behavior exhibited by many polymeric materials. In its simplest form, it is the so-called thermorheologically simple behavior, wherein time and temperature affect mechanical behavior essentially through one parameter, reduced time. With so-called *physical aging*, the aging is usually characterized quite well at low stresses by using a modified form of reduced time (e.g. Struik 1978, Sullivan 1990 and McKenna 1994). In the linear viscoelastic range of behavior these observations lead to the use of so-called *master* creep or relaxation functions in predicting long-term strain or stress response; such functions may be found from short-time tests conducted over a range of temperatures. Chemical aging may be significant in some applications (e.g. Skontorp 1995); but it is beyond the scope of this paper.

Although not established nearly as well, certain types of simplicity exist at moderate-to-high stresses. They are used here along with nonequilibrium thermodynamics to achieve the form of the nonlinear constitutive equations described in this paper.

In Section 2 the thermodynamic theory is first used to provide the framework for incorporating the idea of master creep functions in a three-dimensional constitutive model for anisotropic media. Specific constitutive equations are then developed. Section 3 is concerned with additional types of simplicity in stress and time effects, including specialization to transverse isotropy. Section 4 discusses specific physical models for describing microcracking effects and for interpreting various material functions using free volume and thermal activation considerations.

The thermodynamic constitutive theory discussed in this paper is an updated version of a model proposed and applied many years ago (Schapery 1969a, 1969b, and Lou and Schapery 1971), and used by many investigators over the years. This early work did not include an explicit model for nonlinear viscoplastic (nonrecoverable) strain. Here this topic receives considerable attention in view of the growing use of structural thermoplastics along with recently available experimental information and new models for these materials.

Mechanical characterization of time-dependent, nonlinear behavior of high-modulus (e.g. glassy) polymers is usually simpler using stresses, rather than strains, for independent variables, as done here (Schapery, 1969b). The opposite seems to be true for low modulus (e.g. rubbery) polymers; strains were used as independent state variables in Schapery's (1966) nonlinear constitutive theory.

2 THERMODYNAMIC CONSTITUTIVE THEORY

2.1 State variables and notation

The thermodynamic system is taken as a representative composite material element having unit volume in the reference (initial) state. In the usual way microstructural details are suppressed, and the material is modeled as an anisotropic, homogeneous continuum with spatially uniform temperature, moisture and other plasticizers, if any; for gradient effects see Weitsman (1990). This homogenized material is acted upon by stresses σ_i and undergoes deformation as defined by the strains ϵ_i. For notational simplicity, standard single index notation is used for stresses and strains. If geometric nonlinearities are to be accounted for, then ϵ_i should be interpreted as Green's strains and σ_i as their work-conjugates, the Kirchoff stresses (Fung 1965).

The thermodynamic state of the material is assumed to be fully defined by the stresses, σ_i, three different sets of internal state variables (ISVs), η_i, ξ_m, and S_q, as well as absolute temperature T and moisture content M (and possibly other plasticizers.) Using boldface in the text to designate a complete set of components, the quantities ϵ and η will always have an i, j or k subscript, in which

i, j, $k = 1, ...6$. The subscripts m and n (with the range m, $n = 7, ...N$) will be reserved for use with $\boldsymbol{\xi}$, while subscripts q and r will be used only on $\boldsymbol{S}$ (with the range q, $r = N + 1, ...Q$). A reference equilibrium state is defined such that $\boldsymbol{\sigma} = \boldsymbol{\epsilon} = \boldsymbol{\eta} = \boldsymbol{\xi} = \boldsymbol{S} = 0$, $T = T_R$ and $M = M_R$. The tensorial order of the ISVs is discussed in Section 3.5.

2.2 Gibbs free energy and entropy production

Three sets of ISVs have been introduced to account separately for viscoplastic ($\boldsymbol{\eta}$) and viscoelastic ($\boldsymbol{\xi}$) effects, as well as damage and other high-energy structural changes ($\boldsymbol{S}$) such as microcracking and shear banding. It is assumed that $\boldsymbol{\xi}$ is associated with energy changes that are small enough to permit the use of a Gibbs free energy which is at most of second order in these ISVs. In order that $\boldsymbol{\eta}$ represent viscoplastic (nonrecoverable) strains the free energy is linear in these ISVs.

Specifically, the Gibbs energy is written in the form

$$G = G_0 - \sigma_i \eta_i - A_m \xi_m + \frac{1}{2} B_{mn} \xi_m \xi_n \qquad (1)$$

where, in general, G_0, A_m and B_{mn} are functions of $\boldsymbol{\sigma}$, $\mathbf{S}$, T and M; time t may be included to account for aging. Throughout this paper, the summation convention for repeated indices is used, unless indicated otherwise.

The strain is obtained from the standard equation,

$$\epsilon_i = -\partial G / \partial \sigma_i \qquad (2)$$

so that

$$\epsilon_i = -\frac{\partial G_0}{\partial \sigma_i} + \eta_i + \Delta \epsilon_i \qquad (3a)$$

where

$$\Delta \epsilon_i \equiv \frac{\partial A_m}{\partial \sigma_i} \xi_m \qquad (3b)$$

after neglecting terms of second order in $\boldsymbol{\xi}$. Later, it will be seen that $\boldsymbol{\eta}$ and $\boldsymbol{\xi}$ increase continuously, starting at zero in the reference state, even when a large stress is suddenly applied. Thus, at least for a limited period of time, one can expect the expansion (1) to be valid.

We shall think of $\boldsymbol{\sigma}$, T and M as specified functions of time in formulating the constitutive equations. In order to predict the strains in eqn(3) additional equations are needed for predicting changes in all ISVs. These so-called evolution equations will be selected so that they satisfy thermodynamic and stability conditions, and lead to constitutive equations that are con-

22

sistent with a large amount of experimental data on polymers and their composites.

The First and Second Laws of thermodynamics lead to the inequality (e.g., Rice 1971),

$$Ts \equiv f_i \dot{\eta}_i + f_m \dot{\xi}_m + f_q \dot{S}_q \geq 0 \tag{4}$$

where s is the *entropy production rate* and the f's are the so-called thermodynamic forces,

$$
\begin{aligned}
f_i &\equiv -\partial G/\partial \eta_i \\
f_m &\equiv -\partial G/\partial \xi_m \\
f_q &\equiv -\partial G/\partial S_q
\end{aligned}
\tag{5}
$$

The evolution equations are relationships between each ISV set and the conjugate set of thermodynamic forces.

2.3 Equations for predicting $\boldsymbol{\xi}$ and $\Delta \epsilon$

For now, linear relationships between f_m and the time rate of change $\dot{\xi}_m$ are assumed

$$\dot{\xi}_m = c_{mn} f_n / a_1 \tag{6}$$

where c_{mn} is a constant, positive definite, symmetric matrix and a_1 is a positive scalar quantity, which may be a function of $\boldsymbol{\sigma}$, T, M and possibly other quantities, such as the history of the state variables. So-called physical aging may be accounted for by introducing explicit dependence of a_1 on time, t. The contribution of $f_m \dot{\xi}_m$ to the entropy production rate s in eqn(4) is seen to be non-negative. The symmetry of c_{mn} comes from Onsager's principle (Fung 1965). Observe that eqn(6) may be written in the form

$$\dot{\xi}_m = \frac{1}{a_1} \frac{\partial F_v}{\partial f_m} \tag{7}$$

where

$$F_v \equiv \frac{1}{2} c_{mn} f_m f_n \tag{8}$$

Equations for predicting $\boldsymbol{\xi}$ are obtained by first using eqns(1) and (5) to derive the thermodynamics forces. Thus,

$$f_m = A_m - B_{mn} \xi_n \tag{9}$$

which, together with the inverse form of eqn(6), yields

$$a_1 \hat{c}_{mn} \dot{\xi}_n + B_{mn} \xi_n = A_m \tag{10}$$

where $\hat{c}_{mn} = [c_{mn}]^{-1}$. A simplifying assumption which leads to equations that are consistent with observed behavior is

$$B_{mn} = a_2 C_{mn} \tag{11}$$

where C_{mn} is a constant, symmetric matrix and a_2 is a positive scalar function that may depend on $\boldsymbol{\sigma}$, T, M and t time; thermodynamic stability requires that C_{mn} be positive definite. Equation (10) may then be written in the form

$$\hat{c}_{mn} \frac{d\xi_n}{d\psi} + C_{mn} \xi_n = \frac{A_m}{a_2} \tag{12}$$

where

$$\psi \equiv \int_0^t \frac{a_2}{a_1} dt' \tag{13}$$

is *reduced time*. It is always possible to find another set of ISVs that are related linearly to $\boldsymbol{\xi}$ and simultaneously diagonalizes $\hat{c}_{mn}$ and C_{mn} (Fung 1965). Without loss in generality we may suppose that $\boldsymbol{\xi}$ is this set, and thus eqn(12) can be written as an uncoupled set of equations (m not summed),

$$\frac{d\xi_m}{d\psi} + c_m C_m \xi_m = \frac{c_m A_m}{a_2} \tag{14}$$

where c_m and C_m replace c_{mn} and C_{mn}, respectively, as a result of the diagonalization. The general solution of eqn(14) is simply (m not summed)

$$\xi_m = \frac{1}{C_m} \int_0^\psi [1 - e^{-(\psi - \psi')/\tau_m}] \frac{d}{d\psi'} \left(\frac{A_m}{a_2} \right) d\psi' \tag{15}$$

where $\tau_m \equiv (c_m C_m)^{-1}$ is positive and is a so-called *retardation time*. It is assumed that the system is in its reference state for $t < 0$.

This result for $\boldsymbol{\xi}$ may be used in eqn(3b) to obtain $\Delta \epsilon$. However, it is helpful to introduce first another significant, realistic simplification which, in the nonlinear range, is needed to preserve the underlying time-dependence of the viscoelastic creep response; i.e. to enable characterization in terms of *master* creep functions. Specifically, let

$$A_m = C_{mj} \hat{\sigma}_j + \alpha_m \phi \tag{16}$$

where C_{mj} and α_m are constants. Also, $\hat{\sigma}$ and ϕ may depend on $\boldsymbol{\sigma}$, T, M, and t. In the linear viscoelastic range $\hat{\sigma} = \sigma$. Also, ϕ accounts for thermal and moisture expansion effects; thus $\phi = 0$ when $T = T_R$ and $M = M_R$.

Substitution of eqn(16) into (3b) and use of eqn(15) yields

$$\Delta \epsilon_i = \frac{\partial \hat{\sigma}_j}{\partial \sigma_i} (\Delta \hat{\epsilon}_j + \Delta \hat{\alpha}_j) \tag{17}$$

where

$$\Delta \hat{\epsilon}_j \equiv \int_0^\psi \Delta S_{jk}(\psi - \psi') \frac{d}{d\psi'} \left(\frac{\hat{\sigma}_k}{a_2} \right) d\psi' \tag{18}$$

and

$$\Delta\hat\alpha_j \equiv \int_0^\psi \Delta\alpha_j(\psi - \psi') \frac{d}{d\psi'}\left(\frac{\phi}{a_2}\right) d\psi' \qquad (19)$$

Also,

$$\Delta S_{jk}(\psi) \equiv \sum_m \frac{C_{mj}C_{mk}}{C_m}(1 - e^{-\psi/\tau_m}) \qquad (20)$$

$$\Delta\alpha_j(\psi) \equiv \sum_m \frac{C_{mj}\alpha_m}{C_m}(1 - e^{-\psi/\tau_m}) \qquad (21)$$

Observe that ΔS_{jk} is symmetric and positive definite as a result of the way in which the C's appear in eqn(20).

The functions in eqns(20) and (21) are independent of all variables except for reduced time. Because they are independent of stress, and give the strain response to unit values of $\hat\sigma_k/a_2$ and ϕ/a_2, it is appropriate to call them *master* creep compliances or creep functions.

If the thermodynamic system is neutrally stable, then B_{mn} will be positive semi-definite rather than positive-definite. In this case one or more C_m in eqn(14) will vanish (giving $\tau_m = \infty$) and the equation for each associated ξ_m becomes

$$\xi_m = c_m \int_o^t \frac{A_m}{a_1^\infty} dt' \qquad (22)$$

(m not summed)

The contribution of these ISVs to $\Delta\epsilon$ is found by substituting this result into eqn(3b). In general, this contribution does not eventually vanish after removal of the stresses, and thus it is appropriate to call it *viscoplastic strain* and denote it by $\Delta\epsilon^p$. The physical sources of this strain may be sufficiently different from those for which $\tau_m < \infty$ that the scalar factor a_1^∞ is different from a_1. Similarly, even if the form of A_m in eqn(16) is applicable, the stress functions $\hat\sigma_j$ may be different, and thus will be designated here as $\hat\sigma^\infty$. Also, as shown previously (Schapery 1964) for stable behavior when $\sigma \equiv 0$, the expansion coefficients α_m associated with the ISVs for which $\tau_m = \infty$ must vanish. The viscoplastic strains then become

$$\Delta\epsilon_i^p = \frac{\partial\hat\sigma_j^\infty}{\partial\sigma_i}K_{jk}\int_o^t \frac{\hat\sigma_k^\infty}{a_1^\infty} dt' \qquad (23)$$

where

$$K_{jk} \equiv \Sigma_m\, C_{mj}\, C_{mk}\, c_m \qquad (24)$$

is seen to be a symmetric, positive semi-definite matrix; the summation is extended over only those ISVs for which $\tau_m = \infty$. It should be noted that eqn(23) may be obtained directly from eqns(17), (18) and (20) by using a limiting process with $\tau_m \to \infty$ and then introducing functions $\hat\sigma^\infty$ and a_1^∞ in place of $\hat\sigma$ and a_1.

Consider, for example, a special case in which $\hat\sigma^\infty = \sigma$ and a_1^∞ is a function of only the current stress state. Then, if the stresses are constant in time, $\Delta\epsilon^p \sim t$. However, some experimental data on a fiber composite (Tuttle et al. 1995) and a semi-crystalline polymer (Lai 1995) indicate that the viscoplastic strain is not proportional to the time under load. Rather, $\Delta\epsilon^p \sim t^\beta$ where $0 < \beta < 1$, which is usually called *primary creep* in studies of metals. This type of behavior results if a_1^∞ obeys a power law in the viscoplastic strain, such as discussed in the next section. If $\hat\sigma^\infty$ is a nonlinear function of stress, then $\Delta\epsilon^p$ changes discontinuously in time in response to discontinuous changes in σ_j, and may decrease even if the stress does not change sign. This type of behavior is not traditionally used in models of viscoplastic strain, and, at this time, there does not appear to be any experimental data that exhibit this behavior. With this motivation, let us assume $\hat\sigma^\infty$ is linear in stress. Then, without any further loss in generality in the form of $\Delta\epsilon^p$, we may simply use $\hat\sigma^\infty = \sigma$ in eqn(23) to obtain

$$\Delta\dot\epsilon_i^p = K_{ij}\frac{\sigma_j}{a_1^\infty} \qquad (25)$$

Thus, because K_{ij} is symmetric, the viscoplastic strain rate may be expressed as a gradient of a potential function of σ_j/a_1^∞.

It was shown by Schapery (1969a) that the strain, eqn(3), may be represented schematically by a mechanical analogue consisting of nonlinear springs and dashpots, as illustrated in Fig. 1. Each ξ_m, eqn(15), is proportional to the strain in one Voigt unit (consisting of one spring with modulus E_m and one dashpot with viscosity η_m in parallel). Thermal expansion of the springs contributes to $\Delta\alpha$. The strain of the spring E_o represents the term $-\partial G_o/\partial\sigma$ and the strain of the dashpot η_o accounts for the $\tau_m = \infty$ case or, equivalently, for the viscoplastic strain discussed next.

2.4 *Equations for predicting* η

In the preceding section, viscoplastic strains, eqns(23) and (25), arose as a special case of viscoelastic behavior through the ξ ISVs. Here, as previously described (Schapery 1994) the η ISVs

24

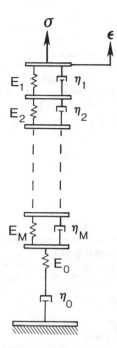

Figure 1. Generalized Voigt Model

are used for comparison purposes to account for viscoplastic strains which are analogous to what is used in the classical strain-hardening theory of metals (e.g. Hult 1966), after allowing for anisotropy. In this case, the evolution laws are nonlinear in the thermodynamic forces f_i, in contrast to eqn(6). However, it is traditional to use a special case in which $\dot{\boldsymbol{\eta}}$ is proportional to the derivative of a potential F_p,

$$\dot{\eta}_i = \frac{1}{a_3} \frac{\partial F_p}{\partial \sigma_i} \qquad (26)$$

where a_3 is a positive scalar function of a scalar measure of the viscoplastic strains, h_η, and we have used the fact that $f_i = \sigma_i$ in view of eqn(5). The quantity h_η is defined through the integral,

$$h_\eta \equiv \int_0^t (\gamma_{ij} \dot{\eta}_i \dot{\eta}_j / 2)^{1/2} \, dt \qquad (27)$$

although it is sometimes defined in terms of total viscoplastic strains (Hult 1966) whether or not there is proportional straining. It is similarly assumed that F_p depends on stresses through a quadratic function of stresses h_σ,

$$h_\sigma \equiv \frac{1}{2}\beta_{ij} \sigma_i \sigma_j \qquad (28)$$

It is supposed for now that a_3 is independent of

stress and F_p is independent of strain, but these functions may depend on other parameters, such as T, M and t. The matrices γ_{ij} and β_{ij} are specified as positive-definite. Without loss in generality they may be taken as symmetric since antisymmetric components have no effect on quadratic functions; although these matrices are usually assumed constant in metal applications, we may allow for dependence on T, M, t and $\mathbf{S}$. Equations (26) and (28) yield

$$\dot{\eta}_i = \frac{1}{a_3} \frac{dF_p}{dh_\sigma} \beta_{ij} \, \sigma_j \qquad (29)$$

A total rather than partial derivative is shown since, by assumption, h_σ is the only mechanical variable that F_p depends on. The entropy production inequality in eqn(4) is clearly satisfied if $dF_p/dh_\sigma \geq 0$. Observe that eqn(29) is equivalent to eqn(25) if we let $K_{ij} = \beta_{ij}$ and

$$a_1^\infty = a_3(dF_p/dh_\sigma)^{-1} \qquad (30)$$

In other words, the classical strain-hardening creep theory is a special case of the thermodynamic theory based on linear relationships between forces f_i and "fluxes" $\dot{\eta}_i$ and on Onsager's principle (which gives the symmetry of K_{ij}.)

In the classical creep theory it is usually assumed there is a relationship between scalars h_σ, h_η and dh_η/dt. This condition is met by assuming $\gamma_{ij} = [\beta_{ij}]^{-1}$. Then, after multiplying eqn(29) by $\gamma_{ij} \dot{\eta}_j/2$ it is found that

$$\frac{dh_\eta}{dt} = \frac{h_\sigma^{1/2}}{a_3} \frac{dF_p}{dh_\sigma} \qquad (31a)$$

From eqn(30),

$$\frac{dh_n}{dt} = \frac{h_\sigma^{1/2}}{a_1^\infty} \qquad (31b)$$

It is commonly found for metals that a_3 obeys a power law in h_η,

$$a_3 = k_3 h_\eta^\gamma \qquad (32)$$

where γ is a constant and k_3 is independent of h_η, but may vary with time; when a_3 increases with h_η, the material is said to exhibit *strain hardening*. Equations (31) and (32) yield

$$h_\eta = \left[\beta^{-1} \int_o^t \frac{h_\sigma^{1/2}}{k_3} \frac{dF_p}{dh_\sigma} \, dt' \right]^\beta \qquad (33)$$

where

$$\beta = (\gamma + 1)^{-1} \qquad (34)$$

Hence, when all stresses are constant in time (and if there are no sources of time-dependence in k_3 and F_p),

$$h_\eta \sim t^\beta \qquad (35)$$

(Observe that $\beta = 1$ without strain hardening.) Equation (35), together with (29), (32) and (34), give $\dot\eta \sim t^{-\gamma/(\gamma+1)}$ or, equivalently, $\eta \sim t^\beta$. As discussed in the previous section, some experimental data on polymers exhibit this so-called primary creep behavior, corresponding to $0 < \beta < 1$.

Finally, observe that eqn(25) is more general than eqn(29) because a_1^∞ is not limited in its dependence on mechanical scalar variables; if eqn(29) is used, then eqn(30) shows that a_1^∞ is limited to a product of functions of h_η and h_σ. If we use eqn(25) it may, for example, be acceptable to assume a_1^∞ depends on h_η and h_σ (where $\beta_{ij} = K_{ij} = [\gamma_{ij}]^{-1}$) but without the limitation of eqn(30); one can verify that eqn(31b) is applicable in this case. Thus, for generality it is preferable to use eqn(25) instead of eqn(29) for characterizing the viscoplastic strain component of ϵ in eqn(3a).

2.5 Equations for predicting S

The ISVs S define high-energy structural changes and, as such, the Gibbs free energy is not expressed as a truncated power series in them. In the absence of significant viscoelastic/viscoplastic effects(i.e. $\eta \simeq \xi \simeq 0$), we have found for fiber composites that the stress (strain) is not sensitive to many details of the strain (stress) history (Lamborn and Schapery 1993). This has lead to the use of the evolution law for each $\dot S_q \neq 0$ (Schapery 1990a),

$$f_q = \partial W_s / \partial S_q \qquad (36)$$

where W_s is the work of structural change, which is a function of S and possibly T and M. The left side may be viewed as the force available to change S_q, while the right side is the required force. For the special case in which S_q is a change in crack surface area, the left side is the energy release rate and the right side is the critical energy release rate. There are many processes for which one or more $f_q < \partial W_s/\partial S_q$; then $\dot S_q = 0$ for this subset of S.

It was shown in earlier work (Schapery 1990a) that the total applied work and complementary work are independent of strain and stress path, respectively, for any set of processes (i.e. strain or stress histories) for which the same subset (or full set) of S changes. Moreover, for stable global response, the total work and complementary work are minimized. With stability, for each S_q that

changes, eqn(36) can always be solved to express changes in S_q in terms of changes in strain or stress (Lamborn and Schapery 1993).

Rate-dependence may be introduced while retaining the essential features of path-independent work. This was demonstrated by Schapery (1991) for the case in which the S_q define changes in crack geometry and obey a power law in energy release rate with a large exponent. Whether the S_q define changes in crack geometry or in other local structural changes, let us suppose the evolution law for S_q is of the form (q not summed),

$$\dot S_q = a_4 F_q f_q^{\lambda q} \qquad (37)$$

where thermodynamic force $f_q > 0$, and $F_q = F_q(S_q)$; also, a_4 is a positive, dimensionless, scalar function that may depend on T, M and t. The calculation of f_q for viscoelastic materials is discussed later in this subsection. Now, from eqn(37)(q not summed),

$$\hat F_q(S_q) \equiv \left[\int_0^\cdot \frac{dS_q}{F_q} \right]^{1/\lambda_q} = \left[\int_0^{\hat t} f_q^{\lambda_q} d\hat t' \right]^{1/\lambda_q} \qquad (38)$$

where $\hat t$ is a reduced time,

$$\hat t \equiv \int_0^t a_4(t') dt' \qquad (39)$$

As long as f_q does not decrease, the right side of eqn(38) is approximately $k_q^{-1} f_q \hat t^{1/\lambda_q}$ if $\lambda_q >> 1$, where k_q is a constant and $k_q \to 1$ as $\lambda_q \to \infty$. Thus,

$$f_q = k_q \hat F_q / \hat t^{1/\lambda_q} \qquad (40)$$

which replaces eqn(36). It is seen that if the contribution to W_s associated with those S_q which obey eqn(37) is taken as

$$\Delta W_s = \sum_q k_q \int_0^\cdot \hat F_q dS_q / \hat t^{1/\lambda_q} \qquad (41)$$

then essential results from the original time-independent theory apply. Whenever f_q decreases, the last of eqn(38) is practically constant, and is given by its value at the time f_q started to decrease.

Generally, for evolution laws of the form in eqn(37)(with the same or different exponents for each S_q); the behavior is like that of a time or rate-independent body except the work of structural change is time-dependent. Consequently, the total work or complementary work is like that for an aging elastic material. This result is expected to apply when the evolution law for $\dot{S}_q$ is a strong function of f_q, even if it is not a power law. With cyclic loading a result similar to eqn(41) is found, except the number of cycles appears in place of time (Schapery, 1990a).

When $\boldsymbol{\xi} \neq \mathbf{0}$, the thermodynamic force $-\partial G/\partial S_q$ in general depends on these ISVs. If, however, the A_m in eqn(1) are essentially independent of S_q, then

$$f_q = -\partial G_o/\partial S_q \qquad (42)$$

after dropping second-order terms in $\boldsymbol{\xi}$ (if the B_{mn} depend on S_q). When eqn(42) is used in eqn(36) or eqn(40) to predict S_q, it is seen that S_q is independent of $\boldsymbol{\eta}$ and $\boldsymbol{\xi}$. This type of simplicity is predicted from a mechanics-based microcracking model (cf. Section 4.1).

3 SIMPLIFICATIONS IN STRESS AND TIME EFFECTS

3.1 Quadratic stress functions

There is a considerable amount of experimental data that indicates the stress-dependent functions depend on quadratic scalar functions of the stress components (e.g. eqn(28)) and, to a lesser extent, on a linear function of stress, such as the mean stress. We shall initially use quadratic functions in order to further simplify the constitutive theory.

First, following Brouwer (1986), assume

$$\hat{\sigma}_j = a_5 \sigma_j \qquad (43)$$

where a_5 is a function of one mechanical variable,

$$h_1 = \frac{1}{2}\rho_{ij}\sigma_i\sigma_j \qquad (44)$$

and possibly T, M and t; here, ρ_{ij} is a symmetric matrix. Then,

$$\frac{\partial \hat{\sigma}_j}{\partial \sigma_i} = a_5\delta_{ij} + \frac{da_5}{dh_1}\rho_{ik}\sigma_k\sigma_j \qquad (45)$$

where δ_{ij} is the Kronecker delta ($\delta_{ij} = 1$ if $i = j$ and $\delta_{ij} = 0$ if $i \neq j$). Equation (17) becomes

$$\Delta\epsilon_i = a_5\left(\Delta\hat{\epsilon}_i + \Delta\hat{\alpha}_i\right) + \frac{da_5}{dh_1}\rho_{ik}\sigma_k\sigma_j\left(\Delta\hat{\epsilon}_j + \Delta\hat{\alpha}_j\right) \qquad (46)$$

It is seen that the ith viscoelastic strain compo-

nent may depend on the hereditary integrals associated with more than one strain component unless $da_5/dh_1 = 0$. In earlier work (Schapery 1969b) it was shown (for a uniaxial stress state) that a_5 is not a function of stress if the elastic strain during sudden loading is equal to that during sudden unloading, as in a creep and recovery test.

In a more recent paper, Schapery (1992) suggested the form

$$\hat{\sigma}_i = \frac{\partial \hat{F}}{\partial \sigma_i} \qquad (47)$$

where $\hat{F}$ is a function of a second order stress invariant, such as h_1. Then,

$$\hat{\sigma}_i = \frac{d\hat{F}}{dh_1}\rho_{ij}\sigma_j \qquad (48)$$

When this result is substituted into eqn(16) we find

$$A_m = C_{mj}\rho_{jk}\frac{d\hat{F}}{dh_1}\sigma_k + \alpha_m\phi \qquad (49)$$

On the other hand, eqn(43) gives

$$A_m = C_{mk}a_5\sigma_k + \alpha_m\phi \qquad (50)$$

Clearly, eqns(49) and (50) lead to equivalent constitutive equations since we may set $a_5 = d\hat{F}/dh_1$ and replace $C_{mj}\rho_{jk}$ by another constant matrix. Here, we shall continue to use eqn(43) in further discussion. Observe also that the simplification in eqn(50), wherein all nonlinearity enters through a scalar factor, is analogous to that used in eqns(6) and (11).

3.2 Time effects

At this point it is helpful to introduce abreviated notation for the convolution integrals in eqns(18) and (19). Specifically, braces are used in place of the integral so that we write

$$\Delta\hat{\epsilon}_j = \left\{\Delta S_{jk}d\left(\frac{\hat{\sigma}_k}{a_2}\right)\right\} \qquad (51)$$

$$\Delta\hat{\alpha}_j = \left\{\Delta\alpha_j d\left(\frac{\phi}{a_2}\right)\right\} \qquad (52)$$

A very good approximation for many high modulus, unidirectional, fiber composites and their resin matrices is

$$\Delta S_{jk} = k_{jk}\Delta S \qquad (53)$$

where k_{jk} is a constant symmetric, positive-definite matrix. That a single, time-dependent function is sufficient to characterize approximately all elements of the linear viscoelastic creep compliance matrix stems from the usually weak time-dependence of the resin's Poisson's ratio and the small ratio of resin modulus to fiber moduli (Schapery 1974).

By substituting eqns(43) and (53) into (51) we find

$$\Delta \hat{\epsilon}_j = \left\{ \Delta S d \left(\frac{a_5}{a_2} \frac{\partial h_2}{\partial \sigma_j} \right) \right\} \qquad (54)$$

where

$$h_2 \equiv \frac{1}{2} k_{ij} \sigma_i \sigma_j \qquad (55)$$

If we further assume

$$\rho_{ij} = k_{ij} \qquad (56)$$

so that both a_2 and a_5 depend on stress through only the one function h_2, then

$$\Delta \hat{\epsilon}_j = \left\{ \Delta S d \frac{\partial F_2}{\partial \sigma_j} \right\} \qquad (57)$$

where

$$F_2 \equiv \int_o^{h_2} \frac{a_5}{a_2} dh_2 \qquad (58)$$

Consider next the case in which there are no sources of time-dependence in $\hat{\sigma}$ and a_2 other than through stress. Then, if σ is applied at $t = 0$ and held constant thereafter (as in a creep test), eqn(51) reduces to

$$\Delta \hat{\epsilon}_j = \Delta S_{jk}(\psi) \hat{\sigma}_k / a_2 \qquad (59)$$

Upon use of eqn(53) for the compliances, eqn(59) reduces to

$$\Delta \hat{\epsilon}_j = \frac{\Delta S}{a_2} k_{ji} \hat{\sigma}_i \qquad (60)$$

The viscoelastic component of strain in eqn(17) becomes

$$\Delta \epsilon_j = \frac{\Delta S}{a_2} \frac{\partial \hat{h}}{\partial \sigma_j} \qquad (61)$$

where

$$\hat{h} \equiv \frac{1}{2} k_{ij} \hat{\sigma}_i \hat{\sigma}_j \qquad (62)$$

and, for simplicity, we have assumed either $\Delta \alpha_j = 0$ or $\phi = 0$. If, in addition, eqn(43) for $\hat{\sigma}$ is used, then

$$\Delta \epsilon_j = \frac{\Delta S}{a_2} \frac{\partial (h_2 a_5^2)}{\partial \sigma_j} \qquad (63)$$

Neither eqn(61) nor (63) is based on eqn(56). If, however, this latter simplification is used (so that both a_2 and a_5 depend on stress only through h_2) and we define h_3 as

$$h_3 \equiv h_2 a_5^2 \qquad (64)$$

then

$$\Delta \epsilon_j = \Delta S \frac{\partial F_3}{\partial \sigma_j} \qquad (65)$$

where F_3 is the potential,

$$F_3 \equiv \int_o^{h_2} \frac{1}{a_2} \frac{dh_3}{dh_2} dh_2 \qquad (66)$$

If a_5 is constant, then without loss in generality we may use $a_5 = 1$, and eqn(64) gives $h_3 = h_2$.

Turning to the effect of ψ, let us suppose a_1 is constant when stress is constant. Then eqn(13) yields

$$\psi = t a_2 / a_1 \qquad (67)$$

Further, let us introduce the common observation that ΔS obeys a power law in ψ,

$$\Delta S = k \psi^z \qquad (68)$$

where k and z are positive constants. If a_1 depends on stress only through h_2 (just as a_2) then eqn(65) simplifies to

$$\Delta \epsilon_j = t^z \frac{\partial F_4}{\partial \sigma_j} \qquad (69)$$

in which

$$F_4 \equiv k \int_o^{h_2} \left(\frac{a_2}{a_1} \right)^z \frac{dF_3}{dh_2} dh_2$$
$$= k \int_o^{h_2} \left(\frac{a_2}{a_1} \right)^z \frac{1}{a_2} \frac{dh_3}{dh_2} dh_2 \qquad (70)$$

where $F_3 = F_3(h_2)$ through eqn(66). The experimental data reported by Lou and Schapery (1971) and by Mignery and Schapery (1991) on unidirectional glass/epoxy and carbon/epoxy composites, respectively, are characterized quite well by eqn(69).

So far, the discussion starting with eqn(59) has been based on the use of timewise constant stresses. However, it is not necessary for the stresses to be constant to obtain eqn(59), at least as a very good approximation. When the curvature $d^2 f/(logt)^2$ or $d^2 log f/d(log\, t)^2$ is small, where f represents $\Delta\hat{\epsilon}_j$, $\hat{\sigma}_k/a_2$ and ΔS_{jk}, then the error in eqn(59) is small (Schapery 1974); this equation is called a *quasi-elastic approximation*. Consequently, results like those in eqn(65) and (69), wherein the viscoelastic strain is a gradient of a potential, are *not* limited to constant stress.

Often the viscoplastic strain response to constant stress may also be expressed as a gradient of a potential function. Assuming that F_p and k_3 are independent of time when the stresses are constant, eqns(26), (32) and (33) yield

$$\eta_i = t^\beta \frac{\partial F_5}{\partial \sigma_i} \qquad (71)$$

where

$$F_5 \equiv \frac{k_3^{-\beta}}{\beta^{\gamma+1}} \int_o^{g_\sigma} \left(g_\sigma^{-\gamma/2} \frac{dF_p}{dg_\sigma} \right)^\beta dg_\sigma \qquad (72)$$

If k_3 varies with time due to, for example, transient temperature, moisture or aging, then

$$\eta_i = \psi_p^\beta \frac{\partial F_5}{\partial \sigma_i} \qquad (73)$$

where another reduced time,

$$\psi_p \equiv \int_o^t dt'/k_3 \qquad (74)$$

has been introduced. Equation (72) still applies to eqn(73), but without the factor $k_3^{-\beta}$. Equations (71) and (73) for viscoplastic strain are seen to be similar to that for viscoelastic strain, eqn(69).

Up to this point we have not explicitly introduced the specialized form of the matrices ρ_{ij} and k_{ij} that accompany the mechanical behavior underlying eqn(53). Let us do this now by considering a unidirectional composite in which elastic fibers are oriented in the x_1 direction. We assume the axial stress σ_1 is supported primarily by the fibers, so that the strain ϵ_1 is unaffected by stress history. This behavior, together with the symmetry of k_{ij}, implies,

$$k_{j1} = k_{1j} = 0 \text{ for } j = 1, \ldots 6 \qquad (75)$$

Similarly, the scalar functions which affect nonlinear behavior of the viscoelastic and viscoplastic strains, $\Delta\epsilon$ and η, respectively , will not depend on σ_1. Hence,

$$\rho_{1j} = \rho_{j1} = \beta_{1j} = \beta_{j1} = 0 \text{ for } j = 1, \ldots 6 \qquad (76)$$

An orthotropic material is characterized by nine independent compliances. Recognizing also that we may always use a normalization in which $k_{22} = 1$, it follows for this case that there are only $9 - 3 - 1 = 5$ independent elements of k_{ij}.

Further simplification is possible in the stress-dependent functions as well by introducing material symmetry. In Section 3.6 we discuss this simplification for the common situation of transverse isotropy. First, however, we shall discuss two specific models of viscoplastic strain for polymers.

3.3 *Lai's model for viscoplastic strain*

In a recent study of high density polyethylene (HDPE), Lai (1995) successfully characterized the nonlinear viscoelastic and viscoplastic behavior under constant and time-varying uniaxial load. This material is really a composite, consisting of amorphous and crystalline phases, with approximately 68 per cent crystallinity. An isotropic representation accounting for multiaxial stress states was also proposed, but the discussion here on Lai's model is limited to uniaxial stress. Essentially eqn(3) was used, in which recoverable viscoelastic strain was characterized by a hereditary integral like that in eqn(18), as detailed previously by Schapery (1969b). It should be added, however, that Lai referred to the viscoplastic or plastic component of strain as *pseudo* plastic since it recovered fully, but seven decades later than the viscoelastic component. He first used the sum of two integrals like eqn(18) to account for the viscoelastic and pseudo plastic strain components, each with different nonlinearities; but then, as a practical matter, replaced the second integral by the viscoplastic strain discussed next.

Using η here to denote the axial viscoplastic component of strain, Lai represented the stress history in terms of a piecewise constant stress history, in which a change from one constant stress to another is made at time $t_\ell(\ell = 1, 2, \ldots L)$. Then, the strain at time t during any constant-stress interval, $t_\ell < t < t_{\ell+1}$, is specified as

$$\eta = P\left(\sigma_\ell, \hat{t}_\ell + t - t_\ell \right) \qquad (77)$$

where σ_ℓ is the current, constant value of the axial stress and $\hat{t}_\ell$ is the *effective time*; $t_{\ell+1} - t_\ell$ is *not* necessarily small. The function $P(\sigma, t)$ is,

29

by definition, the viscoplastic strain in a creep test, i.e. when a constant stress σ is first applied at $t = 0$ and subsequently held constant for $t > 0$. It is assumed that η is a continuous function of time, with continuity existing even at stress-discontinuity times, t_ℓ. The effective time $\hat{t}_\ell$ is defined by the equation

$$\eta_\ell = P\left(\sigma_\ell, \hat{t}_\ell\right) \qquad (78)$$

where $\eta_\ell \equiv \eta(t_\ell)$; notice that when $t = t_\ell$ in eqn(77), then eqn(78) is recovered. By starting with experimental values of $P(\sigma, t)$ from creep tests, one may construct the entire viscoplastic strain history using eqns (77) and (78). Observe from eqn(77) that the only effect of prior strain history is through a time-shift, $\hat{t}_\ell$, which depends on the value of η at the beginning of the current time interval, $t = t_\ell$. This formulation is not only easy to use, but it also predicts very well Lai's measured viscoplastic strain under multiple-step and cyclic loading. Application to continuously varying stress is, of course, achieved by using small time increments, $t_{\ell+1} - t_\ell$.

Here we shall relate this formulation to the uniaxial version of eqn(26) and, through this, determine if the two models are equivalent. First, it is to be observed that, apart from a time-scale shift, the strain history during each constant-stress interval is the same as in a single-step creep test. This simplicity distinguishes the behavior from that predicted for viscoelastic behavior, as in eqn(18), where the effect of stress history is far more complex. Now, solve eqn(77) for the time argument, which we may write in the form,

$$\hat{t}_\ell + t - t_\ell = Q\left(\sigma_i, \eta\right) \qquad (79)$$

Differentiation of eqn(79) with respect to time over $t_\ell < t < t_{\ell+1}$ yields the strain rate,

$$\dot{\eta} = R\left(\sigma, \eta\right) \qquad (80)$$

where $R \equiv (\partial Q / \partial \eta)^{-1}$. This representation is equivalent to eqns(77) and (78) and to eqn(25) for uniaxial stress; note that boundedness of R assures continuity of η. A mechanical model analog of eqn(80) is a dashpot in which the viscosity is a function of stress and strain.

Equation (26), specialized to uniaxial stress, is a special case of eqn(80) because the dependence on σ and η enters as separate factors. Let us pursue this matter by examining the creep behavior of HDPE reported by Lai. Specifically, his tensile creep data obey the equation

$$P(\sigma, t) = f(\sigma) t^m \qquad (81)$$

where m is constant ($=0.5$) over most of the stress range studied, $0 < \sigma < 12\ MPa$, and $dm/d\sigma > 0$ for $\sigma > 12\ MPa$. Also, f is a positive, nonlinear function of stress for $\sigma > 0$. Thus,

$$Q = (\eta/f)^{1/m} \qquad (82)$$

and then

$$\dot{\eta} = \eta^{(1-1/m)} m f^{1/m} \qquad (83)$$

Referring to eqn(26), $a_3 \sim \eta^{1/m-1}$ and $\partial F_p / \partial \sigma \sim f^{1/m}$. Equation(26) is seen to be fully consistent with Lai's data when $\sigma < 12\ MPa$; for higher stresses, m is a function of stress. Thus, to account for this behavior under uniaxial or multiaxial stresses, a_3 should include dependence on one or more scalar measures of stress (e.g. eqn (28)) as well as on viscoplastic strain, eqn(27). With this interpretation of a_3, eqn(26) provides a generalization of Lai's model to anisotropic behavior and multiaxial stresses. Certainly, when necessary, one could also allow for dependence of F_p on h_η. Alternatively, eqn(25) provides an immediate generalization.

Finally, assuming m is constant, it is of interest to integrate eqn(83) for a general, time-dependent stress. If $\eta(0) = 0$, then

$$\eta = \left[\int_o^t f^{1/m} dt'\right]^m \qquad (84)$$

showing that the strain is a Lebsegue norm of f of order $1/m$ (Reddy and Rasmussen 1982). This result is simply the uniaxial version of eqn(33) with $\beta = m$.

3.4 Zapas and Crissman's model for viscoplastic strain

In studies of polyethylene under uniaxial tensile stress, Zapas and Crissman (1984) and Crissman (1986) proposed and successfully applied a viscoplastic strain representation of the form

$$\eta = \emptyset \left(\int_o^t g(\sigma) dt'\right) \qquad (85)$$

where $\emptyset$ is a function of the indicated integral of a function $g(\sigma)$. It was found that a power law for $\emptyset$ provided a good fit to the data, and thus eqn(84) applies with $g = f^{1/m}$. Similar to what Lai (1995) did recently for HDPE, Tuttle et al. (1995) characterized the mechanical behavior of carbon fiber/bismaleimide under uniaxial stress by adding Schapery's (1969b) single integral constitutive equation for viscoelastic strain (i.e. eqn(17) for uniaxial stress) to the Zapas-Crissman (1984) viscoplastic strain; they also found that η could be expressed in the power law form, eqn(84).

We may therefore conclude that the viscoplastic strain for the polymers studied under uniaxial stress is well-described by the strain-hardening model originally used for metals. This model for multiaxial loading is given by eqn(26); the author is not aware of any experimental investigations of its validity for unreinforced and reinforced plastics.

3.5 Tensorial order of the ISVs

So far it has not been necessary to specify explicitly the tensorial order of the ISVs. However, some consideration of this question is now required as background to the next section, which is concerned with material symmetry. Clearly η must be a second order tensor for eqn(3a) to be correct (when expressed using double index notation for stresses and strains). In constrast, there is no such requirement for ξ. For example, ξ may be simply a set of M-6 scalars (zeroth order tensors), or ξ may be (M-6)/6 sets of second order tensors. The tensorial order of the coefficients A_m must of course be the same as ξ, so that the resulting strain, eqn(3b), is a second order tensor. Furthermore, as previously discussed, without loss in generality we could omit η entirely and use instead a subset of ξ with infinite retardation times. By assuming the corresponding subset of the coefficients A_m is linear in stress, the resulting viscoplastic strains would be like η, but expressed as a sum of contributions from the subset of ξ (cf. eqn 3b).

Thus, let us suppose both viscoelastic and viscoplastic strains are expressed in terms of scalars ξ_m, and imagine that a change in each represents a specific, local molecular or microstructural rearrangement (cf. Section 4.2). The combined effect of all local changes on the strain tensor is expressed by eqn(3), in which the effect of material symmetry is contained in the quantities G_o, A_m and B_{mn} that appear in the Gibbs free energy, eqn(1), and in c_{mn} and a_1 eqn(6). The ISVs $\mathbf{S}$ may account for damage and other changes which are not small enough to use a second order expansion of G in $\mathbf{S}$. It is common practice to use second and higher order tensors to account for damage (e.g. Onat and Leckie 1988). However, another approach that has proven successful is the use of scalars, each of which accounts for a different family of damage; each may affect the material symmetry differently, as expressed in terms of changes in moduli or compliances. Park and Schapery (1994) and Schapery and Sicking (1995) used this approach for particle-filled rubber and fiber-reinforced plastic, respectively, wherein each S_q is a measure of the cumulative work required to produce the damage. In the next section we shall assume $\mathbf{S}$ is a set of scalars. This simplifies the characterization of nonlinear effects of stress, in that it may be done through invariants of only the stress tensor; otherwise, joint invariants of $\boldsymbol{\sigma}$ and $\mathbf{S}$ would also be needed (Spencer 1971).

3.6 Invariants for transverse isotropy

The various scalar functions which appear in the preceding sections are limited in their dependence on stresses and strains to dependence on invariants which, in turn, are determined by material symmetry. As is well known, for isotropic materials there are at most three invariants of stress or strain. For transversely isotropic materials, such as many unidirectional, fiber composites, there are an additional two stress or strain invariants. Rather than using the traditional invariants, for our purposes it is helpful to use instead the equivalent set of five invariants employed by Hashin (1980) in characterizing failure. In terms of single index notation for stresses, they are

$$H_1 = \sigma_1, \ H_2 = \sigma_2 + \sigma_3$$
$$H_3 = \sigma_4^2 - \sigma_2\sigma_3, \ H_4 = \sigma_5^2 + \sigma_6^2$$
$$H_5 = 2\sigma_4\sigma_5\sigma_6 - \sigma_2\sigma_5^2 - \sigma_3\sigma_6^2 \qquad (86)$$

where x_1, is the isotropy axis. The standard association with double index stresses has been used,

$$\sigma_1 = \sigma_{11}, \ \sigma_2 = \sigma_{22}, \ \sigma_3 = \sigma_{33}$$
$$\sigma_4 = \sigma_{23}, \ \sigma_5 = \sigma_{13}, \ \sigma_6 = \sigma_{12} \qquad (87)$$

The material functions introduced in earlier sections may depend on stress only through the five invariants $H_s(s = 1, 2, \ldots 5)$. Consider a scalar function h, say, that is at most of second order in stesses, and assume the fibers are elastic and are so relatively stiff in the x_1 direction that σ_1 has no effect on the viscoelastic or viscoplastic behavior. Then, the most general form of h that can affect this behavior is

$$h = \lambda_1 H_2 + \lambda_2 H_2^2 + \lambda_3 H_3 + \lambda_4 H_4 \qquad (88)$$

where the λ's are free constants. Explicity, in terms of stresses,

$$h = \lambda_1(\sigma_2 + \sigma_3) + \lambda_2(\sigma_2^2 + \sigma_3^2) +$$
$$(2\lambda_2 - \lambda_3)\sigma_2\sigma_3 + \lambda_3\sigma_4^2 + \lambda_4(\sigma_5^2 + \sigma_6^2) \qquad (89)$$

Taking $h = h_\sigma$, we may compare term-by-term

eqn(89) with eqn(28) to find $\lambda_1 = 0$ and

$$\beta_{22} = \beta_{33} = 2\lambda_2, \ \beta_{44} = 2\lambda_3$$
$$\beta_{23} = 2\lambda_2 - \lambda_3, \ \beta_{55} = \beta_{66} = 2\lambda_4 \qquad (90)$$

All other elements of β_{ij} vanish. We may always normalize h so that one coefficient is unity, say $\beta_{22} = 1$; in effect, this coefficient is absorbed into the material function, such as F_p in eqn(26). Hence, there are only *two* free constants, which we may take as the shear coefficients β_{44} and β_{55}. The coupling coefficient is then $\beta_{23} = 1 - \beta_{44}/2$.

The function h_σ, eqn(28), becomes for a three-dimensional stress state,

$$h_\sigma = \frac{1}{2}\Big[\sigma_2^2 + \sigma_3^2 + \beta_{44}\sigma_4^2 + $$
$$\beta_{55}(\sigma_5^2 + \sigma_6^2) + (2 - \beta_{44})\sigma_2\sigma_3\Big] \qquad (91)$$

Recalling that $[\gamma_{ij}] = [\beta_{ij}]^{-1}$ we find

$$\gamma_{44} = \beta_{44}^{-1}$$
$$\gamma_{55} = \gamma_{66} = \beta_{55}^{-1}$$
$$\gamma_{22} = \gamma_{33} = (1 - \beta^2)^{-1}$$
$$\gamma_{23} = \gamma_{32} = -\beta(1 - \beta^2)^{-1} \qquad (92)$$

where $\beta \equiv 1 - \beta_{44}/2$; all other elements of γ_{ij} vanish. For the case of plane stress ($\sigma_3 = \sigma_4 = \sigma_5 = 0$) there is only one free constant,

$$h_\sigma = \frac{1}{2}(\sigma_2^2 + \beta_{55}\sigma_6^2) \qquad (93)$$

Other quadratic stress functions, such as h_1 in eqn(44) and h_2 in eqn(55), must reduce similarly. It is simply a matter of replacing β_{ij} by ρ_{ij} or k_{ij}. Recall that the k_{ij} are coefficients that relate the transient components of the linear viscoelastic creep compliances to a single scalar function ΔS, eqn(53). By comparing eqns(55) and (91), we conclude that the only non-zero components are

$$k_{22} = k_{33} = 1, \ k_{44}, \ k_{55} = k_{66}$$
$$k_{23} = 1 - k_{44}/2 \qquad (94)$$

Although there are only two independent coefficients, the time-dependence is defined by a third quantity, ΔS, which is the transverse creep compliance ΔS_{22} since $k_{22} = 1$. It may be easily verified that the last of eqn(94) can be derived directly from the standard equation relating trans-

verse normal and shear compliances of linear media.

The form of the quadratic functions (e.g. eqn(91)) may be compared with that used by Mignery and Schapery (1991). Specifically, by starting with the assumption that nonlinear viscoelastic behavior may be characterized in terms of an invariant that is proportional to the square of the average octahedral shear stress in the matrix of a carbon fiber/rubber-toughened epoxy composite, they arrived at a scalar function expressed in their notation as,

$$\tau_s = \sigma_2^2 + \sigma_3^2 + 2c_1\sigma_2\sigma_3 + c_2\sigma_4^2 + c_3(\sigma_5^2 + \sigma_6^2) \quad (95)$$

(The c's in this equation are not to be confused with the c's used in Section 2.3.) For the elementary parallel-element model used to represent the microstructure,

$$c_2 = c_3 = c, \ c_1 = 1 - c/2 \qquad (96)$$

where

$$c = 3/(1 - \nu_e + \nu_e^2) \qquad (97)$$

and ν_e is an *effective* Poisson's ratio, defined as the ratio of the axial stress in the matrix to the transverse applied stress sum ($\sigma_2 + \sigma_3$); for an isotropic linear elastic matrix, ν_e is the Poisson's ratio; experimental results for a plane stress state ($\sigma_3 = \sigma_4 = \sigma_5 = 0$) yielded $c = 4$ (corresponding to $\nu_e = 0.5$) and $c_3 = 4.2$. In an earlier study of glass/epoxy in plane stress, Lou and Schapery (1971) used the octahedral stress model to obtain the form $\tau_s = \sigma_2^2 + c_3\sigma_6^2$, with $c_3 = 3.9$ (corresponding to $\nu_e = 0.35$). These two values of c_3 may be compared to the measured values of k_{66} in order to determine how well eqn(56) is satisfied, where $2h_1 \equiv \tau_s$ and thus $\rho_{66} \equiv c_3$. From Mignery and Schapery (1991) $k_{66} = 3.4$ (compared to $c_3 = 4.2$) and from Lou and Schapery (1971) $k_{66} = 4.4$ (compared to $c_3 = 3.9$); thus, at least as a crude approximation, eqn(56) is satisfied for plane stress.

It should be emphasized that the invariant in eqn(95) does not have to be associated with the octahedral shear stress in the matrix. Indeed, this form is that of eqn(89) in which, for a three-dimensional state of stress, there are two free parameters. In view of equation (90), the coefficients in eqn(95) must satisfy the condition,

$$c_1 = 1 - c_2/2 \qquad (98)$$

As noted previously, in plane stress there is only one free coefficient, c_3. Sun et al. used

32

such a quadratic stress function in studies of off-axis, unidirectional composites employing a plasticity model (Sun and Chen 1989, Sun and Rui 1990) and a viscoplasticity model (Yoon and Sun 1991 and Gates and Sun 1991); they used a so-called *overstress* viscoplasticity representation, which has a three-element mechanical analogue consisting of a dashpot in parallel with a slider element, both of which are attached to spring element E_o (cf. Fig. 1). The values of c_3 ranged from 2.5 for a carbon/epoxy composite to 4.0 for a boron/alumimum composite. Schapery (1995) reported a value of $c_3 = 3.7$ for time-independent behavior of a carbon/epoxy composite.

These and other plane-stress studies of unidirectional fiber composites support the use of quadratic stress invariants. However, experimential three-dimensional nonlinear behavior of fiber composites appears to be a relatively untouched area. It is known that plastics exhibit some sensitivity to pressure (e.g. Boyce et al. 1988), and therefore it may be necessary to allow for explicit dependence on the mean stress in composites. Within the context of eqn(88), mean stress or pressure appears through H_2; the axial fiber modulus is assumed to be relatively large, and therefore σ_1 docs not appear in pressure-sensitive terms. Even if $\lambda_1 = 0$, pressure effects enter the invariant h through H_2^2; however, H_2 may be needed either as a linear term in h or as an entirely separate argument of material functions, such as a_1, a_2, F_p etc. Terms of third and higher order in stress may be needed in some cases. Furthermore, the coefficients in the stress invariants need not be constant. They could depend on viscoplastic strain, temperature, etc., although most characterizations use constant values.

4 PHYSICAL MODELS

4.1 *Effect of distributed microcracking*

The problem of explicitly incorporating microcracking in a nonlinear viscoelastic, homogenized, constitutive model is a very difficult one. However, with a suitably simplified constitutive equation for the uncracked material, some explicit analytical results may be developed. Here, we shall refer to a micromechanical model developed by Schapery (1981) which uses a constitutive model that allows for a J-like integral in characterizing crack growth (Schapery 1984, 1990b). What is of interest here is to identify the material functions in the thermodynamic-based constitutive model in Section 2 that make it equivalent to the previously developed mechanics-based constitutive model; this comparison should be useful in developing more general models with microcracking. For simplic-

ity, thermal, moisture and aging effects are omitted here, although they were taken into account in the earlier work.

The constitutive equation without microcracking (Schapery 1981) may be obtained from eqn(3) by assuming $\eta = 0$, $a_1 = a_2$ (so that $\psi = t$), $a_5 = 1$ (so that $\Delta \epsilon = \Delta \hat{\epsilon}$) and then using eqn(57) for $\Delta \hat{\epsilon}$. Further, it was assumed that G_o is proportional to F_2. Then, eqn(3) becomes

$$\epsilon_i = \left\{ Sd\frac{\partial F_2}{\partial \sigma_i} \right\} \qquad (99)$$

for the uncracked composite, where $S \equiv S_o + \Delta S$ and S_o is an initial compliance. It was shown by Schapery (1981) that the effect of quasi-static crack growth on overall strain ϵ is taken into account by an equation that has the same form as eqn(99), but with F_2 depending on the instantaneous geometry of the cracks.

Let us now use the ISVs, designated by $\boldsymbol{S}$ in Section 2.5, to represent the effect of cracks. Referring to eqn(58), it is seen that all effects of microcracking enter through the one scalar, a_2; inasmuch as G_o is proportional to F_2, G_o is also a function of $\boldsymbol{S}$. The evolution eqn(37), with eqn(42) for f_q, is equivalent to that used previously (Schapery 1981) when crack speed obeys a power law in the viscoelastic J integral. With S_q taken as a change in crack surface area, this J integral is proportional to $-\partial G_o/\partial S_q$, and thus is proportional to the crack driving force f_q. The mechanics-based theory indicates that when J is expressed in terms of stresses (rather than strains) it is independent of viscoelastic functions. This behavior is consistent with that for f_q when eqn(42) is used. Recall that, in general, $f_q = -\partial G/\partial S_q$; but since $a_5 = 1$, viscoelastic effects enter f_q through only second order terms in $\boldsymbol{\xi}$, which are neglected in the context of the thermodynamic model of Section 2. Thus, the simplified thermodynamic constitutive equation is fully consistent with the mechanics-based model with microcracking.

4.2 *Molecular models*

Some molecular models for nonlinear viscoelastic behavior of polymers exist; two will be mentioned here. Although model-to-model details are significantly different, there seems to be general agreement that changes in the so-called free volume, which are voids associated with molecular packing irregularities (Ferry 1980), is a major source of nonlinearity in polymers below their glass transition temperature. Knauss and Emri (1987) and Losi and Knauss (1992) accounted for nonlinear behavior of glassy polymers through the dependence of the time-scale factor on free volume, where the latter was assumed to be linear in the continuum dilatation. Their three-dimensional constitutive equations are identical to those for linear viscoelastic behavior of isotropic materials except for the time scale factor (which is a_1 in eqns(6) and (13)). Emri and Pavsek (1992) studied moisture effects in the context of this free volume model. Schapery (1969a) observed that volume creep and recovery of several plastics subjected to pressure loading (Findley et al. 1967) are consistent with the free volume theory, in that the data imply a_1 is an increasing function of pressure; see also Fillers and Tschoegl (1977) for a study on the effect of pressure on reduced time. Hasan and Boyce (1995) developed a nonlinear viscoelastic/viscoplastic constitutive model for glassy polymers by accounting for an evolving distribution of activation energies; conceptually, they associated a local increase in free volume with a decrease in the local activation energy barrier to a local inelastic shear transformation or rearrangement.

Here we shall incorporate the free energy eqn(1) into a thermal activation model for evolution of ISVs. This free energy and the ISVs are different from those used by Hasan and Boyce (1995), and thus the constitutive equations are different. On the other hand, through appropriate simplifications and choice for a_1, the Knauss-Emri-Losi model may be recovered. It is intended in this brief study to illustrate the connection between the phenomenological thermodynamic model of Section 2 and results from the fundamental notion that local molecular rearrangements are thermally activated (e.g. McClintock and Argon 1966).

The molecular rearrangements responsible for inelastic behavior at low-to-moderate strains were assumed by Boyce et al. (1988) to be segment rotations. In more recent work Hasan and Boyce (1995) refer more generally to local shear transformations. There is no need here to identify specific local rearrangements or structural changes. Rather, we simply employ the thermal activation theory in a form that is similar to the starting point used by Hasan and Boyce (1995),

$$\dot{\xi} = \omega \left[e^{-\Delta G_f/kT} - e^{-\Delta G_b/kT} \right] \qquad (100)$$

where ω is proportional to the attempt frequency and to the number of sites that contribute to $\dot{\xi}$, k is the Boltzmann constant and T is absolute temperature. Also, ΔG_f and ΔG_b are the activation free energies for forward and backward rearrangements, respectively. These quantities are the thermal energies needed for the rearrangements. For notational simplicity we have omitted the index m; eqn(100) should be interpreted as applying to each ξ_m. The activation energies are

$$\Delta G_f = G_e - f\Delta\xi_f v$$
$$\Delta G_b = G_e - f\Delta\xi_b v \qquad (101)$$

where $f \equiv -\partial G/\partial\xi$ is the thermodynamic force, $\Delta\xi_f$ and $\Delta\xi_b$ are the very small changes in ξ associated with each local rearrangement step. Also, v is the local volume of material involved in the step, G_e is the activation energy when $f = 0$, and $\Delta\xi_f = -\Delta\xi_b \equiv \Delta\xi$, where $\Delta\xi$ is a very small positive constant. The quantity $f\Delta\xi$ is the decrease in Gibbs free energy that accompanies the forward rearrangement, thus decreasing the required thermal energy. When the material is locally in an (essentially) equilibrium state, i.e. when $f \simeq 0$ for one or more ISV, rearrangements occur whenever the local thermal energy exceeds the *equilibrium* value G_e; but in this case forward and backward rearrangements are equally likely, and thus $\dot{\xi} \simeq 0$.

Upon introducing the index m and using eqn (101), we obtain (m not summed),

$$\dot{\xi}_m = W_m\sinh(\lambda_m f_m) \qquad (102)$$

where

$$W_m \equiv 2\omega_m e^{-G_{em}/kT}$$
$$\lambda_m \equiv v_m\Delta\xi_m/kT \qquad (103)$$

This result is to be compared with eqn(6), which is linear in f_m. Thus, in the context of the theory in Section 2.3, we assume f_m is small enough to permit replacement of the *sinh* function by its argument, so that (m not summed),

$$\dot{\xi}_m = W_m\lambda_m f_m \qquad (104)$$

The phenomenological eqn(6) and eqn(104) are brought into full agreement by using in place of c_{mn} a diagonal matrix with positive, constant elements c_m, and (m not summed),

$$\frac{c_m}{a_1} = 2\omega_m v_m\Delta\xi_m e^{-G_{em}/kT}/kT \qquad (105)$$

Observe that Onsager's principle does not have to be invoked to achieve symmtry of c_{mn}. Guided by experimental results we want this equation to be satisfied; it leads to thermorheological simplicity and related simplicity in nonlinear and aging behavior. Obviously, then, any variation in the right side must be the same for all ISVs, as expressed in terms of changes in a_1. For example, if we make the reasonable assumption that the product $\omega_m v_m\Delta\xi_m$ is constant, all ISVs must have the same equilibrium activation energy G_e, and

$$a_1 = kTe^{G_e/kT} \qquad (106)$$

where a_1 may be normalized so that it is unity at a reference state. Any experimentally measured dependence of a_1 on stress invariants, age, temperature, etc. (which, at least in part, may be due to changes in the free volume) must be reflected in changes in the equilibrium activation energy G_e. Alternatively, if a molecular model provides a specific function G_e, eqn(106) enables a_1 to be found.

In Section 2.3 it was assumed that a_1 is independent of $\boldsymbol{\xi}$. Now we can show that this independence is implied by eqn(105). Consider the thermodynamic force f_m, which is in eqn(9). If we suppose that $\boldsymbol{\xi}$ is the ISV set that diagonalizes both c_{mn} and B_{mn}, then upon using eqn(11) (m not summed),

$$f_m = A_m - a_2 C_m \xi_m \qquad (107)$$

where the C_m are positive constants. The equilibrium condition $f_m = 0$ gives

$$\xi_m = A_m / a_2 C_m \qquad (108)$$

Thus, even if the equilibium activation energy is given as a function of $\boldsymbol{\xi}$, eqn(108) enables $\boldsymbol{\xi}$ to be eliminated in favor of stress, temperature, etc.

It should be emphasized that the scalar G_e may depend on invariants of not only stress, but also on invariants of strain and strain history as well as temperature, temperature history, etc. Indeed, mechanical and thermal histories may affect the instantaneous free volume, and thus affect G_e; recall that G_e is the activation energy for those ISVs in which the conjugate f_m vanishes, and not all may vanish at any one time. In order to recover the Knauss-Emri-Losi free volume model, we would express a_1 (or equivalently G_e) in terms of macroscopic dilatation and temperature history.

Hasan and Boyce (1995) introduced a probability density function for a distribution of (equilibrium) activation energies in order to characterize nonlinear viscoelastic behavior. This function was expressed in terms of evolving ISVs. The global shear strain rate was then expressed as a sum of local strain rates over all activation energies. Here we have used a single equilibrium energy G_e, together with a second order Gibbs function and eqn(3), to provide the macroscopic strains as a sum over all ISVs. They used one thermodynamic force, and assumed it is proportional to the applied shear stress.

The present model uses many local forces f_m, each of which is a work conjugate of each local rearrangement variable ξ_m. These forces may vanish even when the body is under external stress $\boldsymbol{\sigma}$,

giving eqn(108). Conversely, when $\boldsymbol{\sigma} \equiv \mathbf{0}$ these forces are not necessarily zero; note that eqn (16) gives $A_m = \alpha_m \phi$ when $\boldsymbol{\sigma} = \mathbf{0}$, so that the resulting f_m may be interpreted as a local, internal force or stress due to thermal or moisture expansion.

These observations may be easily illustrated by using the mechanical model in Fig. 1. The force f_m is that acting on the m th dashpot, and ξ_m is (proportional to) the strain of the m th Voigt unit. Equation (108) is simply this strain when this unit's spring carries the entire axial load on the model. In terms of a local transformation site in polymers, the spring represents the elastic resistance of surrounding material to increases in ξ_m; the force on the spring is the so-called backstress, as given by the term $a_2 C_m \xi_m$ in eqn(107). Inasmuch as each spring exhibits thermal and moisture expansion characteristics, f_m may not vanish when $\sigma = 0$. On the other hand the equilibrium state of the free dashpot η_o (which provides the viscoplastic strain) exists only when $\sigma = 0$. The equilibrium activation energy associated with the material represented by this dashpot may depend on viscoplastic strain, which is accounted for in eqns(25) and (26) by changes in a_1^∞ and a_3, respectively; in this case an equation like (108) is not available to eliminate the ISV because the C value is zero.

Finally, we observe that eqn(102) provides a means for generalizing the constitutive model to allow for large thermodynamic forces. This equation is qualitatively similar to that used for $\dot{S}_q$, eqn(37), which was motivated by crack growth behavior. Let us briefly pursue this generalization by supposing there is a subset of $\boldsymbol{\xi}$ for which the thermodynamic force is $f_m = A_m$; i.e., there is no backstress, and thus these ISVs contribute to the viscoplastic strain, say $\Delta \epsilon^p$, as in the linearized version, eqn(25). Suppose further that $\Delta \epsilon^p$ is continuous in time even when $\boldsymbol{\sigma}$, T and M are discontinuous. Then, $A_m = C_{mj} \sigma_j$ for each ξ_m (where the C_{mj} are constants), and eqns(3b) and (102) yield

$$\Delta \dot{\epsilon}_i^p = \frac{\partial A_m}{\partial \sigma_i} \dot{\xi}_m = \sum_m \frac{\partial A_m}{\partial \sigma_i} W_m \sinh\left(\lambda_m A_m\right) \qquad (109)$$

Supposing further that $G_{em} \equiv \hat{G}_e$ (where $\hat{G}_e$ is the equilibrium activation energy for the present subset ξ_m) and that ω_m and $v_m \Delta \xi_m$ are constant, eqn(109) may be written in the same form as eqn(26); viz.,

$$\Delta \dot{\epsilon}_i^p = \frac{1}{a_3} \frac{\partial F_p}{\partial \sigma_i} \qquad (110)$$

where now

$$F_p = \sum_m \frac{2\omega_m}{v_m \Delta \xi_m} [\cosh \ (\lambda_m A_m) - 1] \qquad (111)$$

and

$$a_3 \equiv e^{\hat{G}_e/kT}/kT \qquad (112)$$

Of course we can always normalize a_3 so that it is unity at some reference state. It is seen that the nonlinear thermal activation model expresses the viscoplastic potential F_p as a sum of contributions from local rearrangements, and also relates the co-efficient a_3 to physically-based parameters just as for a_1 in eqn(106).

5 CONCLUSIONS

Constitutive equations for nonlinear viscoelastic and viscoplastic behavior of polymers and poly-meric composites have been discussed. Consider-able attention was given to various types of re-alistic simplifications to make them practical for experimental characterization and structural anal-ysis purposes. Specifically, the hereditary charac-teristics appear through the use of master creep functions in convolution integrals, eqns(18) and (19). All nonlinear, multiaxial stress effects enter only through scalar functions. There is a consider-able amount of experimental data which indicates that these functions depend primarily on an in-variant that is quadratic in stresses; a second in-variant, the mean stress, may be needed in some cases to account for pressure effects. Moreover, for simple stress histories the strains often may be expressed in terms of gradients of stress po-tentials (e.g. eqns(26), (61), (69), and (99). The constitutive equations are consistent with certain physical models for microcracking and molecular-scale inelastic rearrangements or transformations, as discussed in Section 4.

Nevertheless, existing models of time-dependent behavior (including chemical and physical aging) and experimental data are still too limited to have real confidence in making predictions of behavior at high stress levels many years into the future, given a characterization based on data collected over a year or less. One must therefore use less than optimum structural designs to make up for the lack of understanding and validated models. It is encouraging that this problem is well-recognized and that there are now many researchers address-ing and making significant progress on important issues involved in durability analysis.

ACKNOWLEDGMENT

Sponsorship of this research by the Office of Naval Research, Ship Structures & Systems, S & T Divi-sion, and the National Science Foundation through the Offshore Technology Research Center, is grate-fully acknowledged.

REFERENCES

Boyce M.C., Parks, D.M. and Argon, A.S. 1988. Large inelastic deformation of glassy polymers. Part I. Rate dependent constitutive model. *Mech. Mat.* 7:15-33.

Brouwer, R. 1986. Nonlinear viscoelastic characterization of transversely isotropic fibrous composites under biaxial loading. Ph.D. Thesis. Free University of Brussels.

Crissman, J.M. 1986. Creep and recovery behavior of a linear high density polythelene and an ethylene-hexene copolymer in the region of small uniaxial deformations. *Polymer Eng. Sci.* 26:1050-1059.

Emri, I. and Pavsek, V. 1992. On the influence of moisture on the mechanical properties of polymers. *Materials Forum.* 16:123-131.

Ferry, J.D. 1980. *Viscoelastic properties of polymers.* New York: John Wiley & Sons, Inc. Third Ed.

Fillers, R.W. and Tschoegl, N.W. 1977. The effect of pressure on the mechanical properties of polymers. *Trans. Soc. Rheol.* 21(1):51-100

Findley, W.N., Reed, R.M. and Stern, P. 1967. Hydrostatic creep of solid plastics. *J. App. Mech.* 34:895-904.

Fung, Y.C. 1965. *Fundamentals of solid mechanics.* Englewood Cliffs, NJ. Prentice-Hall, Inc.

Gates, T.S. and Sun, C.T. 1991. Elastic/viscoplastic constitutive model for fiber reinforced thermoplastic composites. *AIAA J.* 29: 457-463.

Hasan, O.A. and Boyce, M.C. 1995. A constitutive model for the nonlinear viscoelastic viscoplastic behavior of glassy polymers. *Polymer Eng. Sci.* 35:331-344.

Hashin, Z. 1980. Failure criteria for unidirectional fiber composites. *J. Appl. Mech.* 47:329-334.

Hult, J.A.H. 1966. *Creep in engineering structures*. Waltham Ma: Blaisdell.

Knauss, W.G. and Emri, I. 1987. Volume change and the nonlinearly thermoviscoelastic constitution of polymers. *Polymer Eng. Sci.* 27:86-100.

Lai, J. 1995. Non-linear time-dependent deformation behavior of high density polythylene. Ph.D. Thesis. Delft: Delft University Press.

Lamborn, M.J. and Schapery, R.A. 1993. An investigation of the existence of a work potential for fiber-reinforced plastic. *J. Comp. Mat.* 27(4): 352-382.

Losi, G.U. and Knauss, W.G. 1992. Free volume theory and non-linear thermoviscoelasticity. *Polymer Eng. Sci.* 32(8): 542-557

Lou, Y.C. and Schapery, R.A. 1971. Viscoelastic characterization of a nonlinear fiber-reinforced plastic. *J. Comp. Mat.* 5: 208-234.

McClintock, F.A. and Argon, A.S. 1966. *Mechanical behavior of materials*. Reading Ma: Addison-Wesley.

McKenna, G.B. 1994. On the physics required for prediction of long term performance of polymers and their composites. *J. Research of the National Institute of Standards and Technology*. 99: 169-189.

Mignery, L.A. and Schapery, R.A. 1991. Viscoelastic and nonlinear adherend effects in bonded composite joints. *J. Adhesion*. 34:17-40.

Onat, E.T. and Leckie, F.A. 1988. Representation of mechanical behavior in the presence of changing internal structure. *J. Appl. Mech.* 55:1-10.

Park, S. and Schapery, R.A. 1994. A thermoviscoelastic constitutive equation for particulate composites with damage growth. *Proc. 1994 Jannaf Structures and Mechanical Behavior Subcommitte Meeting*. Chem. Prop. Info. Agency Pub. 617. p.131-144.

Reddy, J.N. and Rasmussen, M.L. 1982. *Advanced engineering analyis*. New York: John Wiley & Sons.

Rice, J.R. 1971. Inelastic contitutive relations for solids: an internal variable theory and its application to metal plasticity. *J. Mech. Phys. Solids*. 19:433-455.

Schapery, R.A. 1964. Application of thermodynamics to thermomechanical, fracture and birefringent phenomena in viscoelastic media. *J. Appl. Phys.* 35:1451-1465.

Schapery, R.A. 1966. A theory of nonlinear thermoviscoelasticity based on irreversible thermodynamics. *Proc. 5th U.S. Nat. Cong. Appl. Mech.* New York: ASME. 511-530.

Schapery, R.A. 1969a. Further development of a thermodynamic constitutive theory: stress formulation. *Purdue Univ. Report No. AA & ES*. 69-2.

Schapery, R.A. 1969b. On the characterization of nonlinear viscoelastic materials. *Polymer Eng. Sci.* 9:295-310.

Schapery, R.A. 1974. Viscoelastic behavior and analysis of composite materials. *Mechanics of Composite Materials*. Edited by G.P. Sendeckyj. Academic Press: 85-168.

Schapery, R.A. 1981. On viscoelastic deformation and failure behavior of composite materials with distributed flaws. *Advances in Aerospace Structures and Materials*. New York: ASME. AD-01: p.5-20.

Schapery, R.A. 1984. Correspondence principles and a generalized J integral for large deformation and fracture analysis of viscoelastic media. *Int. J. Fracture*. 25: 195-223.

Schapery, R.A. 1990a. A theory of mechanical behavior of elastic media with growing damage and other changes in structure. *J. Mech. Phys. Solids*. 38(2): 215-253.

Schapery, R.A. 1990b. On some path independent integrals and their use in fracture of nonlinear viscoelastic media. *Int. J. Fracture*. 42:189-207.

Schapery, R.A. 1991. Simplifications in the behavior of viscoelastic composites with growing damage. *IUTAM Symposium on Inelastic Deformation of Composite Materials*. G.J. Dvorak (ed). New York: Springer-Verlag. p. 193-214.

Schapery, R.A. 1992. On nonlinear viscoelastic constitutive equations for composite materials. *Proceedings of VII International Congress on Experimental Mechanics*. Las Vegas. p.9-21.

Schapery, R.A. 1994. Nonlinear viscoelastic constitutive equations for composites based on work potentials. *Mechanics USA*. A.S. Kobayashi (ed). *Appl. Mech. Rev.* 47:S269-275.

Schapery, R.A. 1995. Prediction of compressive strength and kink bands in composites using a work potential. *Int. J. Solids Structures.* 32(6/7): 739-765.

Schapery, R.A. and Sicking, D.L. 1995. On nonlinear constitutive equations for elastic and viscoelastic composites with growing damage. *Mechanical Behavior of Materials.* A. Bakker (ed). Delft: Delft University Press. p.45-76.

Skontorp, A. 1995. Isothermal high-temperature oxidation, aging and creep of carbon-fiber/polyimide composites. Ph.D. Dissertation, Dept. of Mech. Eng., Univ. of Houston.

Spencer, A.J.M. 1971. Theory of invariants. *Continuum Physics Vol. I: Mathematics.* A.C. Eringen (ed). New York: Academic Press.

Struick, L.C.E. 1978. *Physical aging in amorphous polymers and other materials.* Amsterdam: Elsevier.

Sullivan, J.L. 1990. Creep and physical aging of composites. *Comp. Sci. Technol.* 39: 207-232.

Sun, C.T. and Chen, I.L. 1989. A simple flow rule for characterizing nonlinear behavior of fiber composites. *J. Comp. Mat.* 23: 1009-1020.

Sun, C.T. and Rui, Y. 1990. Orthotropic elasto-plastic behavior of AS4/PEEK thermoplastic composite in compression. *Mech. Mat.* 10: 117-125.

Tuttle, M.E., Pasricha, A. and Emery, A. 1995. The nonlinear viscoelastic-viscoplastic behavior of IM7/5260 composite subjected to cyclic loading. *J. Comp. Mat.* In press.

Weitsman, Y. 1990. A continuum diffusion model for viscoelastic materials. *J. Phys. Chem.* 94(2): 961-968.

Yoon, K.J. and Sun, C.T. 1991. Charn of elastic-viscoplastic properties of an AS4/PEEK thermoplastic composite. *J. Comp. Mat.* 25:1277-1226.

Zapas, L.J. and Crissman, J.M. 1984. Creep and recovery behavior of ultra-high molecular weight polyethylene in the region of small uniaxial deformations. *Polymer.* 25:57-62.

Progress in Durability Analysis of Composite Systems, Cardon, Fukuda & Reifsnider (eds)
© *1996 Balkema, Rotterdam. ISBN 90 5410 809 6*

Creep and failure of fabric reinforced thermoplastics

O.S.Brüller
Technical University of Munich, Germany

ABSTRACT: Isothermal uniaxial creep experiments conducted under constant environmental conditions on glass or carbon fabric reinforced thermoplastics have shown that in the principal directions, warp and woof, the behavior of the material is linear viscoelastic. Under 45° a very strong nonlinearity has been observed. A theory of strength based on the constancy of the time-dependent strain energy seems to describe with good accuracy the failure of the investigated materials under uniaxial tensile loading.

1 INTRODUCTION

Fabric reinforced thermoplastics are relatively new materials with many useful properties which make them suitable for applications in almost domains of engineering. Their most important property is the high resistance, comparable with that of composite materials assembled from unidirectionally reinforced layers. Usually they are available in form of plates or sheets of relativ small thickness. Differing from the plates made of unidirectionally reinforced layers, due to the geometry of the woven reinforced fabrics, they exhibit time dependent properties even in the principal directions, warp and woof. Accordingly, their behavior is viscoelastic and, their failure is - as expected - time-dependent.

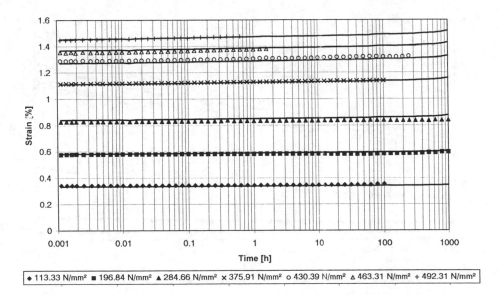

Fig. 1: Creep curves of PA6 / 48% vol. glass linen (warp)

2 VISCOELASTIC CHARACTERIZATION

Glass or carbon fibers usually used for the reinforcement of polymeric materials generally do not exhibit any time-dependent properties; their mechanical behavior is pure elastic. Consequently, unidirectionally reinforced layers practically behave also elastically if they are loaded in the direction of reinforcement.

Due to the viscoelastic properties of the poymeric matrix, in any other loading direction which does not coincide with the direction of the reinforcing fibers, the behavior of composites is time-dependent. In addition, various factors such as temperature, humidity and loading level play a very important role in the characterization of the

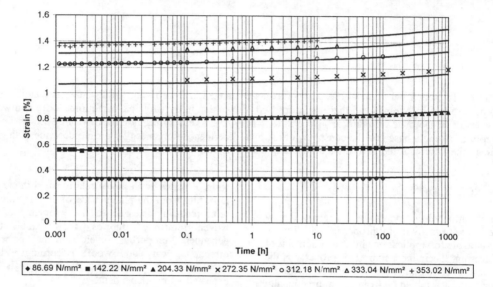

Fig. 2: Creep curves of PPS / 47% vol. glass linen (warp)

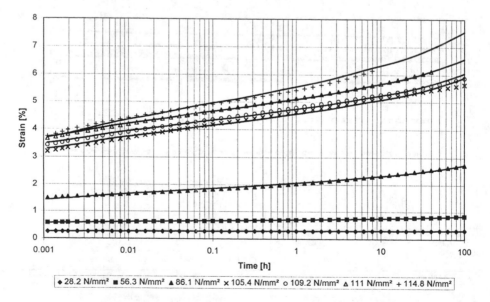

Fig. 3: Creep curves of PA6 / 48% vol. glass linen (45°)

mechanical response of such materials. Even at constant temperature, under non-changing environmental conditions, nonlinear (load-dependent) viscoelastic constitutive equations are necessary for the material characterization.

With woven fabric reinforced plastics, due to their structure, the situation is different. Here, the fiber threads are no more straight, they have the shape of a wave in both principal directions, warp and woof. Consequently, a constant load applied suddenly to a tensile specimen cut in one of the principal directions leads to an imediate elastic material response, followed by a transient, time-dependent increase of the strain, the creep response. While many papers deal with the elastic mechanical properties of such reinforced plastics, neglecting their time-dependent properties, only relatively few papers do present results including viscoelastic (time-dependent) effects.

The knowledge of the viscoelastic properties of woven fabric reinforced plastics, as exhibited for instance under creep loading, is absolutely necessary for a correct design of strucural parts made of such materials. As shown later, the transient part of the strain obtained under constant load represents a non-negligible part of the total strain. Investigation of the tensile creep compliance obtained at different stress levels in warp and woof direction, yields the important conclusion that the material response to the sudden loading step is linear elastic, while the transient behavior is linear viscoelastic, i.e. the creep compliance does not depend upon the level of the applied stress.

Such a kind of linear viscoelastic behavior can easily be described mathematically by using exponential series with a limited number of terms (Prony-Dirichlet series) for the characterization of the linear creep compliance $J(t)$ (Schapery 1969, Brüller 1987).

One has:

$$J(t) = J_0 + \sum_{i=1}^{m} J_i \, [\, 1 - exp\,(-t/\tau_i)\,] \qquad (1)$$

where J_0 = instantaneous compliance; J_i = linear parameters; τ_i = discrete relaxation times chosen one per decade in the experimental range.

As a first example, creep curves of a fabric reinforced PA6 / 48% vol. glass linen are shown in Fig. 1. The very good agreement between experimental values (symbols) and curves computed on the basis of a linear creep compliance can be seen. An explanation for such a kind of behavior could be based on the fact that with fabrics, the creep behavior in fiber direction is caused mainly by a geometrical change of the shape of the woven threads in loading direction. This very small change leads to the observed linear

viscoelastic behavior. Similar results have been obtained with some other glass or carbon fibers reinforced thermoplastics. A second example is depicted in Fig. 2 where creep curves of PPS / 47% vol. glass linen in warp direction are shown.

With specimens cut under 45° to the warp direction the situation is quite different. Tensile creep experiments, conducted on such specimens have shown a very strong nonlinearity of both, the instantaneous material response, and the time-dependent (transient) part of the creep compliance. For the mathematical description of this kind of behavior on can use the characterization method based on the introduction of some nonlinearity factors (Schapery 1969). One could find out, that only two stress-dependent nonlinearity functions were necessary to obtain an accurate mathematical description of the nonlinear creep curves (Brüller 1987). Hence, the relation for the nonlinear creep compliance is:

$$J(t, \sigma_0) = g_0(\sigma_0)\, J_0$$
$$+ g_t(\sigma_0) \sum_{i=1}^{m} J_i \, [\, 1 - exp\,(-t/\tau_i)\,] \qquad (2)$$

where $g_0(\sigma_0)$ and $g_t(\sigma_0)$ are the nonlinearity functions of the instantaneous response and transient creep function, respectively.

For the 'at once' determination of the linear parameters and nonlinearity function of the material a mixed iteration procedure on the basis of least square differences has been proposed by the author (Brüller 1987).

Fig. 3 shows creep curves of 45° specimens machined from the material PA6 / 48% vol. glass linen, while in Fig. 4 the results for the nonlinearity factor of the instantaneous response $g_0(\sigma_0)$ and those of the transient creep behavior $g_t(\sigma_0)$ are depicted as a function of the applied strain. Similar data are shown for PPS / 47% vol. glass linen (45°) in Fig. 5 and Fig. 6, respectively.

3 TIME-DEPENDENT FAILURE

The study of the failure of the investigated materials has shown that the time to fracture under creep loading depends on the level of the applied (constant) stress: the higher the stress, the shorter is the time to fracture.

Conventional strength theories define some function of strain or stress, the limiting value of which is determined empirically. When this limiting value is exceeded, it is assumed that the associated failure takes place. All these theories are generally time-independent and therefore not applicable to viscoelastic materials such as plastics.

Consequently, with time-dependent materials failure depends upon the loading history and one must asses the effect of loading history on the nature of a deformation process. Thus, one must consider the relative participation of the two primary mechanisms, viscous flow and elastic deformation. The participation of these two processes varies with the rate of deformation (Eirich 1966). The viscous processes require appreciable time whereas the elastic ones are quasi-instantaneous, therefore the higher the rate, the greater the elastic part. For this reason, attempts made for predicting failure must take into consideration the whole loading history of the

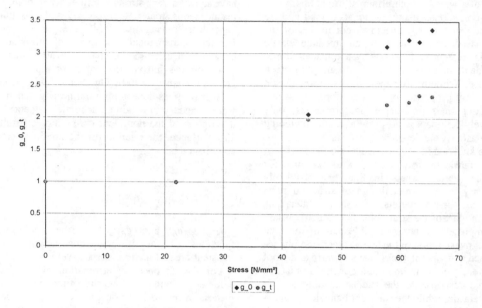

Fig. 4: Nonlinearity factors of PA6 / 48% vol. glass linen (45°)

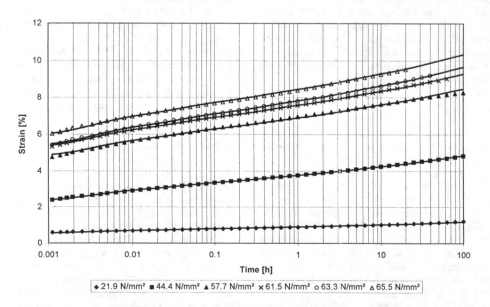

Fig. 5: Creep curves of PPS / 47% vol. glass linen (45°)

material. It is obvious, that the limits obtained are no more "strain limits" or "stress limits", but *energetical limits*, depending upon both, strain and stress. The point at which a certain limit of the material is exceeded, depends upon the loading history.

One of the first theories of strength in which time is also considered was developed by Reiner and Weissenberg. Following them (Reiner and Weissenberg 1939), the work done during loading by the external forces on a viscoelastic material is converted into a stored part (potential energy) and a dissipated part (loss energy), each of which may be divided into two other parts: the isotropic one, connected with volume changes and the deviatoric one, associated with shape changes. According to the Reiner and Weissenberg theory of strength, failure occurs when the stored deviatoric strain energy reaches a certain value, defined by Reiner as "resilience" and assumed to be a material constant.

Trying to explain the time-dependent failure of viscoelastic materials under stress relaxation, one can conclude that the Reiner-Weissenberg theory of strength is not able to explain this phenomenon, since during stress relaxation strain energy is *dissipated* and not stored.

Under creep loading conditions the situation is different. During the transient phase of creep, the time-dependent energy supplied by the external forces is transformed into a stored part and a dissipated part; one can show, that those two parts

of the transient creep strain energy are always equal (Brüller 1993). Investigations conducted by the author on several non-reinforced thermoplastics under creep loading have shown that failure occures when the time-dependent stored strain energy (which is equal - as stated - to the dissipated strain energy) reaches a certain value, which also can be assumed to be a material constant.

Thus, the total time-dependent strain energy in creep W_{tot} is equal to the sum of the stored strain energy at failure W_{sto} and the dissipated strain energy at failure W_{dis}, i.e.:

$$W_{tot} = W_{sto} + W_{dis} \tag{3}$$

where:

$$W_{sto} = W_{dis} \tag{4}$$

On the other hand the total time-dependent strain energy at failure is given by:

$$W_{tot} = \sigma_0 \, (\epsilon_f - \epsilon_{el}) \tag{5}$$

where ϵ_f = strain at failure, ϵ_{el} = strain after the instantaneous loading step, and σ_0 = acting constant stress.

Assuming that the total time-dependent strain energy at failure is a material constant, one obtains the strain at which failure is supposed to occur:

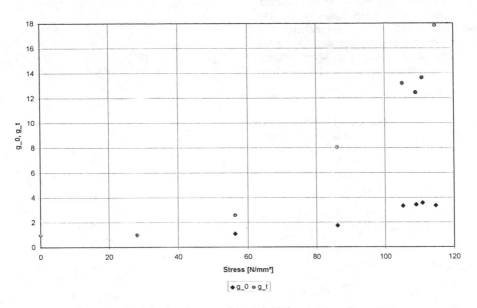

Fig. 6: Nonlinearity factors of PPS / 47% vol. glass linen (45°)

43

$$\epsilon_f = W_{tot}/\sigma_0 + \epsilon_{el} \tag{6}$$

This relation enables the prediction of the expected strain at which failure will occur under a given creep load. Knowing the failure strain ϵ_f and the corresponding creep curve, the time-to-fail can be easily predicted.

4 EXAMPLES

PPS / 47% vol. glass linen in warp direction is considered as an example. Fig. 7 shows the creep curves with the average fracture points.

Computation of the dissipated time-dependent strain energy at fracture (which is, as stated, half of the total time-dependent strain energy at fracture) for the three upper curves have led to the results $.105; .100; .105$ Nmm/mm^3, respectively. Consequently, the average time-dependent dissipated strain energy at fracture under creep loading conditions is for this material:

$$W_f = .103 \ Nmm/mm^3 \tag{7}$$

With this value one can construct the fracture curve under creep load of the material by connecting the fracture points of the individual creep curves. On the basis of the fracture curve one can predict the time-to fail under any desired creep load. For instance, the average time-to-fail under a load of 272.35 N/mm^2 was obtained as about 1000 h (by extrapolation of the creep curve).

Similar results have been obtained for 45° specimens machined from the same material. The average value of the time-dependent dissipated strain energy at fracture under creep loading conditions was here:

$$W_f = 1.44 \ Nmm/mm^3 \tag{8}$$

Similar results were obtained with other fabric reinforced thermoplastics, for instance with PA6 / 48% vol. glass linen in warp and 45° direction.

ACKNOWLEDGEMENT

The author wishes to acknowledge the contributions of M. Bau and H. Hagenbucher.

REFERENCES

Reiner, M. and K. Weissenberg 1939. *Rheol. Leafl.* 10:12-20.

Eirich, F.R. 1966. *High Speed Testing* 5:271-300

Schapery, R.A. 1969. *Polym. Eng. Sci.* 4:295-310

Brüller. O.S. 1987. *Polym. Eng. Sci.* 4:144-148

Brüller, O.S. 1993. On the Time-Dependent Failure of Polymers, pres. Europ. Symp. on Impact and Dynamic Failure of Polym. and Comp., Porto Cervo, (Sardinia), Italy.

Progress in Durability Analysis of Composite Systems, Cardon, Fukuda & Reifsnider (eds)
© *1996 Balkema, Rotterdam. ISBN 90 5410 809 6*

Prediction of delamination growth under cyclic loading using fracture mechanics

M. König & R. Krüger
ISD University of Stuttgart, Germany

M. Gädke
DLR Braunschweig, Germany

ABSTRACT: Results are presented which show that criteria based on elastic fracture mechanics are an appropriate tool for predicting delamination growth in carbon fibre reinforced epoxy laminates caused by fatigue loading. To check the validity of the approach local energy release rates along experimentally determined delamination fronts have been determined using the virtual crack closure method, implemented in a finite element analysis. Plots of computed energy release rate versus measured delamination progression per load cycle have been included in a Paris Law diagram as obtained experimentally using simple specimens for material characterization. The results lie well within the scatter band of the experimentally determined Paris Law.

1 INTRODUCTION

Delamination - the disbond of two adjacent fibre reinforced layers of a laminate - is a prevalent state of damage for fibre reinforced materials such as CFRE (carbon fibre reinforced epoxy). A recent survey on problems concerning composite parts of civil aircrafts [Miller 1994] shows that delamination, mainly caused by impact, presents 60% of all kind of damage observed. Up to now, empirically determined maximum strains arc used as design limit in the layout of components made of CFRE. This prevents failure due to delamination but for an optimal utilisation of the potential offered by this material as well as for the determination of inspection intervals during service, it is essential to be able to predict delamination growth.

Criteria based on elastic fracture mechanics are used to describe the delamination failure. Propagation therefore is to be expected when a function of the mixed mode energy release rates G_I, G_{II}, G_{III} along the delamination front locally exceeds a certain value G_c. We suppose that G_c is a material property independent from geometry and stacking sequence, however, dependent on the ply orientations of the layers adjacent to the plane of delamination. The goal of the investigations presented is to obtain information on the dependence of delamination growth on the mixed mode energy

release rates for cyclic loading and to verify the approach considered.

The specimen shown in Fig. 1, made of prepreg material (Ciba Geigy T300/914C), was designed to simulate a specific state of damage. The specimen with $[0_2/+45/0_2/-45/0/90]_S$ layup contains an artificial $10 \times 10\,mm$ square delamination at interface 3/4 . An additionally introduced ply cut serves to simulate fibre fracture. The specimen is one of several types of specimens that have been tested under cyclic loading to provide input to set up a predictive model for delamination onset and growth [Consortium 1994].

2 COMPUTATIONAL TOOLS

In order to achieve correct results for mode separation when computing energy release rates the use

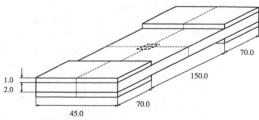

Figure 1: CFRE Specimen

of three-dimensional FE (Finite Element) models is required. A newly developed *layered element with eight nodes* formulated according to a continuum based three-dimensional (3D) shell theory [Parisch 1995] has been used for the three-dimensional models of the specimen. This element has been employed in order to reduce computing time, to prevent the elements from locking for small element thickness to element length ratios and to assure complete compatibility with the contactor and target elements that are used to avoid structural overlapping [Parisch 1989].

The most significant step for the current approach is the accurate computation of the distribution of the mixed mode energy release rates along arbitrarily shaped delamination fronts. It has been found that the virtual crack closure method is most favourable for the computation of energy release rates because the separation of the total energy release rate into the contributions by the different crack opening modes is possible in a straight forward manner [Rybicki 1977]. When using this method only one FE-computation is necessary for a given delamination front which is beneficial especially when solving large geometrically non-linear problems. Preliminary investigations employing DCB and ENF specimens assured the reliability of the technique used for computation of energy release rate distributions along straight and curved delamination fronts [Krüger 1993]. In addition, convergence studies for ENF and SLB specimens [Krüger 1994b] did not show the non-convergence of the virtual crack closure method associated with the oscillatory singularity as reported in the literature, e. g. in [Raju 1988, Hwu 1992]. This may be due to very small bimaterial constants of the considered interfaces [Gao 1992] or due to finite element meshes that are in the range of relatively constant results for G_I, G_{II} and G_{III} [Hwu 1992]. The issue will be investigated further.

3 EXPERIMENTAL PROGRAMME

In preparation of the experimental part of the program, specimens - as shown in Fig. 1 - have been cut from prefabricated plates made of prepreg material (T300/914C) with a stacking sequence of $[0_2/ + 45/0_2/ - 45/0/90]_S$. Artificial $10 \times 10\,mm$ square delaminations have been introduced at interface 3/4 during manufacturing by embedding a double foil of release film. For the R=0.1 fatigue loading a frequency of 3 Hz was selected in order to avoid any heat generation which might alter the material behaviour. Stress maxima were chosen to be 30% of the ultimate tensile strength (UTS). For this first setup delamination growth could not be observed even after several hundred thousand load cycles.

In order to induce even more severe damage and to force the delamination to grow the [0]-ply directly beneath the deliberate delamination was damaged by a $10\,mm$ cut through the fibres. Using C-scan for the detection, delamination growth was then observed at ply interface 4/5 while the deliberate delamination at interface 3/4 remained uneffected and did not start growing before the new delamination beneath had reached the same size and shape. At a stress level of 30% UTS no further delamination growth could be observed at all. Increasing the stress level to 40% and 50% UTS caused both delaminations to grow simultaneously.

4 ENERGY RELEASE RATES

Seven delamination fronts - illustrated in Fig. 2 - have been selected for 3D-analysis, i.e. a starter front s_0 with an assumed delamination length of $0.2\,mm$ and the fronts s_1, s_2, s_3, s_4 and s_7 which developed after 1,000 up to 7,000 load cycles respectively, further front s_∞ which had developed after several 10,000 load cycles (after the growing delamination had reached the same size and shape as the artificial one above).

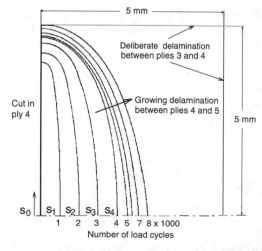

Figure 2: Delamination growth obtained from C-scans (only one quarter depicted)

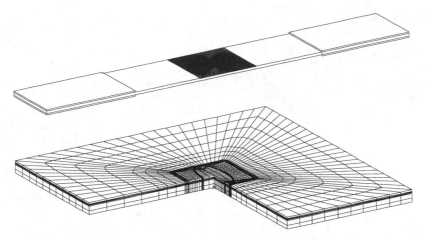

Figure 3: The specimen and FE-model for front s_4

Due to the structural unsymmetry of the geometrically symmetric specimen caused by the $\pm45°$ plies, models extending over the entire width of the specimen are necessary as shown in Fig. 3. The two top [0]-layers of the sublaminate have been grouped into one 3D-shell element, followed by one element which models the [45]-ply adjacent to the artificial square delamination. Two elements over the thickness were used for the cut [0]-layer, one for the [0]-ply to follow which is adjacent to the new growing delamination. The remaining 11 layers of the base laminate were joined into two elements of $[-45/0/90_2/0/-45]$ and $[0_2/+45/0_2]$ layup respectively. This configuration was used for the entire 3D-investigation, i.e. for all fronts.

First the behaviour of the assumed initial front s_0, which might be existing prior to any loading, having a delamination length of 0.1 mm to the left and right of the initial cut is considered in the present discussion. The deformed geometry is shown in Fig. 4 and an interpenetration of the cut [0]-ply with the ply above is noticeable. In all following computations this was prevented by using a contact processor that utilizes a contactor target concept applying the penalty method [Parisch 1989]. Results show that the contact between the delaminated surfaces occurs only locally (Fig. 5), resulting in a mode I opening which had been neglegted in previous 2D and 3D models [Krüger 1994a].

The distribution of the energy release rate along the front of the deliberate delamination (interface 3/4) is shown in Fig. 6. Along the entire front the energy releases rate components are far below any threshold values for delamination growth (see section 5) so that the contour is expected to remain constant. Significant peaks are being observed in the immediate vincinity of the ply cut ($s = 0.25$ and $s = 0.75$) due to the stress concentration caused by the cut. For the growing delamination (interface 4/5) a significant mode I contribution caused by a crack opening is noticeable as shown in Fig. 7. The total energy release rate $G_{tot} = G_I + G_{II} + G_{III}$ and the contributing modes are fairly constant along the major part of the front, their values being considerably higher than the experimentally obtained threshold values for delamination growth (G_{Ith}, G_{IIth}) so that propagation is to be expected along the entire front. Computed results are dropping towards the cut ($s = 0.25$ and $s = 0.75$) which indicates that crack propagation will be slower in this area. These results are in excellent agreement with experimental observations.

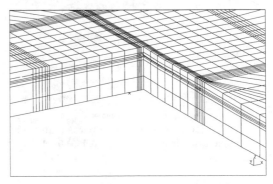

Figure 4: Detail of deformed geometry for front s_0 without contact processor

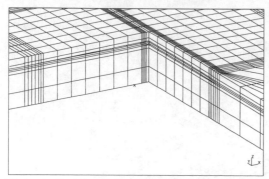

Figure 5: Detail of deformed geometry for front s_0 using contact processor

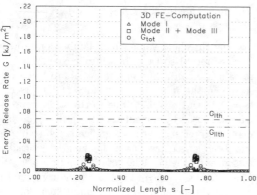

Figure 6: Distribution of energy release rate along the square delamination (interface 3/4) for 1st load cycles

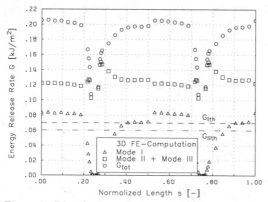

Figure 7: Distribution of energy release rate along the propagating front s_0 (interface 4/5) for 1st load cycles

Carefully observing the numerically predicted displacement behaviour for the growing delamination we notice that the cut [0]-ply first bends upward contacting the upper [45]-layer, however with increasing delamination growth it also tends to bend downward. For front s_7 and s_∞ therefore contact analysis becomes necessary also at interface 4/5 complicating the total FE-analysis by requireing carefully selected small load increments and larger computing time for the non-linear computation. The global (in effect linear) load deflection behaviour remains uneffected by the local non-linearities. As a typical example the deformed geometry for front s_4 is shown in Fig. 8.

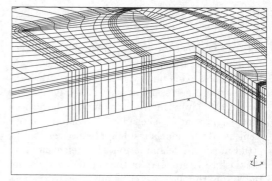

Figure 8: Detail of deformed geometry for front s_4

Looking at the distribution of the energy release rate along the artificial square delamination we notice a slight increase for higher number of load cycles (compare Fig. 9 with Fig. 6). Theshold values however are not reached until after some 10,000 load cycles when the growing delamination has reached the same size as the artificial one (see Fig. 12), explaining the fact that the deliberate delamination remains constant over the entire test sequence. For the propagating delamination front at interface 4/5 a significant mode I contribution caused by a crack opening is noticeable. The total energy release rate $G_{tot} = G_I + G_{II} + G_{III}$ is fairly constant along the entire front, its value being considerably higher than the experimentally obtained threshold values for delamination growth ($G_{I\,th}$, $G_{II\,th}$), as seen e.g. in Fig. 10, so that delamination growth is to be expected along the entire front. Irregularities occur only in the immediate vincinity of the cut ($s = 0.25$ and $s = 0.75$) caused by the extremely complex stress state resulting from the cut. The 3D-FE mesh used in this area is not fine enough to give accurate information so that val-

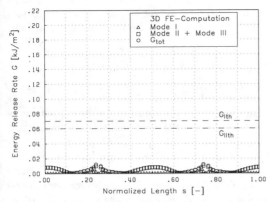

Figure 9: Distribution of energy release rate along the square delamination (interface 3/4) after 4,000 load cycles

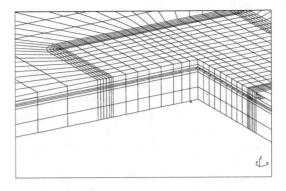

Figure 11: Detail of deformed geometry for front s_∞

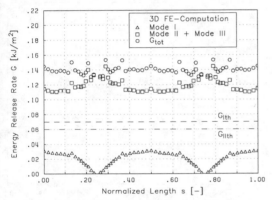

Figure 10: Distribution of energy release rate along the propagating front s_4 (interface 4/5) after 4,000 load cycles

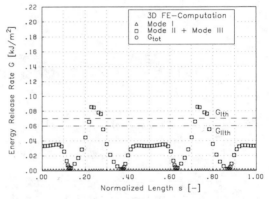

Figure 12: Distribution of energy release rate along the square delamination (interface 3/4) after several 10,000 load cycles

ues computed at the location of the cut and in its immediate vincinity should be disregarded, which has been done for the plots presented.

Taking a look at the deformed geometry for front s_∞ (Fig. 11) we notice that the cut [0]-ply is in multiple contact with the upper [45]-layer as well as with the [0]-ply underneath. Crack opening leading to mode I contributions is no longer observed. Corresponding distributions of the energy release rates show that the mode I values have dropped to zero (Fig. 12 and 13). Comparing with the threshold values a minor growth of front s_∞ should still be possible, this however could not be verified by the experiments. Increasing the load level up to 40% or 50% UTS raises the computed energy release rates well above the threshold values for both delaminations, yielding a prediction of simultaneous growth as observed experimentally.

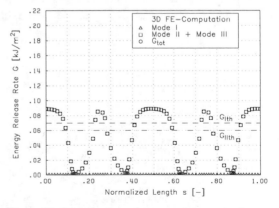

Figure 13: Distribution of energy release rate along the propagating front s_∞ (interface 4/5) after several 10,000 load cycles

In Fig. 14 the total energy release rates (G_{tot}) for the propagating front, i.e. along s_1, s_2, s_3, s_4 and s_7 are presented. It is obvious that (G_{tot}) is fairly constant along the individual fronts, dropping progessively with growing delamination which coincides with the decreasing growth rate observed in Fig. 2. As illustrated in Fig. 15 the mode I contribution decreases until it finally becomes negligible for front s_∞ (Fig. 13). Comparing the above results to those obtained using a quarter model [Krüger 1994a] we notice a remarkable difference, concluding that the unsymmetry caused by the ±45° plies has to be taken into account, yielding the necessity of a full width model. Additionally it becomes obvious that simply suppressing any mode I contribution to avoid layer interpenetra-

tion is insufficient and contact analysis becomes essential to catch the local deformation behaviour.

5 COMPARISON WITH PARIS LAW

Mode I and mode II failure under cyclic loading have been characterized for the considered material in [Prinz 1991]. For mode I characterization unidirectional DCB specimens were tested under R=0.1 loading. The results are shown in Fig. 16. Commonly known unidirectional ENF specimens and TCT (Transverse Crack Tension) specimens developed at DLR were used for mode II characterization under R=0.1 loading yielding the information shown in Fig. 17.

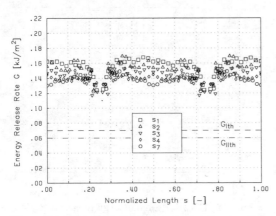

Figure 14: Distribution of energy release rates G_{tot} along propagating fronts s_1, s_2, s_3, s_4 and s_7

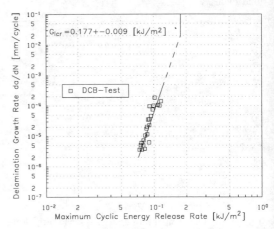

Figure 16: Paris Law for mode I, obtained using DCB-specimens (R=0.1)

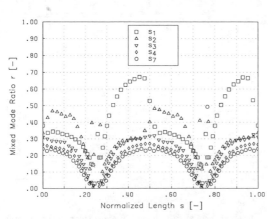

Figure 15: Distribution of mixed mode ratio $r = G_I/(G_{II} + G_{III})$ along the propagating fronts s_1, s_2, s_3, s_4 and s_7

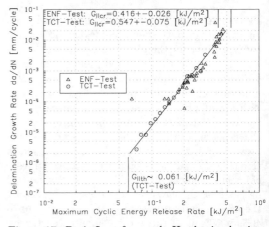

Figure 17: Paris Law for mode II, obtained using ENF- and TCT-specimens (R=0.1)

Threshold values for delamination growth have only be determined for mode II. Therefore, threshold energy release rates in both cases were defined as those values which would cause the delamination to grow 1 mm during one million load cycles $(da/dN = 10^{-6})$. These values have also been used in all of the previous diagrams. The results for mode I and mode II characterization were combined into one plot yielding Fig. 18.

Due to the fact that for the specimen discussed delamination growth occurs between two 0°-plies, plots of measured delamination progression per load cycle (da/dN-values) versus computed total energy release rates along the delamination fronts can be included in the above Paris Law diagram, as shown in Fig. 19.

For a front progressing from s_k to s_n the energy release rates used in the Paris Law plot were those obtained for the starter front s_k. Another possibility would be taking the average value from the computed energy release rates for front s_k and front s_n. Computational results lie well between the experimentally determined straight lines for mode I and mode II failure, the points with small mixed mode ratio r being closer to the failure line for mode II. The influence of the mixed mode ratio r on the location of points between the two Paris Law lines has to be investigated further.

6 CONCLUSIONS AND OUTLOOK

The application of elastic fracture mechanics criteria as a possible tool for predicting delamination growth under tension-tension fatigue loading has been tested. This approach requires the accurate knowledge of the distribution of the local energy release rate along the contour of an arbitrarily grown delamination. The virtual crack closure method has been found most favourable for computation of mixed mode energy release rates in connection with finite element modeling employing a layered 3D-shell element and elastic contact analysis.

Experiments with cyclic loading show that measured delamination progression fits into a Paris Law as obtained experimentally using simple specimens for material characterization.

Due to the observed mixed mode conditions, Mixed Mode Bending (MMB) tests become necessary to determine Paris Law parameters for a final verification of the approach. In addition, a test program is necessary for characterization of fracture toughness for interfaces with adjacent non-unidirectional layers [Davidson 1994].

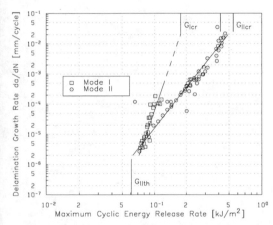

Figure 18: Paris Law from material tests (R=0.1)

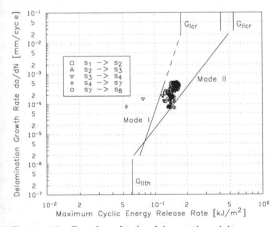

Figure 19: Results obtained in section 4 in comparison to Paris Law from material tests

REFERENCES

Project Consortium. *Brite Euram Project BE3444, Analytical and Experimental Approach to Cumulative Damage and Residual Strength Prediction for CFRE Composites.* CEC, Sept. 1990 – Aug. 1994.

Davidson, B.D., R. Krüger and M. König. Three Dimensional Analysis and Resulting Design Recommendations for Unidirectional and Multidirectional End-Notched Flexure Tests. Technical Report No. MAME-94-401, Department of Mechanical, Aerospace and Manufacturing Engineering, Syracuse University, 1994.

Gao, H., M. Abbudi and D.M. Barnett. Interfacial Crack-Tip Field in Anisotropic Elastic Solids. *J. Mech. Phys. Solids*, 40(2):393–416, 1992.

Hwu, C. and J. Hu. Stress Intensity Factors and Energy Release Rates of Delaminations in Composite Laminates. *Eng. Fracture Mech.*, 42(6):977–988, 1992.

Krüger, R. and M. König. Computation of energy release rates: A tool for predicting delamination growth under cyclic loading ? In David Hui, editor, *Proceedings of International Conference on Composites Engineering, ICCE/1,* New Orleans, pages 805–806, 1994.

Krüger, R., M. König, and T. Schneider. Computation of local energy release rates along straight and curved delamination fronts of unidirectionally laminated DCB- and ENF-specimens, AIAA-93-1457-CP.
In *The 34th SSDM Conference,* La Jolla, California, pages 1332–1342, 1993.

Krüger, R. Three Dimensional Finite Element Analysis of Multidirectional Composite DCB, SLB and ENF Specimens. ISD-Report No. 94/2, Institute for Statics und Dynamics of Aerospace Structures, University of Stuttgart, 1994.

Miller, A.G., D.T. Lovell and J.C. Seferis. The evolution of an aerospace material: Influence of design, manufacturing and in-service performance. *Composite Structures*, 27:193–206, 1994.

Parisch, H. A consistent tangent stiffness matrix for three-dimensional non-linear contact analysis. *Int. J. Num. Meth. Eng.*, 28:1803–1812, 1989.

Parisch, H. A continuum-based shell theory for nonlinear applications. *Int. J. Num. Meth. Eng.*, 35, 1995.

Prinz, R. and M. Gädke. Characterization of interlaminar mode I and mode II fracture in CFRP laminates. In *Proc. Int. Conf.: Spacecraft Structures and Mechanical Testing*, pages 97–102. ESA SP-321, 1991.

Raju, I.S., J.H. Crews and M.A. Aminpour. Convergence of Strain Energy Release Rate Components for Edge-Delaminated Composite Laminates. *Eng. Fracture Mech.*, 30(3):383–396, 1988.

Rybicki, E.F. and M.F. Kanninen. A finite element calculation of stress intensity factors by a modified crack closure integral. *Eng. Fracture Mech.*, 9:931–938, 1977.

Progress in Durability Analysis of Composite Systems, Cardon, Fukuda & Reifsnider (eds)
© *1996 Balkema, Rotterdam. ISBN 90 5410 809 6*

Frequency dependent electromagnetic sensing (FDEMS) for life monitoring of polymers in structural composites during use

D. Kranbuehl, D. Hood, L. McCullough, H. Aandahl, N. Haralampus & W. Newby
Departments of Chemistry & Applied Science, College of William and Mary, Williamsburg, Va., USA

M. Eriksen
Robit as, Billingstadt, Norway

ABSTRACT: Frequency dependent electromagnetic sensors, FDEMS, provide a sensitive, convenient, automated means for monitoring how polymers are influenced by their environment during use. FDEMS in situ sensors can be designed and calibrated to monitor changes in mechanical service life properties of polymer materials as for example in flexible pipes and other types of composites used in the production and transport of oil and gas. With a proper understanding of the type of polymer and the composition, pressure and temperature of the use environment, the FDEMS sensor output can be related to changes in modulus, maximum load and elongation at break. The technique has the following advantages over other monitoring techniques: nondestructive; increased accuracy/reproductivity; ultra sensitivity; in situ capability; remote sensing; automated.

1 INTRODUCTION

With the future expanded use of composites in extended use structures, such as oil pipelines bridges and aircraft with life times of 20 to 40 years, there is a need to develop an insitu health monitoring sensor capability. These polymer health sensing techniques need to involve small scale, short time test methods capable of monitoring large scale long time aging of performance properties. This area is already acute in the oil industry as it strives for more cost effective solutions, especially offshore. For the oil industry the use of polymer materials in various composite structures is an enabling technology making reduced weight and lower cost solutions possible. One of the already existing uses of polymer materials is as the fluid gas barrier in flexible pipes for transport of oil and gas from the ocean floor to the surface on offshore platforms. A critical need in using polymers in flexible pipes and in extending the application of polymers to offshore load bearing structures is a sensor method that can verify the suitability of polymer materials under realistic aging conditions. Even more important is the ability to monitor the change in the polymer materials performance properties insitu in the field under variable, unpredictable, degradation use conditions. Thus the ultimate objective is to develop effective sensing methods both for efficient

evaluation, selection and qualification of polymer materials for a particular extended use environment and for monitoring in situ the remaining life, rate of aging and projected replacement of the polymer structure during use in the variable, field environment. In this paper, we discuss the use of frequency dependent electromagnetic sensors, FDEMS, to monitor aging of the polyamide thermoplastic, nylon 11, in its commercial form Rilsan BESNO P40TL. For over a decade, this material has been used as the oil, gas barrier in flexible pipes on offshore platforms throughout the world and particularly in the North Sea. This paper addresses the ability of the FDEMS sensing technique to monitor degradation in mechanical stress-strain performance properties such as load at break, elongation at break as well as the breakdown in the polymers molecular weight.

2 BACKGROUND

In the past, frequency dependent electromagnetic sensing techniques made over many decades of frequency, Hz-MHz, have provided a sensitive, convenient automated means for characterizing the processing properties of thermosets and thermoplastics [1-6]. Through the frequency dependence of the impedance, the FDEMS sensing technique using a universal wafer thin Dek Dyne sensor, has been able to monitor chemical and physical changes throughout the entire cure process for a range of polymeric materials. FDEMS sensing techniques have the advantage of monitoring in situ the changes that occur in processing properties in the tool during fabrication.

The FDEMS technique and related lower frequency work has been shown to be effective for monitoring a variety of resin cure processing properties such as reaction onset, point of maximum flow, extent of reaction, T_g and reaction completion as well as the variability in processing properties due to resin age and exposure to moisture. The FDEMS technique has been shown to be able to monitor similar processing properties in thermoplastics such as T_g, T_m, recrystallization and solvent-moisture out-gassing. The FDEMS technique has the particular advantage over other chemical characterization measurements for being able to monitor these processing properties continuously and in situ as the resin changes from a resin of varying viscosity to a cross linked insoluble solid. Another advantage is that measurements can be made simultaneously on multiple samples or at multiple positions in a complex part and on polymers used as composites, adhesives, and coatings.

The advantage of FDEMS is that the FDEMS sensor can detect and monitor the process of "cure in reverse" or degradation. The FDEMS sensor is able to monitor the cure process of a polymer resin in situ during manufacture by detecting, on a molecular level, the changes in the ionic and dipolar mobility. During polymerization, the changes in the processing properties such as viscosity and the build-up in end-use properties are due to changes in molecular structure and the decreasing ability of the molecular ions and dipolar groups to move. Similarly irreversible degradative processes such as embrittlement (that is, increasing modulus and decreasing elongation-strain at break or max load) or cleavage of the crosslinking structure (that is, a decrease in modulus or load carrying capability) are both due to changes in the molecular state of the polymer resin. As in cure, these changes in molecular structure and mechanical properties are normally due to changes in the mobility of ions and dipoles at the molecular level. Thus, FDEMS sensing has the potential to be a highly sensitive means for in situ monitoring of the changes in mechanical and physical properties of polymeric materials due to aging.

Normally, irreversible degradation will take place in an environment that will also produce changes in the polymer that affect the FDEMS measurements. In the case of polymers used in flexible pipes these aging effects include uptake of hydrocarbons, uptake of water and loss of plasticizer, all of which will influence the FDEMS measurements. Use of FDEMS to quantify changes in mechanical properties does require a thorough understanding of such influences through test and calibrations in representative environments.

3 FDEMS THEORY

Frequency dependent measurements of the sensor's dielectric impedance are characterized by its equivalent capacitance, C, and conductance, G. C and G are used to calculate the complex permittivity, $\varepsilon^*(\omega)=\varepsilon'-\iota\varepsilon''$, where $\omega = 2\pi f$, f is the measurement frequency and C_o is the air replaceable capacitance of the sensor.

This calculation is possible when using a sensor whose geometry is invariant over all measurement conditions. Both the real and the imaginary parts of ε^*, the complex permittivity, can have an ionic and a dipolar component.

$$\varepsilon'(\omega)=\frac{C(\omega)\ material}{C_o}$$

$$\varepsilon''(\omega)=\frac{G(\omega)\ material}{\omega C_o}$$

$$\varepsilon' = \varepsilon'_d + \varepsilon'_i$$

$$\varepsilon'' = \varepsilon''_d + \varepsilon''_i$$

The dipolar component ε_d arises from rotational diffusion of molecular dipole moments or bound charge. The ionic component ε_i arises from the translational diffusion of charge. The dipolar term is generally the major component of the dielectric signal at high frequencies and in highly viscous media. The ionic compound dominates ε'' at low frequencies, low viscosities and/or higher temperatures.

Plots of the product of frequency (ω) multiplied by the imaginary component of the complex permittivity $\varepsilon''(\omega)$ make it relatively easy to visually determine when the low frequency magnitude of ε'' is dominated by the mobility of ions in the resin and when at higher frequencies the rotational mobility of bound charge dominates ε''. A detailed description of the frequency dependence of $\varepsilon^*(\omega)$ due to ionic, dipolar and charge polarization effects has been previously described [1,2].

4 INSTRUMENTATION

Frequency dependent complex impedance measurements are made using an Impedance Analyzer controlled by a microcomputer. Measurements are taken continuously throughout the entire aging process at regular intervals and converted to the complex permittivity, $\varepsilon^* = \varepsilon' - i\varepsilon''$. Measurements are made with a geometry independent Dek Dyne micro-sensor. The sensor is planar, 2.54 x 1.27 cm in area and 0.5 mm thick. This single sensor-bridge microcomputer assembly as shown in figure 1 is able to make continuous uninterrupted measurements of both ε' and ε'' over 10 decades in magnitude at all frequencies. The sensor is inert and has been used at temperatures of 400°C.

5 LIFE MONITORING OF MECHANICAL PROPERTIES

Fundamentally, mechanical and dielectric measurements are rooted in the same equation, that of a spring, force = constant times displacement. The proportionality between the force and the displacement, k for a spring, is the constant of interest, *i.e.*, F = kx. The modulus is a form of that proportionably constant. In dielectric experiments the same question is asked only the force is an electrical force E rather than a mechanical force F and the displacement is the change in position of dipolar molecules and ions. The proportionality constant between how far these molecular groups move for a given electric force is called the polarizability and given the symbol α where α is analogous to k. The dielectric permittivity ε is the material property form of this proportionably constant, α.

Dielectric permittivity, ε, is analogous to the mechanical modulus in its various uses. The fundamental difference between dielectric and mechanical force displacement measurement (in addition to the obvious fact that one uses a mechanical force and the other an electric force) is the fact that mechanical measurements look at displacements over macroscopic, visible distances. Dielectric measurements look at displacements over

molecular distances, ≈ 1 to 1000 Angstroms or 10^{-10} to 10^{-7} meters. *All* macroscopic dimensional changes *originate* from changes in the position of molecules over molecular distances, i.e., Angstroms 10^{-8} cm. Hence, all changes in the mechanical properties of a material must originate in changes in the ability of molecules to move on a molecular level. That is the first of several reasons why FDEMS sensing is so sensitive, in detecting and monitoring aging (or cure) of a polymeric material.

The second important aspect to recognize is that mechanical measurements are destructive. You can not remeasure the maximum load, extension or the load-extension at yield several times on the same sample. You must make a second sample. Each tensile test sample must be made in a machine shop. Precise dimensions and total reproducability are essential. Each sample must be mounted in an identical fashion in the apparatus. Exact reproducability of alignment is essential. This is only partially achieved to varying degrees with each operator. Finally, unless the samples are exactly identical in molecular structure on the surface as well as inside the distribution of the force within the material will vary. As a result, so will the results. In practice the results vary for all of these reasons. Generally, 3 to 5 mechanical samples must be destructively tested to get an *average estimate* of the materials' force-displacement properties.

In a dielectric FDEMS sensor experiment, in contrast, the same sample can be looked at over and over again at the exact same point from day to day to month or year. There is no variation due to sample machining, sample mounting. The measurement can be made at many different frequencies, rates of change in displacement-force. Differences between one sample and another can be detected because each sensor continues to look at the same piece material over all time. Finally, since the material is the same and not destroyed, as many measurements as required can be made and averaged on any day to produce the desired precision, and thereby have the ability to detect long time changes over short periods insitu in the use environment.

None of these important advantages are possible with mechanical measurements. They are all required in order to monitor aging of a material when the number of measurements and length of time needed to monitor cannot be determined. FDEMS sensor's continuous nondestructive in situ measurement is needed to monitor the state of the material as time elapses in an unpredictable, continuously varying, hostile environment.

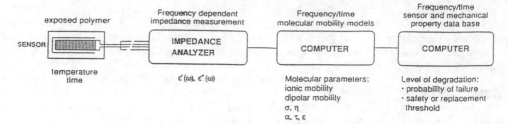

Figure 1   Schematic at FDEMS equipment

Health Sensors

· Allow in–situ measurement
· Allow repeated measurement of undisturbed samples
· Capability for measurement in field

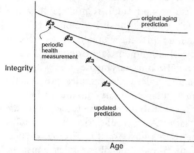

· Aging model alone is not completely reliable because it must predict an uncertain future.
· Measurements by health sensors keep predictions on track.

Figure 2   Institu. health monitoring sensor concept

In an unpredictable environment, Figure 2 shows how the FDEMS sensor can be used to revise and update predictions on the expected life of a polymer structure. By taking measurements over a short period of time, the state and rate of change in state of the polymer's properties can be used to determine the current state and to predict an updated future aging history. FDEMS sensor measurements taken at later times during use in the field can then be used to continually verify and correct earlier predictions. Figure 2 is a representation of a family of such updated sensors life monitoring curves for a part which ages faster than expected.

6 EXPERIMENTAL

A series of accelerated aging tests have been carried out on Rilsan BESNO P40 to investigate the ability of FDEMS to detect degradation under different conditions. The test we report here was carried out in a mixture of 95% water and 5% volume ASTM grade 3 oil saturated with CO_2. The oil, water, CO_2 mixture was kept at a temperature of 105°C in a closed container producing an equilibrium pressure of about 1.2 bar. One of the main reasons for the temperature was to create aging over a reasonably short time.

Samples were instrumented with embedded FDEMS sensors. The samples were approximately 6mm thick and 40 mm square and the sensors were embedded well inside the material typically around the center in the plane of the square. A large number of premachined "dog-bone" samples for mechanical testing were included together with the instrumented samples. All the samples were cut from a pipe of Rilsan and "thermo-flattened" in a jig to ensure consistent pre-treatment. In the aging bath all the samples were stacked in a small rack to give full exposure on each side.

At pre-determined intervals after the test was initiated the aging bath was opened to carry out FDEMS measurements on all sensors and to retrieve in triplicate samples for mechanical testing. The mechanical testing was carried out according to an ASTM D638 standard and average values reported. The test reported here was completed after 150 days of aging.

7 IN THE FIELD MEASUREMENTS

In practice, for offshore use of flexible pipes, FDEMS sensors can be embedded in "witness" (or coupon) sample of the polymer material in the pipe and be exposed to the same environment and conditions that the material in the pipe itself sees. Typically this will be done in a spool piece either in front or just after the flexible pipe in the production system. In the field, the high sensitivity of FDEMS allows an early verification of compatibility between the material and the production fluid. Since it is an in situ monitoring system it measures the cumulative degradation which is very important in cases where the composition, temperature and pressure of the production medium are continuously changing over time. The FDEMS measurements can, based on

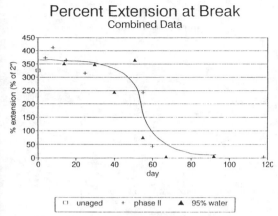

Percent Extension at Break
Combined Data

| □ unaged | + phase II | ▲ 95% water |

Figure 3a   Aging results of 2 studies, % extension

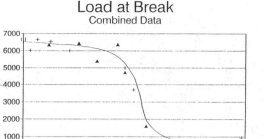

Load at Break
Combined Data

| □ unaged | + phase II | ▲ 95% water |

Figure 3b   Aging results of 2 studies, load

laboratory calibration, serve as an attractive means of characterizing the health of the polymer material used in the pipe. The FDEMS sensors are small and inert to all environments. The sensing equipment has been adopted for use on offshore platforms. A prototype system for continuous FDEMS monitoring of Rilsan BESNO P40TL exposed to crude flow has been developed and has been operated on a North Sea offshore platform since the fall of 1994.

8 RESULTS

In Figures 3a and 3b, we present the results from two mechanical aging studies. Each value is the average of 3 samples. Measurements of elongation and load at yield were also made but are note reported here. The scatter of the mechanical measurements for 3 nominally identical samples indicated a significant variance demonstrating the uncertainties for this kind of tests. After 60 days the elongation at break has dropped to

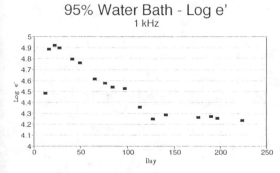

95% Water Bath - Log e'
1 kHz

Figure 4   FDEMS sensor output

approximately 45% compared to 350% for the virgin material. The rate of drop is also large. Significant degradation has taken place. The material would no longer, from the point of ductility be suitable for use in a flexible pipe.

Measurements of the complex permittivity of the FDEMS sensor embedded in the Rilsan and left in the CO_2 saturated water oil tank at 105°C were also made over a frequency range of 100 Hz to 100 kHz. Both the value of ε' and ε'' showed a rapid rise over the initial 20 day period as water diffused into the Rilsan and plasticizer diffused out. These conclusions are supported by weight loss studies on the Rilsan specimens. Figure 4 shows the FDEMS sensor output for ε' measured at 1 kHz. Similar results are observed at other frequencies and for the loss component ε''. For this report we focus only on ε' measured at 1 kHz.

The FDEMS sensor output, has been correlated with the mechanical aging results in Figures 5a and 5b. The FDEMS sensor samples and the mechanical test specimens age at different rates because of differences in sample volume-surface area and the role of diffusion in affecting the extent of aging. In order to make the correlation, the molecular weight of both the mechanical and the FDEMS sensor specimens was made. The data in figure 5 is based on samples which have been aged to the point of equivalent molecular weight. Other techniques including FTIR, Tg and H (crystallinity) were used. These support the result that the molecular weight is decreasing. This results in increased crystallinity. The result is a less ductile lower strength material with age.

Figure 6 shows the FDEMS sensor output normalized relative to its maximum value on day 20, when the water and plasticizer diffusion process

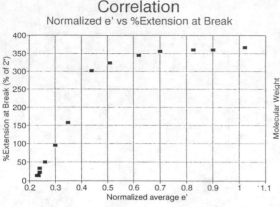

Figure 5a   Correlation with % extension

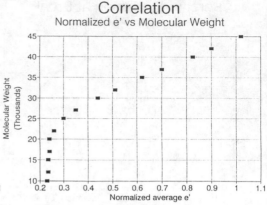

Figure 5b   Correlation with molecular weight

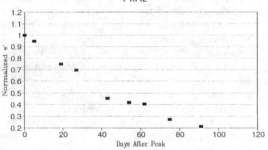

Figure 6   Normalized sensor output

has reached the Rilsan on the sensor. Using the correlation plots of Figures 5a and 5b, FDEMS sensor output shown in Figure 6 is monitoring the effect of aging - degradation with time on the mechanical properties of the Rilsan oil-gas barrier

while it is in the hostile environment. The FDEMS sensor monitored molecular weight and the elongation at break is shown in Figure 7a & 7b. Similarly the FDEMS sensor output can be correlated with other mechanical test criteria for thermosets as well as thermoplastics or directly with molecular information and used to monitor degradation as in the nylon 11 barrier.

9 CONCLUSIONS

FDEMS is an attractive method for laboratory and in the field in situ monitoring of polymer aging in structural composites. Rilsan BESNO P40TL that is used as a barrier material in flexible pipes has been monitored during accelerated aging. It has been shown that FDEMS measurements, under the tested exposure regime, can be correlated and used to measure insitu degradation in mechanical properties.

FDEMS has a number of attractive features. It is

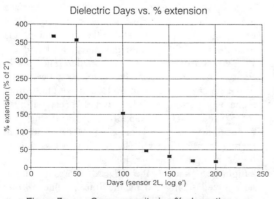

Figure 7a   Sensor monitoring % elongation

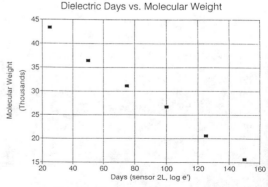

Figure 7b   Sensor monitoring MW

necessary to understand all influences to make a comprehensive interpretation. It is therefore necessary to carry out laboratory tests and calibrations to support the use of the method for field applications. For Rilsan BESNO P40TL we are currently carrying out investigations to quantify the calibration from different hostile constituents in relevant exposure environments. We are also looking at depth dependence of degradation and how degradation is affected by adjacent steel layers in the flexible pipe. These investigations will go both into the design criteria for the offshore monitoring equipment that is under development and for developing offshore oil industry guidelines for the use of various types of flexible pipes.

10 ACKNOWLEDGEMENT

The work reported here is a part of the FPT inspection joint industry program where the participating oil companies are Amerada Hess, BP, Elf, Esso, Norsk Hydro, Saga Petroleum, Shell and Statoil. The authors would like to thank the participants for support to make the work reported here possible and the NSF Center of Excellence at Virginia Polytechnic University, Blacksburg, Virginia grant number NSF-DMR 9120004.

11 REFERENCES

1. Kranbuehl, D. "Cure Monitoring", Int. Encyclopedia of Composites, ed. S. Lee, 531-543 (1990).
2. Kranbuehl, D. "In situ Online Measurement of Composite Cure with Frequency Dependent Electromagnetic Sensors," *Plastics, Rubber, and Composite Processing* 16, 213-219 (1991).
3. Hart, S., D. Kranbuehl, D. Hood, A. Loos, J. Koury, J. Harvey, "FDEMS Sensor-Loss Model QPALS Intelligent Cure Processing of Polyimides," Int. SAMPE Symp. Proc. 39(1), pp. 1641-51 (1994).
4. Kranbuehl, D., P. Kingsley, S. Hart, G. Hasko, B. Dexter, A. C. Loos, "Sensor-Model In situ Monitoring and Intelligent Control of the Resin Transfer Molding Composite Cure Process," Polymer Composites, 15 (4) 297-305 (1994).
5. Wang, Y., M. Argiriadi, W. Limburg, S. Mahoney, D. D. Kranbuehl, D. E. Kranbuehl "Monitoring Polymerization and Associated Physical Properties Using Frequency Dependent Sensing", Polym. Mat. Sci. and Eng. 70, pp. 279-280 (1994).
6. Kranbuehl, D. *Journal of Non-Crystalline Solids* 131, 930-934 (1991).

2 Fatigue

Progress in Durability Analysis of Composite Systems, Cardon, Fukuda & Reifsnider (eds)
© 1996 Balkema, Rotterdam. ISBN 90 5410 809 6

Constant-stress fatigue response and life-prediction for carbon-fibre composites

B. Harris, N. Gathercole, H. Reiter & T. Adam
School of Materials Science, University of Bath, UK

ABSTRACT: A method is presented for predicting the fatigue behaviour of carbon-fibre composites of current interest to the aerospace industry. A description is given of the overall patterns of fatigue response under constant-stress loading, with particular emphasis on materials of $[(\pm45,0_2)_2]_s$ lay-up. A method of analysis of data in terms of constant-life diagrams for life prediction purposes is presented, together with an indication of the level of success that may be achieved in prediction and of the potential hazards.

1 INTRODUCTION

In a recent paper (Gathercole *et al*, 1994) we presented a detailed examination of the constant-stress-amplitude fatigue response of a T800/5245 composite laminate consisting of a high-failure-strain intermediate-modulus carbon fibre in a toughened resin (said to be a bismaleimide) with a $[(\pm45,0_2)_2]_s$ lay-up. Stress/life curves were presented for a range of R ratios from repeated tension to repeated compression, and from these data parametric constant-life curves were derived which provided an empirical relationship between the alternating and mean components of stress which can be used for design purposes. A limited comparison with some results for some early laminate materials, XAS-carbon/epoxy, Kevlar-49/epoxy, and hybrids of the two, in unidirectional lay-up (Adam *et al*, 1986; Fernando *et al*, 1988; Adam *et al*, *1989*), suggested that the procedure may have some general level of validity. In the continuation of this work we have therefore attempted to examine a wider range of composites

with a view to assessing the potential usefulness of the technique and the level of confidence that a designer using it might have in making preliminary predictions of life from limited data sets. In this paper, further results are presented for the older T800/5245 laminate and three other modern carbon-fibre-reinforced plastics (CFRP), all having the same lay-up, *viz.* $[(\pm45,0_2)_2]_s$). The parameters of the constant-life relationship for all four of these laminates are compared and an illustration is given of how the method may be used for life prediction.

2 MATERIALS AND TESTING PROCEDURES

2.1 *Experimental materials*

The three new laminates discussed in this paper are the carbon-fibre composites HTA/913, T800/924 and IM7/977. Details of these materials, together with relevant information for the older T800/5245 (Gathercole *et al*, 1994), are given in table 1. Within the group, it will be seen that there is one representative of the older-established variety of

Table 1. Experimental materials and sources

Material	Fibre	Matrix	Supplier
HTA/913	ENKA high strength, standard modulus Tenax fibre	BSL 913 standard epoxy (low-T cure)	Ciba-Geigy (UK)
T800/5245	Toray intermediate modulus, high failure strain	Narmco epoxy/bis-maleimide	BASF (Europe)
T800/924	Toray intermediate modulus, high failure strain	Ciba 924 high-strain, toughened epoxy (high-T cure)	Ciba-Geigy (UK)
IM7/977	Hercules intermediate modulus, high failure strain	977 modified epoxy	Fiberite (Europe)

composite, HTA/913, which is similar in many respects to the T300/913 that has been the subject of many research programmes. The higher-performance fibres, T800 and IM7, have similar properties but are embedded in quite different resins. Among these four laminates, therefore, we have the opportunity to investigate the specific effects of fibre and resin characteristics on the fatigue performance of the composite and on the applicability of the life-prediction model which we are presenting.

The prepregs were all notionally zero-bleed materials (although manufacturers often appear to differ in their interpretation of this phrase) and the final laminates, autoclaved according to the suppliers' recommendations, were of nominal fibre volume fractions, V_f, between 0.65 and 0.69. Test pieces for strength and fatigue tests were cut to nominal dimensions of 200mm x 20mm with a water-cooled diamond saw, regions near to plate edges and defective areas identified by C-scanning being excluded from the test programme. The cut edges of the samples were lightly polished and, after abrasion of the end surfaces, end tabs of 1.6mm thick soft aluminium were glued on with Ciba Geigy Redux 403 epoxy-resin paste. The adhesive was cured in a dry oven at 40°C for 24hr. The remaining central gauge sections of the test samples were 100mm long.

2.2 Testing procedures

Fatigue tests were carried out in 1300 series servo-hydraulic Instron machines capable of constant load (±100kN) or constant deflexion (±50mm) cycling. A programmable signal generator and analyser (SArGen, by Marandy Instruments, Bath, UK) directs a voltage to the fatigue machine actuator (of ±10V maximum range, tension-compression), bypassing the machine's internal signal generator. Outputs from load cell or strain sensors are returned to the generator, recorded, and compared with preset levels, the output control signal to the actuator then being adjusted, as necessary, to ensure that the two remain coincident. Unlike the Instron internal control system, the SArGen is not limited to constant amplitudes. A Macintosh computer acts as a user interface which supervises SArGen and stores data on hard discs.

In constant-amplitude fatigue testing, the required frequency and maximum and minimum levels of the load cycle are inputs to the Macintosh. A D-A converter in the SArGen converts this information into a sinusoidal voltage signal which is directed to the Instron actuator. The load cell output signal (and up to three additional output signals) are directed to an A-D converter in the SArGen and these signals are converted to 180 data points per cycle. The load maxima and minima are recorded and if the averages of five consecutive maxima and minima are different from preset ones (for example as a result of changing specimen compliance) the output signal to the actuator is adjusted. Converted output signals are directed to the Macintosh where maximum and minimum load values and the number of cycles are written to disc if the extreme values differ from previous values by more than a set amount (say 1% full scale).

Tension and compression strength tests were carried out in the same servo-hydraulic testing machines as were used for the fatigue experiments at approximate strain rates of $1.5 \times 10^{-4} s^{-1}$ (tension) and $1 \times 10^{-4} s^{-1}$ (compression). For compression tests anti-buckling guides of the kind described by Curtis (1988) were used. Elastic modulus values were obtained by means of clip-on strain gauges.

3 MECHANICAL PROPERTIES

The mechanical strength and stiffness characteristics of the four laminates are summarised in table

Table 2. Mechanical properties of experimental CFRP laminates (standard deviations in brackets)

Property	Material			
	HTA/913	T800/5254	T800/924	IM7/977
Tensile strength, GPa	1.27(0.05)	1.67(0.09)	1.42(0.09)	1.43(0.07)
Tensile modulus, GPa	69.8(4.4)	94.0(3.1)	92.0(8.0)	90.2(11.3)
Failure strain, %	1.7(0.1)	1.7(0.1)	1.5(0.1)	1.5(0.1)
Compression strength, GPa	0.97(0.08)	0.88(0.10)	0.90(0.090)	0.90(0.07)
Mode I delamination fracture energy, Jm^{-2}	214(31)	————	107(7)	204(12)

2. It can be seen that the tensile moduli of the T800/5245, T800/924 and IM7/977 materials are all alike, the average value being about 92GPa, whereas that of the HTA/913 composite is only about 75% of this level. The rather surprising difference between the strengths of the two T800 materials must be due, it is supposed, to the different levels of fibre/resin adhesion. The IM7/977 composite, with a reinforcing fibre not unlike the T800, is closer to T800/924 in mechanical behaviour than to the T800/5245.

By contrast, the compression strengths of the two T800 composites and the IM7 material show much more homogeneous behaviour. The HTA/913 material, surprisingly, possesses a slightly higher compression strength than the other materials. Perhaps the most significant aspect of the results in table 2 is the fact that major differences in fibre tensile properties and matrix characteristics exert almost no influence on the compression resistance of these laminates.

It can be seen from table 2 that the IM7/977 has a higher G_{IC} value than the T800/924, with a mean of 204Jm^{-2} compared to 107Jm^{-2}. This would indicate that the likelihood of delamination occur-ring, especially during all tensile and mixed-mode loading (eg. at R = –0.3), is much higher for the T800/924 material than for the IM7/977. The occurrence of fibre bridging accompanied by higher G_{IC} values and greater scatter in the results appears to be related to the fibre/resin adhesion of the material, at least in the case of the IM7/977. This difference in the apparent interlaminar adhesion would be expected to have a bearing on certain aspects of the damage growth in these materials under fatigue loading conditions.

4 CONSTANT-STRESS FATIGUE RESPONSE

4.1 *Fatigue stress/life curves*

Replicate stress/life data for the composite laminates were obtained at five stress ratios, R (defined as $\sigma_{min}/\sigma_{max}$), of +0.1, –0.3, –1.0, –1.5, and +10. Where possible, at least five replicate tests were run at each stress level, although in some instances there were more and from time to time fewer, depending on operational problems and material availability.

Graphical representation of S/logN$_f$ curves depends largely on the purpose for which the data are to be used. A designer may require only the minimum life values or some statistically appropriate parameter, such as a given quantile of the asymptotic extreme-value distribution of minima deduced from a large population of replicate test results. For the purposes of developing our model, however, we have made use of the median lives, m(N$_f$), which are regarded as more useful than mean values for fatigue life data (Johnson, 1964; Little and Jebe, 1975; Spindel and Haibach, 1981).

A typical set of stress/life data for the HTA/913 laminate is shown in figure 1. The diagram shows the actual data points and two sets of fitted curves. The broken curves are third-order polynomial fits to the full set of stress/life data for a given R ratio, while the full curves are polynomial fits to the median points of the replicate stress data only. The significance of this will be discussed later.

A limited comparison of fatigue-life data for the four experimental laminates reveals some interesting similarities, as shown by the data sets for R = 0.1, –1 and 10 plotted in figure 2. It can be seen that the data sets for R = 10 and –1 for all four materials are largely overlapping and, although not shown in figure 2, the same is true for R = –1.5. As we have found with all materials tested in our research programme, the fitted curves

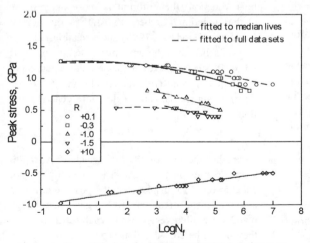

Figure 1. Stress/life data for HTA/913 [(±45,0₂)₂]$_s$ *laminate.*

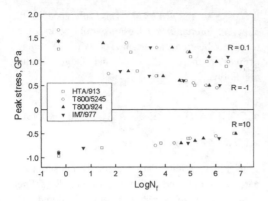

Figure 2. Selected comparison of fatigue data for four CFRP laminates at R ratios of 0.1, –1, and 10.

for repeated tension and repeated compression extrapolate smoothly back to the monotonic tension and compression strengths, respectively. The data for an R ratio of –0.3 only deviate downwards from the curve for R = 0.1 after some 1000 cycles, this curve too extrapolating back to the tension strength, and our earlier work has shown that the same is also true for R = –0.6. Between R = –0.6 and –1, a marked change in failure behaviour occurs, as indicated by the fact that the stress/logN$_f$ curves no longer extrapolate back to the monotonic tension strength, signifying the increasing importance of the compression load component in determining behaviour. The similarity of the stress/life curves for R ratios between –1 and +10 and the relatively small dispersion in the monotonic compression strengths of the four laminates are likely to be related facts. The wide spread of tension strengths of the four laminates, from HTA/913 up to T800/5245, is reflected in the locations of the initial portions of the stress/life curves, but it can be seen that the initial separation of the individual curves is reduced as the fatigue life increases and the expected benefit of higher-performance fibre/resin combinations is apparently not realised in laminates of this kind.

5 CONSTANT-LIFE ANALYSIS

The data in figures 1 and 2 are plotted in terms of peak stress as a function of life. It is apparent, however, that as the compression component of cycling increases (R increasingly negative) the stress range,

$2\sigma_{alt}$, to which the sample is subjected must at first increase and if the data were plotted as stress range versus log life the data points for a stress ratio of –0.3 would then apparently lie above those for R = +0.1. This is a familiar feature of fatigue in fibre composites, and it leads to the notion that some element of compression load in the cycle can apparently improve the fatigue response. It also results in a well-known aspect of composites fatigue, *viz* that master diagrams of the constant life, or Goodman, variety are displaced from symmetry about the R = –1 plane (the alternating-stress axis) (Shütz and Gerharz, 1977; Kim, 1988; Adam *et al*, 1992).

We have shown previously (Adam *et al* 1986; Fernando *et al*, 1988) that for a family of carbon/Kevlar hybrid composites the effects of R ratio could be illustrated by presenting the fatigue data as a normalised constant life diagram by means of the fatigue parameter:

$$f = \frac{a}{(1-m)(c+m)} \quad \text{.................1)}$$

where a = σ_{alt}/σ_t, m = σ_m/σ_t, and c = σ_c/σ_t. In these normalised functions, σ_{alt} is the alternating component of stress, equal to ½($\sigma_{max} - \sigma_{min}$), σ_m is the mean stress, ½($\sigma_{max} + \sigma_{min}$), and σ_t, σ_c are the monotonic tensile and compressive strengths, respectively. For the purposes of this parametric analysis, we keep the sign of σ_c positive, so that the parameter c is also positive. The stress function, *f*, depends on the test material. Since this is a parabolic function, the criterion is more akin to the Gerber function (Gerber, 1874) than the normal Goodman (Goodman, 1899) linear relationship.

As further data became available for other R ratios, it became apparent that a bell-shaped curve was required to describe the results, and that equation 1 was a special case of the more general function:

$$a = f(1-m)^u(c+m)^v \quad \text{.................2)}$$

Another special case, that for which u = v, represents a curve which is symmetric about a plane not equal to R = –1.0 displaced in the positive direction along the mean stress axis. But in general, although u and v are often very close to each other in value, they should be considered as separate parameters in order to increase the generality of the model.

In the first instance, the development of this model was purely empirical: there was no *a priori* reason to suppose that the values of the parameters *f*, *u* and *v* had any special significance relative to the materials structure/properties relationship.

5.1 *Application of the model for life prediction*

Since the acquisition of fatigue data for any new material is an expensive and time-consuming process, and since new composite materials are constantly being introduced, it is of advantage to the designer to have available some conservative means of estimating the potential usefulness of such a new material as soon as possible. Thus, although it is a *sine qua non* that before any new material is put into real service, its fatigue properties must be exhaustively investigated, there is considerable value in having available some means of judging the likely fatigue performance from a relatively small data bank. This is not as hazardous as it sounds, despite what we know about statistical methods of analysis, because of the regular patterns of behaviour that are appearing, as we shall see. In this part of our paper, we shall therefore describe the stages in the analysis, illustrated with results for various materials.

Stage 1: Acquisition of data

In addition to determining the tension and compression strengths of any new material, it is necessary to obtain stress/life data for a series of stress levels that cover to a reasonable extent the whole working range from the monotonic strength level down to any notional 'endurance limit'. Initially, five or so replicate tests at each of four or five stress levels will define a reasonable stress/life curve for a particular R ratio, and data for R ratios from repeated tension to repeated compression will be required. Such small numbers of tests are not in themselves adequate for defining a true probability distribution function (pdf) at any stress level, but the multiplicity of stress levels increases the confidence in fitting a curve, whether it be a polynomial or some other specific function. The same is true whether testing is replicate or random. It has also been shown (Whitney, 1981; Yang and Jones, 1981; Gathercole *et al*, 1994) that pooling fatigue data sets can be used to provide statistical information which can then be applied with greater levels of confidence to the smaller test data sets.

In what follows, then, we use the median as the working life parameter, but we assume that the same principles will apply to some other statistically

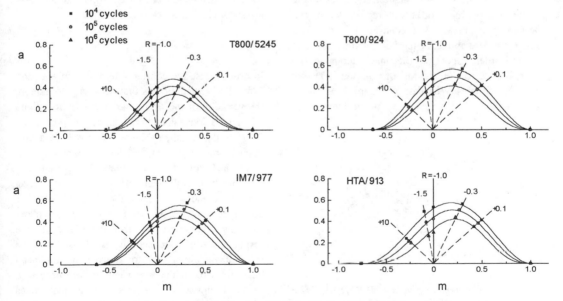

Figure 3. Constant-life plots for four CFRP laminates of $[(\pm45,0_2)_2]_S$ *construction.*

67

derived parameter. We are currently in the process of testing this hypothesis.

Stage 2: Preliminary data analysis

For the plotting of constant-life diagrams, a suitable method of interpolation of the data in figure 1 is required. There are several possibilities, including linear interpolation between the median data points and non-linear interpolation, either between the median data points or along a curve fitted to the whole data set for a given R value. Hitherto, we have used only non-linear fitting to polynomials, usually of third order, but it is by no means certain that this is the most appropriate method. We have observed, however, that for a reasonably well-arranged set of data, polynomial fits to the median points and the full data sets are indistinguishable, as can be seen for the data for HTA/913 in figure 1. One advantage of fitting a curve to the full data set is that the extent to which the fitted curves may be safely extrapolated is somewhat greater than when only the median points are used. This can be seen for the R = −1.5 data in figure 1.

One problem that we have observed in fitting polynomials is that when there is a large gap between the monotonic tensile strength and the shortest life for the next highest stress level, the shape of curve may not be well represented by the best-fit polynomial. A second-order curve will sometimes rise above the tensile strength value, and a third-order curve may become sinuous in the short-life region. It is for this reason that we referred earlier to the data of figure 1 as 'well-arranged'. In order to avoid this problem, it is

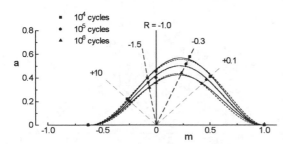

Figure 4. Constant-life plots for a $[(\pm45,0_2)_2]_s$ *IM7/977 laminate. The full curves represent the fitting of equation 2 without constraint, while for the dashed curves, f has been fixed at the mean value of 1.3.*

tempting to ensure that some fatigue tests are done at higher stress levels. But since tests resulting in lives of between 100 and 1000 cycles are often affected by initial machine stability, a better solution may be to choose a fitting function that allows a fairly linear initial portion. Such models have been suggested by Talreja (1981), Curtis (1986), and Harris *et al* (1990).

Our present method of analysis is to plot graphs of the kind shown in figure 1 in the package Origin, by MicroCal Software Inc., which provides a flexible group of curve-fitting routines, including user-defined ones. Once an acceptable level of goodness-of-fit is established, the resulting polynomial coefficients are entered into a spread-sheet (Microsoft Excel), together with values for the monotonic tension and compression strengths of the laminate. Excel then produces a set of data pairs (a,m), as defined earlier, which includes the end points (0,c) and (0,1) representing the monotonic failure conditions. These data sets, for given lives (*eg.* 10^4, 10^5, 10^6 cycles), are then exported back to Origin for plotting in the form of constant-life diagrams. Clearly, the extent to which extrapolation of the polynomial curves beyond the actual experimental data window is permissible or safe is a matter which must be carefully controlled. The results of such an analysis for all four experimental laminates are shown in figure 3.

Stage 3: Life prediction

In order to use the constant-life model of equation 2 for life prediction, it is first necessary to explore the variation of the fitting parameters with fatigue life. This is done, again in Origin, by fitting equation 2 to the data sets, as shown in figure 3. In a preliminary fitting session, all three parameters, *f*, *u* and *v*, are allowed to vary freely. The parameter *f* controls the overall height of the bell-shaped curve, whereas *u* and *v* determine the shapes of the left and right wings of the curve, and therefore allow for any asymmetry in the material's fatigue response. Early experience suggested that the same value of *f* could be used for all lives, and the mean value from the initial fitting is therefore used in a second fitting session to determine the final values of *u* and *v*. Figure 4 shows the relatively insignificant effect of this assumption on the shapes of the fitted curves.

For the fitting processes the results of which are plotted in figure 3, the coefficients of variation for the fitting of u and v were always of the order of 4%.

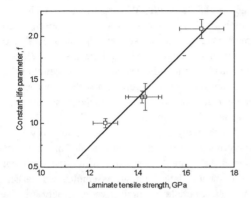

Figure 5. Dependence of the constant-life parameter f on the laminate tensile strength.

As we have already remarked, the constant-life plots in figure 3 represent relationships between the normalised parameters a and m. It appears, nonetheless, from the results that we have obtained so far that the scaling parameter, f, is related to the laminate tensile strength, as illustrated in figure 5.

This is unexpected, and requires more detailed examination, but the potential value of finding a generally valid linear relationship of this kind is that it would provide a useful initiation point for the fitting procedure, based on known information. It follows from the derivation of the constant-life plots that the higher the values of f, the better the fatigue performance at any given life, since f, overall, determines the relative 'height' of the curve, and the higher the curve, the greater the alternating stress that can be tolerated for a given mean stress at a given life. This feature corroborates to some extent the commonly accepted 'stress/life equal-rank' assumption. Clearly, the linear regression line in figure 5 does not extrapolate to the origin.

The parameters u and v both depend on the fatigue life, as is apparent from figure 3. The exact nature of this dependence for the four experimental laminates is illustrated in figure 6. The error bars in this figure represent the standard deviations for the fits shown in figure 3. It can be seen that the relationships are again linear over the range of lives

studied here, and extrapolation to 10^7 cycles would probably not be too hazardous.

For the T800/5245 and IM7/977 laminates, the slopes of the $u,v/\log N_f$ lines are all identical, but while the u and v values for the latter material are also almost identical, indicating a highly symmetric constant-life curve, there is a larger difference between the u and v values for the T800/5245 material than for any of the others. The slopes of the $u,v/\log N_f$ lines for T800/924 and IM7/977 are slightly different, the former being somewhat higher, but the actual values for these two laminates are very close indeed, certainly to within the sensitivity of the fit. Thus, in terms of the relationships:

$$u,v = A + B\log N_f \dots\dots\dots\dots 3)$$

it can be said that B is, to all intents and purposes the same for all three of the higher-performance laminates, while A shows some differences from laminate to laminate and is different for u and v.

Likewise, the higher the values of u and v, the poorer the fatigue performance because the further u and v rise above unity (the parabolic 'special case' of the generalised constant-life relationship) the more the 'wings' of the curve are pulled down-

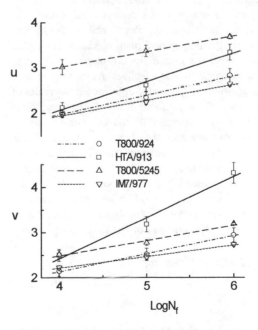

Figure 6. Dependence of the constant-life parameters u and v on life, for four $[(\pm45,0_2)_2]_s$ CFRP laminates.

69

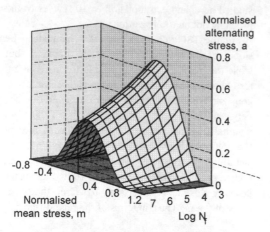

Normalised
alternating
stress, a

Normalised
mean stress, m

Log N_f

Figure 7. Three-dimensional surface plot of the a,m,logN$_f$ constant-life relationship defined by the first of equations 4 for the $[(\pm45,0_2)_2]_s$ IM7/977 laminate.

wards and the more bell-shaped the curve becomes, so *reducing* the level of alternating stress that can be tolerated for a given mean stress at a given life. It rather appears that the high *v* values for HTA/913 are associated with the high ratio of the monotonic compression and tensile strengths, defined as the parameter *c*, σ_c/σ_t, since, as we have already noted, the HTA/913 laminate stands out from the others in this respect. And, finally, the greater the slopes, du/dlogN$_f$ and dv/dlogN$_f$, the poorer the fatigue performance because the higher the slope, the greater the downward deviation of the S/logN$_f$ curve at long lives. As we have already suggested, the greater the difference in the values of *u* and *v* for a particular material, the greater the degree of asymmetry of the constant-life curve. This may influence the choice of material if it is known that for a given application a relative degree of compression or tension loading will predominate.

Having established the parameters of the relationships in equations 3, It is now possible to predict σ/logN$_f$ curves for any desired R ratio. This is done by solving the pair of simultaneous equations:

$$a = f(1-m)^{u(N_f)}(c+m)^{v(N_f)}$$

$$\dots\dots\dots\dots\dots\dots 4)$$

$$a = m\left(\frac{1-R}{1+R}\right)$$

The first of these is the constant-life equation, equation 2, modified to include information about the life-dependence of the two exponents, u(N$_f$) and v(N$_f$), as established from the form of equations 3. The second equation is derived from the conventional definition of the stress ratio. Solution of these two equations is easily carried out in a package like the MathSoft MathCAD programme which will graph or tabulate S/logN$_f$ curves for a chosen range of R values. The output can be in one of two forms, depending on the requirements of the user. The first is a three-dimensional generalisation of the constant life plots of figure 3, a surface plot showing the full variation of (a,m,logN$_f$), as illustrated for the IM7/977 laminate in figure 7.

Alternatively, a family of stress/life curves of conventional form can be produced for ranges of lives that are consistent with the original experimental data window, as emphasised earlier. As an example of predictive capability of the method, we reproduce in figure 8 a set of stress/median-life curves for the IM7/977 composite which were predicted at a time when only the monotonic strengths (σ_t and σ_c) and part of the S/logN$_f$ curve (the first five data points only) for R = 0.1 were available. Choice of values for *f, u* and *v* was made from knowledge of the behaviour of the T800/924 laminate already tested. Superimposed on the plot are the full data sets that were subsequently obtained for the IM7/977 laminate: the initial five points used in the prediction are shown as filled symbols. Predictions

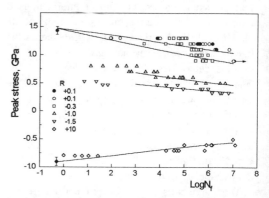

Figure 8. Stress/life curves for the $[(\pm45,0_2)_2]_s$ IM7/977 laminate predicted from an S/logN$_f$ at a single R ratio (R = 0.1) and the monotonic strengths. The filled data points for R = 0.1 are the values used in the prediction. All other data were obtained after the predictions were made.

were made with MathCAD for the range of loglife values from 3 to 7 only, but the full $S/logN_f$ curves for R = +0.1, −0.3 and +10 shown in the diagram were drawn by extrapolating back to the monotonic strength values, in the normal way.

It can be seen that that the level of agreement between the predicted curves and the experimental data for stress ratios of −1.0, −1.5 and +10 is extremely good. The polynomials for R = −1.0 and −1.5 could also have been extrapolated back at least one more decade without danger. The agreement of the predicted curve for R = +0.1 with the full data set is acceptable, although the two run-out values at 10^7 cycles have not been allowed for. The poorest fit in figure 8 is for R = −0.3. For this laminate the results at this R value were somewhat different from what we have observed for all other composites. The point at which the $S/logN_f$ curve begins to deviate downwards from the curve for R = 0.1 is between 10^4 and 10^5 cycles — perhaps a decade later than is usually the case — and the rate of downward deviation is then quite rapid. As a consequence, the predicted $S/logN_f$ curve for an R ratio of −0.3 is too conservative over the greater part of the range. From the designer's point of view, however, this is a safe prediction.

6 CONCLUDING REMARKS

The life-prediction model discussed here was first conceived as an empirical description of fatigue data obtained over a limited range of R values for some unidirectional composites consisting of members of the hybrid family CFRP/KFRP (Adam *et al*, 1986; Fernando *et al* 1988; Adam *et al*, 1989). By inspection, it appeared that a simple parabolic function fitted the normalised (a,m) data for a limited range of R values, and the model was successful in predicting stress/life relationships for a range of compositions within the hybrid family, from plain CFRP to plain KFRP. The potential value of a model which permitted the prediction of fatigue response from a relatively limited experimental data base was emphasised at the time. Since that early work was done, a variety of new high-performance composite materials has been introduced, and the obvious requirement was to investigate the validity of the early model for different composite types and different lay-ups.

It became apparent as experiments were carried out over a wider range of R ratios that the original parabolic model was inadequate, and it was refined, first, to a symmetric power-law relationship (Adam *et al*, 1992) and, subsequently, to an asymmetric power law (Adam *et al*, 1992), as described by equation 2. Fatigue results for four modern CFRP laminates (of varying character but all of the same $[(\pm45,0_2)_2]_s$ lay-up) are now shown to fit this power-law model very well. The original data for unidirectional hybrids fit the new model acceptably since, as we have shown elsewhere (Gathercole *et al*, 1994), (a,m) data pairs for R ratios between +0.1 and −0.6 fit either model equally well.

A somewhat surprising feature of the results that we have obtained is the apparent similarity of the behaviour of the four composites described in this paper. Only the tensile properties and tension-dominated fatigue response appear to show real differences, which is unexpected, given the known differences in interfacial bond strength in some of these materials, a matter upon which we have also commented elsewhere (Gathercole *et al*, 1994). This similarity means that we have not yet been able to demonstrate unambiguously the generality of the power-law model. We hope to extend the current work to include glass-fibre-reinforced plastics in the near future.

We also acknowledge the limitations of a model that has been developed on the basis of median fatigue lives and polynomial curve fits, and we are at present evaluating the likely validity of the model for use with statistically more realistic parameters. We are interested in offering the designer a conservative predictive tool that will work with a minimum of experimental data but which will allow continuous up-dating of the predictions as more data become available. It seems that a possible approach is to develop a stand-alone expert system. A suggested procedure might be as described below.

An initial experimental programme would be carried out to determine the monotonic tension and compression strengths, together with stress/life data for stress ratios of +0.1 and perhaps −1.2 in order to locate curves in the left-hand quadrant of the constant-life diagram. Perhaps three replicate tests might be done initially at three stress levels for both R ratios. By pooling the 15 to 20 fatigue lives so obtained, following the descriptions of Whitney (1981) and of Yang and Jones (1981) a trial value of the Weibull shape parameter, m, could be obtained. A crude estimate could then be made of, say, either the characteristic minimum value (the Weibull scale parameter) or the modal value of the distribution of minima for each stress level on the basis of the extreme-value-theory (Bury, 1975; Castillo, 1988) premise that the distribution of the minima of a Weibull distribution with shape parameter m is also a Weibull distribution with shape parameter m and characteristic minimum value $b/n^{1/m}$, where n is the size of the sample and b is the scale parameter of the replicate test set. The characteristic smallest values (or other appropriate quantile values of life) would then be used to derive trial constant-life curves, with appropriate 'guesses' being made initially for the parameters f, u and v until sufficient data were available to permit realistic curve fitting.

Initially, the statistical validity of the procedure would be open to question, and would require careful testing before serious use, but as further data were acquired, the validity would improve, together with the true predictive capacity of the model.

We are currently evaluating the predictive capability of the constant-life model described here against the possible application of an artificial neural network for the same purpose.

7 ACKNOWLEDGEMENTS

The authors are very grateful to the Defence Research Agency for financial support of this work, and to Mr Matthew Hiley of the DRA, Farnborough, for his support and guidance. The work on artificial neural networks mentioned above is being funded by the Engineering & Physical Sciences Research Council of the UK.

REFERENCES

Adam T, Dickson RF, Jones CJ, Reiter H and Harris B, 1986, Proc Inst Mech Engrs: Mech Eng Sci, **200**, C3, 155-166.

Adam T, Fernando G, Dickson RF, Reiter H and Harris B, 1989, Int J Fatigue, **11**, 233-237.

Adam T, Gathercole N, Reiter H and Harris B, 1992, Advanced Composites Letters, **1**, 23-26.

Bury KV, 1975, *Statistical Models in Applied Science*, (J Wiley & Sons London).

Castillo E, 1988, *Extreme Value Theory In Engineering*, (Academic Press, Boston/London).

Curtis PT, 1986, *An investigation of the mechanical properties of improved carbon-fibre composite materials*, RAE (Farnborough) Technical Report 86021, (Ministry of Defence, Farnborough, UK).

Curtis PT, (Editor), 1988, *CRAG Test Methods for the Measurement of Engineering Properties of Fibre-Reinforced Plastics Composites*, RAE (Farnborough), Technical Report TR 88012, Procurement Executive, Ministry of Defence, Farnborough, Hants, February 1988.

Fernando G, Dickson RF, Adam T, Reiter H and Harris B, 1988) J Mater Sci, **23**, 3732-3743.

Gathercole N, Reiter H, Adam T and Harris B, 1994, Int J Fatigue, **16**, 523-532.

Gerber W, 1874, Z Bayer Archit Ing Ver, **6**, 101.

Goodman J, 1899, *Mechanics Applied to Engineering*, (Longman Green, London).

Harris B, Reiter H, Adam T, Dickson RF and Fernando G, 1990, Composites, **21**, 232-242.

Johnson LG, 1964, *The Statistical Treatment of Fatigue Experiments*, Elsevier Publishing Company (Amsterdam, The Netherlands).

Kim, RY, 1988, chapter 19 of *Composites Design* (4th Edition), SW Tsai, Editor, (Think Composites, Dayton, Ohio, USA).

Little RE and Jebe EH (Editors), 1975, *Manual on Statistical Planning and Analysis of Fatigue Experiments*, ASTM STP 588, American Society for Testing and Materials (Philadelphia, USA).

Schütz D and Gerharz JJ, 1977, Composites, **8**, 245-250.

Spindel JE and Haibach E, 1981, in *Statistical Analysis of Fatigue Data*, ASTM STP 744, editors RE Little and JC Ekvall, American Society for Testing and Materials (Philadelphia, USA), 89-113.

Talreja R, 1981, Proc Roy Soc (Lond), **A378**, 461-475.

Whitney JM, 1981, in *Fatigue of Fibrous Composite Materials*, ASTM STP 723, American Society for Testing and Materials (Philadelphia, USA), 133-151.

Yang JN and Jones DL, 1981, in *Fatigue of Fibrous Composite Materials*, ASTM STP 723, American Society for Testing and Materials (Philadelphia, USA), 213-232.

Progress in Durability Analysis of Composite Systems, Cardon, Fukuda & Reifsnider (eds)
© *1996 Balkema, Rotterdam. ISBN 90 5410 809 6*

Comparison of fatigue test methods for research and development of polymers and polymer composites

V. Altstädt, W. Loth & A. Schlarb
BASF AG, Polymer Research Laboratory, Ludwigshafen, Germany

ABSTRACT: This article provides a comparison of different test methods for characterizing the fatigue behaviour of polymers and polymer composites for materials research and development purposes. Fatigue crack growth experiments can be employed as a very fast and effective screening method, while hysteresis measurements provide the capability to describe non linear material behaviour due to fatigue damage accumulation.

1 INTRODUCTION

The fatigue behaviour of polymers and composite systems is generally described by Wöhler-diagrams (S-N curves) in accordance with classical construction materials. A large number of specimens need to be tested in order to generate a Wöhler-diagram. This is often not possible for materials screening, especially in the initial phases of research when only limited amounts of material are available. Despite rigorous testing, measurement of cycles to failure provides no information about structural changes related to the fatigue process.

To gain a more fundamental understanding of the effects of chemistry and processing on fatigue behaviour, two test methods:
• Fatigue crack propagation (FCP)   and
• Hysteresis measurements (HM)
have been compared with regard to their applicability in describing the fatigue of polymer materials.

After a brief description of the experimental methods, the effect of the molecular weight and distribution of Polystyrene (PS) prepared by different methods on the fatigue behaviour in a molecular weight range from 100,000 to 1.5 mio. is presented.

Sheet moulding compounds (SMC) are more fatigue sensitive than glass mat-reinforced polypropylene (GMT) because of the heterogeneous composition of SMC. Both materials are used in load-bearing components for automotive applications. With both materials, the applicability to describe the different aspects of fatigue behaviour is compared.

Because polymer research is complicated and expensive, advances in test methods are important not only for their fundamental scientific value, but also for their ability to make research more efficient and effective.

2 FATIGUE CRACK GROWTH

All fatigue failures in polymers or polymer composites involve one phase in which a defect zone, such as a craze or microcrack initiates, followed by a propagation phase to final fracture (Figure 1).

Based on the assumption, that the fatigue lifetime is determined by the propagation phase, a pre-existing flaw is assumed. The stress state at the tip of the crack is defined by the stress intensity factor range ΔK. For the case of fatigue, Paris 1964 showed, that a linear relationship predicted by a simple power law for a double logarithmic scale exists between the FCP rate da/dN and the applied ΔK (Figure 2).

The linear dependence is frequently observed only over an intermediate range of growth rates. When investigating a wide range of da/dN, deviations from this linear behaviour may be observed, as illustrated schematically in Figure 2. That is, FCP-rates decrease rapidly to vanishingly small values as ΔK approaches the threshold value ΔK_{th}. This ΔK level defines a design criterion that is analogous to the fatigue limit determined from traditional S-N curves (Herzberg 1980, Janzen 1991). FCP rates increase markedly as ΔK approaches K_{cf}, at which unstable fracture occurs within one loading cycle.

From the standpoint of evaluating a materials fatigue resistance, any decrease in FCP rates at a given value of ΔK or, alternately, any increase in ΔK to drive a crack at a given speed is, of course, beneficial.

Frequently in the screening phase of newly synthesized materials, only small amounts of polymer are available for testing. Applying fracture mechanics concepts of FCP has the advantage that with very small amounts of polymer (e.g. 6 g) it is possible to prepare a small compact tension specimen to study the propa-

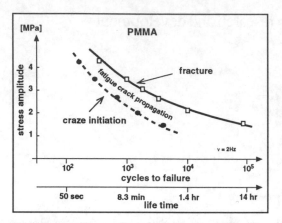

Figure 1. Stress amplitude vs. cycles to craze
initiation and to fracture (Sauer 1990).

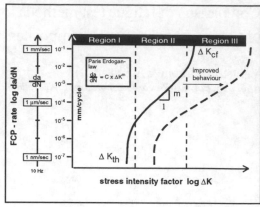

Figure 2. Fatigue crack propagation diagram.

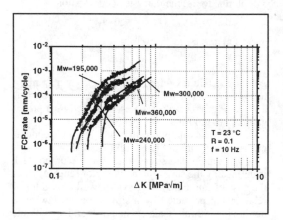

Figure 3. Effect of molecular weight for PS
prepared by free radical polymerization.

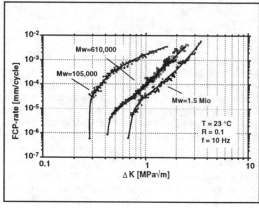

Figure 4. Effect of molecular weight for PS
prepared by anionic polymerization.

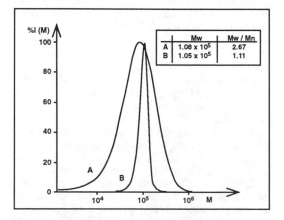

Figure 5. GPC curve of MW-distribution for PS
with M_w of $\approx$100,000

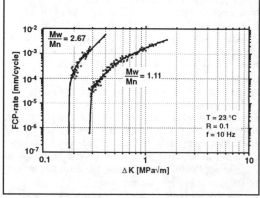

Figure 6. Effect of the polydispersity index, M_w/M_n
for PS with M_w of $\approx$100,000

gation of a fatigue crack in the material using a computer controlled servohydraulic test equipment. In contrast to tensile or impact tests, stable fatigue crack growth within a large volume of the specimen gives a volume information about the interaction between the propagating crack tip, designated as a microscopically small probe, and the polymer material itself. By this it is possible to study the viscoelastic fracture behaviour within a FCP range between 10^{-7} to 10^{-2} mm per cycle. In addition, since the formation of the fracture surface is well defined, SEM studies of the fatigue fracture surface, can provide valuable information about structure property relationships.

The FCP method has been described in detail by Herzberg 1980 and others, e.g.: interchangeability of specimen geometry (Lang 1984), the extent of energy dissipation at the crack tip (Pavsek 1994), interlaminar FCP (Altstädt 1990).

2.1 Effect of molecular -weight and -distribution

Physical properties of a polymer are strongly dependant on the average molecular weight M_w and sequence distribution M_w/M_n, because the presence of molecular entanglements can significantly affect the mechanical behaviour. In contrast to free radical polymerization, anionic polymerization can under proper conditions, produce near-monodisperse molecular weight distributions.

Figure 3 shows the FCP behaviour as a function of M_w for commercial PS systems prepared by free radical polymerization and Figure 4 for model systems of PS prepared by anionic polymerization. While the quasi static fracture properties for PS above a M_w of 150,000 are little effected by M_w, the FCP behaviour is effected almost by a factor of ten. The FCP rate is generally less for PS with higher molecular weight at any given value of the stress intensity factor. This indicates, that disentanglement and chain scission are very much effected by M_w under fatigue conditions and is in agreement with the observation, that lifetime to fracture in a corresponding S-N curve at any given stress level is appreciably greater for PS with a higher molecular weight.

In figure 4, PS with M_w of 100,000 performs much better compared to the commercial systems with M_w in the range between 195,000 to 300,000 (figure 3). This surprising trend was further investigated with model polymer systems with different polydispersity indexes - M_w/M_n and a constant M_w of 100,000. Figure 5 shows the GPC chromatograms of the two different PS systems and figure 6 the corresponding FCP diagrams. The fatigue resistance differs drastically as a function of M_w/M_n. In this instance, the system with $M_w/M_n = 1.11$ exhibited the best fatigue resistance over the entire range of ΔK. In addition to the improved fatigue resistance, the fracture toughness K_{IC} is improved by a factor of 4. This indicates, that to improve

the fatigue performance, the low molecular part of the M_w distribution must be suppressed.

Clark 1993 observed that for PMMA, the effects of molecular weight on the fatigue resistance can be substantially changed by the applied mean stress as a consequence of the competition between chain scission and chain slippage.

2.2 Effect of fiber reinforcement on FCP

Chopped glass strand reinforced thermoset composites like sheet moulding compound (SMC) or glass mat reinforced thermoplastics (GMT) often fail as a result of macroscopic crack propagation. For example, large size parts in automotive applications may tend to fail by crack propagation from stress concentration points.

For these highly anisotropic materials, FCP experiments can also be employed to study the effect of the constituents on the FCP resistance. Figure 7 compares FCP data for GMT and R-SMC at a load ratio R = 0.1. Despite the very different fiber structures, matrix toughness and compositions (R-SMC contains 29 vol% of calcium carbonate filler), the ranking of the materials reflects directly the different glass fiber volume fractions:

PP-GM30 : 30 wt% - 13 vol%
PP-GM40 : 40 wt% - 19 vol%
SMC-R30: 30 wt% - 23 vol%

To better understand this trend, it is necessary to consider all possible fiber effects, mat structure, aspect ratio and fiber-matrix bonding (Karger-Kocsis 1991, Schlarb 1993).

3 HYSTERESIS MEASUREMENTS (HM)

Fatigue crack propagation diagrams provide useful information as to the ability of different materials, or of different variants of a material, to resist crack propagation. Such tests, however, provide no information about the resistance of the polymer to craze or crack initiation. Further, since tests are run at positive R values, they provide little information concerning the possible damaging influences of cyclic compressive stresses (Sauer 1990).

While fracture mechanics investigations require test specimens with pre-cracks, HM are feasible either with or without cracks. Therefore the initiation phase for crazing, shear yielding or debonding and the crack propagation phase can be also investigated.

By monitoring strain and stress during a cyclic loading experiment, the response of the material can be clearly identified by continuously monitoring the change of the hysteresis loop. The signals from the load cell and an axial extensiometer connected to the test specimen are measured simultaneously, amplified, digitized and on-line processed by a computer. The digitized information is used to define strain, stiffness

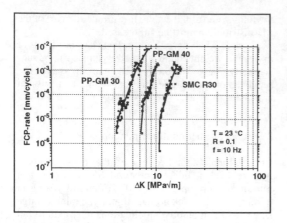

Figure 7. FCP diagram for R-SMC and PP-GMT.

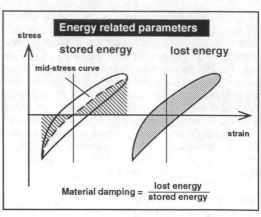

Figure 8. Energy related evaluation of the hysteresis loop.

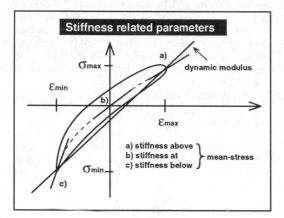

Figure 9. Stiffness related evaluation of the hysteretic loop.

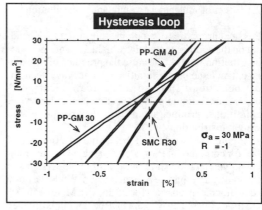

Figure 10. Hysteresis loop for R-SMC and PP-GMT, at 30 MPa, R= -1

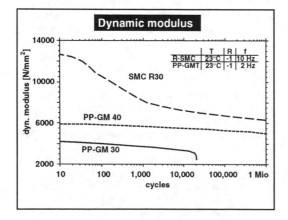

Figure 11. Change in dynamic modulus. Load amplitude 30 MPa, R = -1

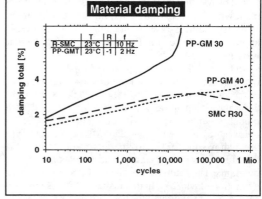

Figure 12. Change in mechanical damping. Load amplitude 30 MPa, R = -1

and energy related characteristics (figures 8 and 9). For the determination of the material damping, a mid-stress curve is fitted numerically to the hysteresis-loop, which permits the definition of energy related characteristics (i.e. lost energy, stored energy, damping) for non-linear viscoelastic material behaviour. In this case the shape of the hysteresis loop is distinctly different from a ellipse, for example a material with a hysteresis loop having unequal tensile and compressive properties, such as observed in SMC. Since the mid-stress curve bisects the hysteresis loop area, as suggested by Lazan 1968, it is consistent with the definitions used for linear elastic materials. The HM method is described in detail by Renz 1988.

3.1 Effect of fiber reinforcement on HM

Cyclic tensile compression loading is the most critical fatigue loading condition for any material. For comparison with the FCP data reported in figure 7, the same materials were tested in load control by HM, but under completely reversed load conditions (R = -1). Specimens of the same geometry (80 x 20 x 4 mm) were tested. To prevent compressive instability failures, the free span was reduced to 25 mm, and to suppress hysteretic heating effects, GMT was tested at 2 Hz.

The different shapes of the hysteretic loops after 10,000 cycles at 30 MPa are shown in figure 10 for each material. The dynamic moduli of the materials with different fiber volume content are reflected in the different slopes of the hysteretic loops. The horizontal shift of the loops in relation to the origin describes the cyclic creep behaviour. The zero point of the hysteresis loop for R-SMC moves during 10,000 load cycles to positive strains. This is accompanied by an intensive microcrack development, visible on the specimen surface, during the tensile loading phase. The loop of GMT moves simultaneously in the opposite direction, indicating plastic deformations effects in the compression loading phase.

For the present example, at a load amplitude of 30 MPa, SMC-R30 is much more fatigue sensitive than PP-GM40. Despite this, no fracture was observed with either material during 10^6 load cycles, although the change in the dynamic stiffness behaviour is quite different. The dynamic modulus (figure 11) of R-SMC decreased by more than 50 % in the range from 10 and 10,000 load cycles. Where as the loss for PP-GM40 was only modest. Surprisingly PP-GM-30 fails after only 20,000 cycles, probably because of a compressive instability failure.

The differences in damage progress can be clearly followed by the observed changes in the material damping (figure 12). PP-GM40 shows only a modest increase in damping, indicating only small changes in the material properties during fatigue at this load level. In contrast, PP-GM30 shows a stronger increase in damping up to 10,000 load cycles. Shortly before failure, the steep increase in damping, proclaims the specimen failure. This observation may explain the early fatigue fracture at this load level. Because of the low fiber volume content (13 vol%), the matrix content is very high. Hysteretic heating effects soften the matrix, reducing the fiber support which leads to microbuckling. Further cyclic heating accelerates the process, resulting in a compressive instability failure.

SMC-R30 shows a maximum in the damping curve at 10,000 cycles. Microscopic investigations by Altstädt 1987 showed that the range with increasing damping is characterized by the development of microcracks between the resin and filler interphase. An increase in mechanical friction as a result of an increase in the microcrack density explains this behaviour. At higher load cycles, the damping decreases. This can be explained by the fact that in this phase of fatigue, the new friction surfaces due to micro-crack formation are reduced and more and more filler particles get into the crack areas, reducing the effective friction area between the boundaries.

4 CONCLUSIONS

In contrast to Wöhler-diagrams, both methods, FCP and HM are able to monitor the development of damage during fatigue life. HM can be used to describe non linear material behaviour. Changes in the material damping during fatigue can be used as a very sensitive integral damage criterion. Strain, stiffness and energy related parameters are valuable for design purposes. Fatigue crack growth experiments are well suited for the investigation of the effect of defects on fatigue behaviour. They can be employed as a very fast and effective screening method especially for the analysis of the rate dependence of fatigue damage mechanisms.

While HM are very sensitive to specimen preparation, fatigue crack growth experiments are specimen geometry independent, with some limitations with regard to the dimensions of the plastic zone size and offers the potential of lifetime computation.

SMC and HM are best suited to the investigation of the debonding of the resin - filler interface and the influence of material composition, such as amount of filler and resin toughness. Fatigue crack growth experiments, however, are more sensitive to fiber and matrix-fibre effects. They provide a method to quantify the effect of different fiber mat-structures in GMT.

ACKNOWLEDGMENTS

The support of Mrs. E. Wagner, Mr. Th. Armbrust in conducting the experiments is gratefully acknowleged.

REFERENCES

Altstädt, V. & R.W. Lang, A. Neu 1990. Influence of Fiber, Resin and Interphase on the Delamination Fatigue Crack Growth of Composites. Durability of Polymer based Composite Systems for Structural Applications. Elsevier Applied Science, pp. 180-189.

Altstädt, V. 1987, Hysteresismessungen zur Charakterisierung der mechanisch-dynamischen Eigenschaften von R-SMC. Ph.D. Diss. Univ. Kassel, Germany.

Clark, T.R. & R.W. Hertzberg, N. Mohammadi 1993. Fatigue mechanisms in poly (methyl methacrylate) at threshold: effects of molecular weight and mean stress. J.of Materials Sci. 28, pp. 5161-5168.

Janzen, W. & G.W. Ehrenstein 1991. Bemessungsgrenzen von glasfaserverstärktem PBT bei schwingender Beanspruchung. Kunststoffe 81-3, p. 231.

Karger-Kocsis, J. 1991. Fatigue Crack Growth in Polypropylene and its Chopped Glass Composites. J. of Polymer Eng., Vol. 10, Nos. 1-3 pp. 99 - 119.

Lang, R.W. 1984. Applicability of Linear Elastic Fracture Mechanics to Fatigue in Polymer and Short - Fiber Composites. Ph.D. Dissertation Lehigh University, Bethlehem, PA, USA.

Herzberg, R.W. & Manson, J.A. 1980. Fatigue of Engineering Polymers, Academic Press, NY.

Lazan, B.J. 1968. Damping of Materials and Members in Structural Mechanics. Pergamon Press Inc., NY.

Paris, P. C. 1964. Proceedings of the 10th. Sagamore Conference. Syracuse Univ. Press, NY, p. 107.

Pavsek, V. & G.W. Ehrenstein 1994. Fatigue Crack Propagation in PBT-GF and SAN-GF. ANTEC'94, pp. 3223-3227.

Renz, R. & V.Altstädt, G.W. Ehrenstein 1988. Hysteresis Measurements for Characterizing the Dynamic Fatigue of R-SMC. Journal of Reinforced Plastics and Composites, Vol. 7, Sept., p. 413.

Sauer, J.A. & M. Hara 1990. Effect of Molecular Variables on Crazing and Fatigue of Polymers. Crazing in Polymers, Vol. 2, Editor H.-H. Kausch, Springer Verlag, pp. 69 - 118.

Schlarb, A & V.Altstädt, H.Baumgartl, R.Drumm 1993. Glasmattenverstärktes Polypropylen - ein recyclingfähiger Werkstoff. Kunststoffe 83-5 p. 377.

Progress in Durability Analysis of Composite Systems, Cardon, Fukuda & Reifsnider (eds)
© *1996 Balkema, Rotterdam. ISBN 90 5410 809 6*

Fatigue characterization of general purpose thermoplastic composites subject to elementary mechanical loadings

N. Himmel, M. Kienert & M. Maier
Institut für Verbundwerkstoffe GmbH, University of Kaiserslautern, Germany

ABSTRACT: The uniaxial fatigue behaviour of E-glass reinforced polyamide 12 (Vestopreg® G101) was investigated. From the raw material [0°/90°]5s plates were consolidated in an autoclave. The fatigue tests under load control were performed at a stress ratio of 0.1 and a sinusoidal loading with a frequency of 10 s⁻¹. The specimens were loaded in the warp direction of the weave and tested at ambient temperature without further conditioning after consolidation. The study included the investigation of unnotched specimens with different widths and the determination of the notch sensitivity of the material. The experimental results could well be fitted to linear S-logN curves depending on the experimental parameters. The relationship of the dynamic modulus and the damping from the load cycle number was inconsistent among the specimens of the various samples. Furthermore, no significant change of these parameters could be observed at the end of the fatigue life prior to final failure.

1 INTRODUCTION

In the past the scientific and technological interest in the fatigue behaviour of composite materials was in the past more or less restricted to the materials and needs of the aerospace industry. Consequently, considerable fatigue data and damage expertise is available for carbon fibre reinforced epoxy resins. Later, due to their advantageous damage tolerance, high performance thermoplastic resins like PEEK again with carbon fibre reinforcement were investigated (Baron 1990).

Some fatigue studies were carried out on glass fibre reinforced polyesters for wind turbine blade applications where particularly large cycle numbers have ve to be faced (Kensche 1995).

For applications in the automotive industry the predominant factor for the choice of materials is cost. In this respect the constituents of composites are chosen under the aspect of
• raw material costs
• production costs of the semifinished product and the final structural part
• automation potential
while the mechanical performance looses its predominant role (as at least formerly was the case in the aerospace industry).

The aim of a research project currently being carried out at the Institute for Composite Materials is to evaluate the potential of cost effective thermoplastic composite materials for fatigue dominated applications mainly directed toward the automotive industry and to investigate the effect of subsequent production processes like thermoforming on the fatigue behaviour of these materials.

One part of this project was the investigation of the uniaxial fatigue behaviour of Vestopreg® G101, an E-glass fabric reinforced polyamide 12. The study included the determination of specimen size and notch effects as structure related aspects. From the prepreg material symmetric [0°/90°]5s "organic sheets" were consolidated in an autoclave. The fatigue behaviour was characterized by the determination of S-N curves from specimens cut out of the laminated sheets in the warp direction. During fatigue testing the damage evolution was continuously monitored by the measurement of the stress-strain-hysteresis, from which the dynamic modulus and the damping of the material were evaluated.

To investigate size effects on the fatigue performance, the width of the unnotched specimens was varied. For each configuration complete S-N curves were measured. Furthermore, the notch sensitivity of the material was tested on specimens with an open hole in the center of the gauge section.

2 EXPERIMENTAL DETAILS

2.1 *Manufacture and preparation of specimens*

The raw material used in this test programme was an E-glass 7 end satin reinforced polyamide 12 with a fibre volume fraction of 52 %. The prepreg developed by Hüls, Germany, under the trade name Vestopreg® G101 is meanwhile licenced to Verseidag, Germany, for production. From the two-sided powder impregnated prepreg of a ply thickness of 0.2 mm "organic sheet" laminates with a nominal thickness of 2.2 mm and a symmetric stacking sequence of [0°/90°]5s were consolidated in an autoclave. Figure 1 shows the temperature and pressure cycle of the autoclave process.

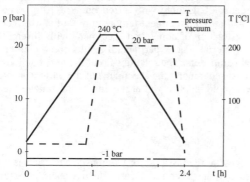

Figure 1: Temperature and pressure cycle of the autoclave process

Nondestructive testing of the laminates was performed by ultrasonic C-scanning which showed a good consolidation quality.

From the sheets plates with the required length were cut parallel to the warp direction of the fabric using a diamond saw. Glass fibre reinforced tabs with ±45° stacking sequence were bonded to the plates by an epoxy adhesive. After curing of the adhesive the rectangular specimens were cut from the plates with the aid of a diamond saw.

Figure 2 shows the geometry of the tested specimens with a free gauge length of 174 mm. The width of the specimens was 10 and 25 mm (unnotched) and 25 mm (notched).

2.2 *Fatigue and tensile strength tests*

For the fatigue tests under load control Schenck servohydraulic test machines (PSA 100, PSB 250) with a load capacity of 100 and 250 kN were used. The specimens were clamped in the center of the tabs using hydraulic grips. The load was measured by the inductive load cell of the test machine. For strain measurement a strain-gauge extensometer (Sandner EXA 50-5 o) with a base length of 50 mm was mounted symmetrically to the center on the wide edge surfaces of the specimen.

The sinusoidal fatigue load was applied at a stress ratio R (minimum stress/maximum stress) equal to 0.1 and a frequency of 10 s^{-1}. To determine the S-N curves for each specimen configuration 4 stress amplitudes with 5 specimens at each stress level were tested. The number N of cycles to failure represents

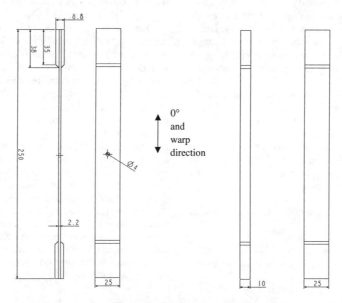

Figure 2: Specimen geometry

the complete fracture of the specimens into two parts.

The tensile strength tests were carried out using the same servohydraulic testing machines as for the cyclic testing. As suggested in the literature (Curtis 1989), the tensile strength was measured on 5 specimens at a strain rate of 0.38 s⁻¹ to apply similar loading conditions as during the fatigue tests. The strain rate of 0.38 s⁻¹ is the maximum strain rate for a sinusoidal loading with an amplitude of 350 MPa which is the static tensile strength of [0°/90°] Vestopreg® G101 as reported by Hüls.

2.3 Statistical analysis of data

In the literature, static strength and fatigue lifetimes are often reported to fit a Weibull distribution (Kim 1982). In this study, a two parameter Weibull distribution was used to describe the life cycles at a given stress level. To estimate the shape and scale parameters of the Weibull distribution a least-square approximation was used based on the life cycle number data of the specimens.

3 RESULTS

3.1 Temperature rise during fatigue loading

During the fatigue tests of the 25 mm wide unnotched specimens an infrared thermography camera was used to measure the temperature rise due to thermal heating in the specimen which was fatigued at a frequency of 10 s⁻¹ at the highest stress amplitude. As a result, a maximum temperature rise of 20 °C occurred from which no evidence was followed for a reduction of the loading frequency.

3.2 S-N curves of the unnotched specimens

The S-N curves of the unnotched specimens are shown in Figure 3. The measured data suggest a linear relationship between the maximum stress S and the (decimal) logarithm of the number N of load cycles to failure according to the following equation:

$$S = m \cdot \log N + b \qquad (1)$$

The parameters m and b, determined from the maximum stress amplitudes and corresponding cycle numbers applying a linear regression analysis are listed in Table 1.

As can be seen from Figure 3 the fatigue life increases with increasing specimen width in the lower cycle range. This difference in the fatigue life is significant at the 5% level for the stress levels 219 and 318 MPa as result of a significance test. The maximum cyclic stress decrease per load cycle decade is between 67 and 76 MPa for both specimen widths. The tensile tests resulted in a strength of 552 ± 12

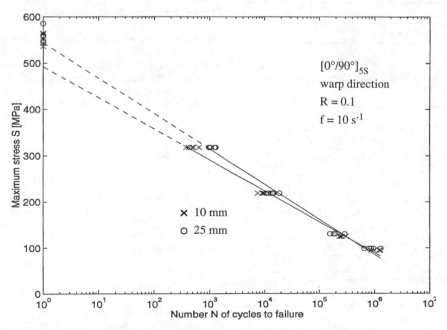

Figure 3: S-N curves of Vestopreg® G101 for unnotched specimens

83

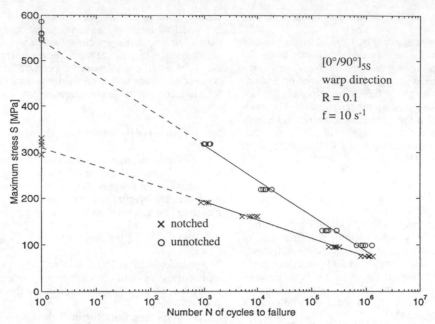

Figure 4: S-N curves of Vestopreg® G101 for notched and unnotched specimens (width 25 mm)

Table 1: Results of linear regression analysis for un-notched specimens

width [mm]	10	25
m [MPa]	-67,1	-76,30
b [MPa]	492	544
correlation coefficient	-0,997	-0,993

Table 2: Estimated Weibull parameters for the un-notched specimens

S [MPa]	α	β	$N_{10\%}$	$N_{50\%}$	$N_{90\%}$
width 10 mm					
95	4,8	$1.13 \cdot 10^6$	$7.04 \cdot 10^5$	$1.05 \cdot 10^6$	$1.35 \cdot 10^6$
125	2,7	$2.70 \cdot 10^5$	$1.19 \cdot 10^5$	$2.36 \cdot 10^5$	$3.66 \cdot 10^5$
219	5,1	$1.03 \cdot 10^4$	6630	9580	$1.03 \cdot 10^4$
318	4,3	532	316	489	646
width 25 mm					
99	3,5	$1.02 \cdot 10^6$	$5.38 \cdot 10^5$	$9.17 \cdot 10^5$	$1.29 \cdot 10^6$
130	3,6	$2.28 \cdot 10^5$	$1.22 \cdot 10^5$	$2.05 \cdot 10^5$	$2.87 \cdot 10^5$
219	4,8	$1.56 \cdot 10^4$	9760	$1.45 \cdot 10^4$	$1.86 \cdot 10^4$
318	6,6	1230	875	1170	1400

and 560 ± 16 (mean ± standard deviation) for the unnotched specimens with a width of 10 and 25 mm, respectively. Comparing this result to the static strength of 350 MPa as given by the supplier of the material a strong loading rate dependence of the tensile strength is obvious as it is well known for glass fibre reinforced plastics (Curtis 1989).

Table 2 gives the estimated shape and scale parameters, α and β, respectively, of the Weibull distribution of the fatigue life for the tested maximum stress levels S of the 10 and 25 mm specimens. From the estimated distributions the cycle numbers to failure corresponding to 10, 50 and 90 % of the specimens were evaluated.

3.3 S-N curves of the notched specimens

Figure 4 shows the S-N data for the notched specimens. As was the case for the unnotched specimens, a linear S-logN relationship can be used for curve fitting the experimental results. A linear regression analysis on the basis of equation (1) provides the following parameters

$$m = -39.4 \text{ MPa}$$
$$b = 311 \text{ MPa}$$

with a correlation coefficient of -0.997. The tensile tests resulted in a strength of 316 ± 15 MPa (5 specimens, strain-rate 0.38 s^{-1}).

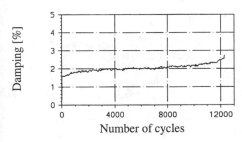

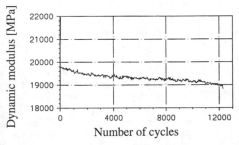

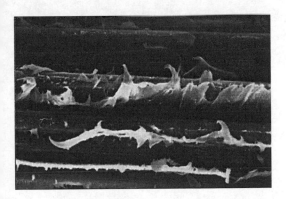

Figure 5: Dependence of damping and dynamic modulus on cycle number (specimen 10 mm wide, unnotched, maximum stress 219 MPa)

Table 3: Estimated Weibull parameters for the notched specimens (width 25 mm)

S [MPa]	α	β	$N_{10\%}$	$N_{50\%}$	$N_{90\%}$
75	4,5	$1.15 \cdot 10^6$	$6.96 \cdot 10^5$	$1.06 \cdot 10^6$	$1.38 \cdot 10^6$
95	5,5	$2.87 \cdot 10^5$	$1.90 \cdot 10^5$	$2.68 \cdot 10^5$	$3.34 \cdot 10^5$
160	3,4	8790	4510	7890	$1.13 \cdot 10^4$
190	4,5	1150	694	1060	1380

Figure 6: SEM photograph of fracture surface, 90°-ply, stress level 99 MPa, 2500 x

Table 3 gives the estimated shape and scale parameters, α and β respectively, of the Weibull distribution of the fatigue life for the tested maximum stress levels S of the notched specimens.

3.4 Fatigue damage characterization

Fatigue induced damage in the material causes changes in the stiffness and damping properties of the fatigue specimen. Thus, monitoring of these properties can provide information for the damage evolution during fatigue loading.

As an example Figure 5 shows the damping and the dynamic modulus plotted across the cycle number for a 10 mm wide unnotched specimen tested at a maximum stress of 219 MPa. Disregarding the first 1000 load cycles approximately a constant 25 % increase of the damping and a 5 % decrease of the dynamic modulus relative to the initial value can be regarded.

However, there was an equal number of specimens within the several samples which exhibited no changes in the dynamic modulus or the damping during fatigue. Generally, no consistent shape of the curves correlating with certain damage processes could be observed. No significant change in these parameters indicating the sudden failure of the tested material arised at the final stage of the fatigue failure process.

As a consequence, it may be stated that for the investigated material stress-strain-hysteresis monitoring does not provide an appropriate method for the characterization of fatigue damage evolution as was suggested for other fibre reinforced plastics materials (Buxbaum, Ehrenstein 1992).

3.5 Fracture surface morphology

The fractured cross section was always perpendicular to the specimen longitudinal axis. Figure 6 shows an SEM photography of the fracture surface of an unnotched fatigue specimen tested at a maximum stress level of 99 MPa. From the matrix drawing it can be observed that the specimen failed in a ductile tension mode of the matrix (Purslow 1987).

4 DISCUSSION

The dependence of the fatigue life of the specimens on their width may arise from two aspects:
• There are testing conditions that change with the specimen width. The clamping force acting on the specimen during fatigue testing was kept constant which results in different clamping pressures for the different specimens. As the fatigue failure usually

occurs near the tab region, it may be influenced by the clamping pressure. The temperature rise in the specimen should also be different for the different geometries as the different surface to volume ratios of the specimens will cause different heat flux resistances.

• There may be a real "size effect" of the specimen, i.e. the observed behaviour arises from aspects that have to do with the specimen size or geometry. This subject was recently discussed by Zweben (1994). At the time one can say that there are no strong implications that there is such a size effect in composites like it is well known to occur for brittle materials like ceramics. In the case discussed, a possible reason could arise from the influence of free edge effects that are supposed to have a different influence for the different specimen sizes.

It should be emphazised that the dependence of the maximum stress on the load cycle number for the investigated specimen configurations as shown in Figure 3 and 4 are statistically approved (on the basis of a sample size of 5 specimens) within 10^3 and 10^6 load cycles. An extrapolation to cycle numbers above the limit cycle number of the current study seems problematic and will need further experimental effort for the determination of the shape of the S-N curves in this region.

5 CONCLUSION

The fatigue behaviour of Vestopreg® G101 subject to uniaxial sinusoidal loading in the warp direction at R = 0.1 and a frequency of 10 s^{-1} under load control was studied. The tests were performed at ambient temperature without any preconditioning of the material after consolidation. During the fatigue tests a maximum temperature rise of 20 °C was found. S-N curves were derived for unnotched and notched specimens which showed a linear dependence of the logarithm of the load cycle numbers to failure and the maximum fatigue stress level. An increase of the unnotched specimen width from 10 mm to 25 mm resulted in a higher fatigue life in the lower cycle range. A strong strain-rate dependence of the tensile strength of the material was found.

To enable the measurement of the stiffness degradation and the damping during fatigue the specimen elongation was monitored. The dependence of the dynamic modulus and the damping for the load cycle number was inconsistent over the range of tests and no significant change in these parameters indicating the final failure of the specimen could be observed.

Future work will include the investigation of the influence of the double belt press consolidation technique and of an additional thermoforming process on the fatigue behaviour of this material.

REFERENCES

Baron, C. 1992. Mechanische Eigenschaften kohlenstoffaserverstärkter Kunststoffe (CFK) bei Variation der Matrixduktilität und der Bruchdehnung der Fasern. *Forschungsbericht DLR-FB 92-06, Deutsche Forschungsanstalt für Luft- und Raumfahrt.*

Buxbaum, O., Ehrenstein, G.W. 1992. Ermüdungsverhalten von Faserverbundkunststoffen und Lebensdauervorhersage für Faserverbundbauteile. *Fachtagung, Würzburg, 11. und 12. Nov. 1992.*

Curtis, P.T. 1989. The fatigue behaviour of fibrous composite materials. *Journal of strain analysis* 24: 235-244.

Kensche, C.W. 1995. Influence of composite fatigue properties on lifetime predictions of sailplanes. *XXIV OSTIV Congress, Omarama, New Zealand.*

Purslow, D. 1987. Matrix fractography of fibre-reinforced thermoplastics. *Composites* 18: 365-374.

Zweben, C. 1994. Is there a size effect in composites? *Composites* 25: 451-453.

Progress in Durability Analysis of Composite Systems, Cardon, Fukuda & Reifsnider (eds)
© *1996 Balkema, Rotterdam. ISBN 90 5410 809 6*

Fatigue life of a fiber composite under spectrum loading

Mohamad S. El-Zein
Deere & Company Technical Center, Moline, Ill., USA

S.C. Max Yen, K.T. Teh & C.Y. Huang
Southern Illinois University at Carbondale, Ill., USA

ABSTRACT: When a structural component is installed, it is anticipated to withstand the actual service loading condition that a structure would experience. Typically, the service loading condition of a structure can be represented by a repeated loading block (also referred to as the spectrum loading). A spectrum loading is a loading history which cannot be characterized through a continuous mathematical function.

In this study, the fatigue life of a composite material under a spectrum loading is presented. The spectrum loading represents the bending stress amplitude developed in the drive shaft of log-skidder during a session of log cutting. Each session of the spectrum loading is approximated 15 minutes in duration with about 8,000 alternation of peak to valley or valley to peak. To reproduce the service condition of the log-skidder, a data recorder was attached to its drive shaft during a session of log cutting. The data record was then translated into the ASCII format acceptable to the program that controls an MTS machine. Upon the development of a normalized spectrum loading data, the MTS machine was programmed to test the composite specimens at different stress levels. Each specimen was subjected to repeated cycles of the spectrum loading until fracture. The S-N curve of the fatigue spectrum life of the composite material can be modeled into a smooth curve. Such an S-N curve also resembles the shape of the S-N curves of the same material under the constant tension/compression fatigue. In all of the tests conducted, it was found that the final fracture always occurred during the spectrum of the maximum tensile stress. Finally, the fatigue life prediction under spectrum loading will be presented.

INTRODUCTION

The response of structural components to fatigue loading has been a major concern since engineers became familiar with strength of materials. For metals, the fatigue process has a distinct crack initiation period followed by crack growth until final fracture occurs. Consequently, there is both a crack initiation and a crack propagation model for the prediction of the nucleation and growth of a single crack. Under uniaxial fatigue loading, models for life prediction of a component and modes of failures are well developed. For biaxial fatigue loading, however, life prediction models are still in the developing stages.

When metal coupons are subjected to long life fatigue testing, the majority of life is spent in initiating a crack while propagation is a small percentage of the overall life. If we measure the change in stiffness of the specimen during the crack initiation step, very little stiffness change will be observed. However, during the crack propagation step, there is rapid and noticeable stiffness degradation. During low cycle fatigue, both initiation and propagation may contribute significantly to the total life before fracture. Figure 1 shows a typical rate of crack propagation for metal components. For a metal component with this type of crack growth, the stiffness reduction could certainly be used as a measure of damage. This concept of stiffness reduction and damage was first used by Kachanov (1986) [4].

On the other hand, in fiber reinforced composites (FRC), cracks develop during the manufacturing process and final failure is the result of several types of damage modes interacting. Damage can be due to matrix cracking, fiber/matrix debonding, fiber breakage, etc. Various modeling schemes have been formulated to predict the life of structural FRC. Currently, the essential parameter monitored and used for predicting life is stiffness. Although stiffness of a component is fundamental for structural integrity, and in some component designs, a criterion for failure (helicopter rotor blades), it is uncertain if it can be used to predict life where stress raisers (cracks) are always present. Moreover, the majority of stiffness degradation in FRC materials occurs toward the very end of life, making this an insensitive measurement parameter through most of the life cycle.

Although damage development on a microscopic

scale is distinctly different between metals and FRC, the author proposes that on the macroscopic scale these materials behave similarly. A stiffness degradation curve for a composite structure is shown in Figure 2. Comparing Figures 1 and 2 reveals that shape of the curves is similar for both the metallic and composite materials.

This paper shows life prediction of fiber reinforced composites subjected to specimen loadings.

EXPERIMENTAL PROGRAM

The material used was thermoset polyurethane with 37% glass. The specimens tested were cut using a diamond plated wheel and machined on a tensile cut machine. The specimens were dog-bone shaped with dimensions of [6/0.5/0.125] inches.

The properties of the material were determined and a typical stress-strain curve is shown in Figure 3. Fatigue testing was performed on a closed loop servo-controlled machine at a frequency of 5 Hz. The S-N curve for this material is shown in Figure 4.

In order to introduce a stress raiser, a 1/8 inch hole was drilled in the middle of the specimen. This type of notch was used for all life predictions investigated in this paper.

THEORETICAL ANALYSIS

The theoretical stress concentration in an infinite plate containing a circular hole is (3), provided the stress applied is in the elastic region. In the linear elastic region, there is no difference between the stress and strain concentration factors. As the linear elastic region is exceeded, the stress concentration decreases while the strain concentration increases. In order to define a fatigue stress concentration factor, Neuber (1961) [6] proposed to consider the geometric mean of the stress and strain concentrations or the square root of the product of the stress and strain concentrations. Neuber's rule has also been used to relate nominal and notch root stress and strain relationships for fatigue analysis as noted by Professor Morrow et al (1969) at the University of Illinois (see Reference [7]). Neuber's rule used in this manner can be written in the following form:

$$K_f^2 S_N e_N = Se$$

(1)

where S_N and e_N are the nominal stress and strain, and S and e are the notched stress and strain values respectively.

If both sides of an equation (1) are divided by a factor n (n depends on the area under the e stress-strain curve), the terms will become energy terms. For linear elastic materials, n is equal to 2. In this model for random fiber composites, a relationship

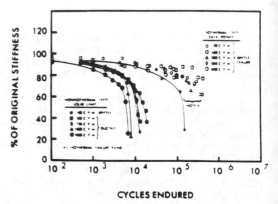

Rate of Crack Growth

Figure 1

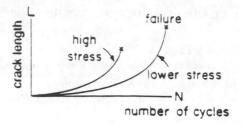

Figure 2. Stiffness Change Versus Number of Cycles (2)

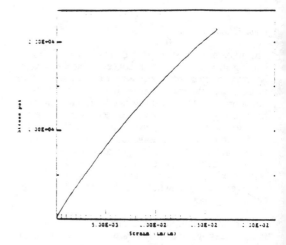

Stress-Strain Curve

Figure 3

88

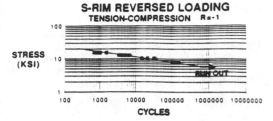

Figure 4

similar to Neuber's equation is written in terms of energy, i.e.:

$$K_f^2 W_N = W_f$$

(2)

where W_N and W_f are the areas under the stress-strain curve (energies) corresponding to the stresses S_N and S_F, and S_F is the fatigue stress of an unnotched specimen required to produce an equivalent life for the notched specimen.

K_f has the same definition as in Neuber's approach:

$$K_f = (K_s K_e)^{\frac{1}{2}}$$

(3)

where K_s and K_e are the stress and strain concentration factors.

For components subjected, the question of damage accumulation in composites has not been addressed thoroughly. Boller (1966) [1], Hofer and Olsen (1967) [3], and Broutman and Sahu (1972) [2] attempted to use Miner's linear damage rule to study damage accumulation. Broutman and Sahu concluded that Miner's rule was not adequate for composites. However, a review of the results presented in [3] revealed that Miner's rule is within reasonable accuracy except for two loading levels.

In this report, Miner's (1945) [8] linear theory is used to study damage accumulation in random fiber composite materials. In mathematical form, Miner's rule is written as:

$$\frac{n_1}{N_1} + \frac{n_2}{N_2} + - - - + \frac{n_m}{N_m} = 1$$

RESULTS AND DISCUSSION

Simple load histories consisting of few cycles were set up and experimental results were obtained. In the meantime, it became apparent that an analytical tool was needed. Hence, a computer program was used to do the rainflow counting and life predictions.

Figures 5, 6 and 7 show some simple load histories along with experimental and prediction results. It is apparent that a good correlation exists between the experimental and the predicted data.

To further verify the model, a long history block is considered as shown in Figure 8. The good correlation between the experimental and predicted data is apparent.

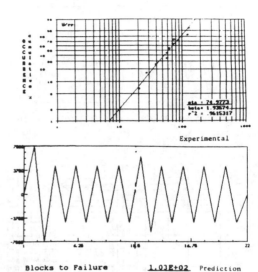

Figure 5

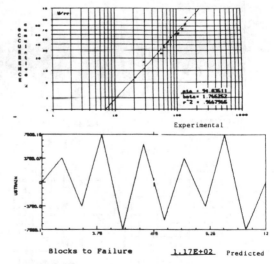

Figure 6

89

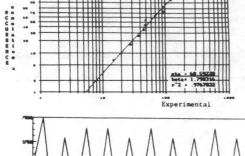

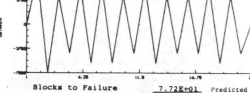

Experimental

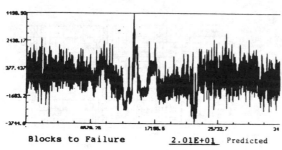

Blocks to Failure   <u>7.72E+01</u> Predicted

Figure 7

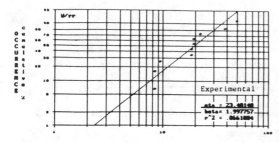

Experimental

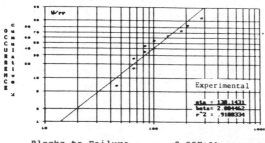

Blocks to Failure   <u>2.01E+01</u> Predicted

Figure 8a

Blocks to Failure   <u>8.28E+01</u> Predicted

Figure 8b

CONCLUSION

The ability to predict fatigue life is essential for acceptance of FRC materials in load bearing designs. The fatigue prediction methodology proposed in this report promises to be an effective and versatile tool in fatigue life analysis.

REFERENCES

Boller, K. H., "Effect of Single Stress Change in Stress on Fatigue Life of Plastic Laminates Reinforced With Unwoven E Glass Fibers," Technical Report AFML-TR-66-220, December 1966.

Broutman, L. J., and Sahu, S., "A New Theory to Predict Cumulative Fatigue Damage in Fiberglass Reinforced Plastics," ASTM STP 497, 1972, p. 170-188.

Hofer, K. E., Jr., and Olsen, E.M., "An Investigation of the Fatigue and Creep Properties of Glass Reinforced Plastics for Primary Aircraft Structures," AD-652415, IIT Research Institute, Chicago, IL, April 1967.

Kachanov, L. M., "Introduction to Continuum Damage Mechanics," Martinus Nihoff Publishers, 1986

Miner, M. A., "Cumulative Damage in Fatigue," Journal of Applied Mechanics, Vol. 12, 1945.

Neuber, H., "Theory of Stress Concentration for Shear Strained Prismatical Bodies with Arbitrary Non-Linear Stress-Strain Law," Journal of Applied Mechanics, ASME, 1961, P. 544-550.

Topper, T. H., Wetzel, R. M., and Morrow, JoDean, "Neuber's Rule Applied to Fatigue of Notched Specimens," Journal of Materials, JMLSA, Vol. 4, No. 1, March 1969, P. 200-209.

Progress in Durability Analysis of Composite Systems, Cardon, Fukuda & Reifsnider (eds)
© *1996 Balkema, Rotterdam. ISBN 90 5410 809 6*

A model for predicting the fatigue damage of filament wound pipes

E. Joseph & D. Perreux
Laboratoire de Mécanique Appliquée R. Chaléat, URA CNRS, Besançon, France

ABSTRACT: This article is mainly devoted to the study of the fatigue performance of [+55,-55] laminate composite pipes. In a first part, the experimental results obtained with biaxial loading are described. These results relate to the lifetime, to the damage and to the effect of frequecy on both these aspects. In the second part, a thermomechanical model is proposed which is capable of describing the results obtained, principally the fatigue/creep interactions responsible for the frequency effects.

1 INTRODUCTION

The aim of this study is to analyse the fatigue behaviour of glass-epoxy laminate pipes obtained by filament winding. This material is of the composite type most commonly used in pipe systems. The good corrosion resistance of composites is one of the main reasons that explain the research effort expended on these materials by the oil industry (Salama ;1994) as well as by the electricity generators (Le Courtois ; 1995). In these applications, the need to understand the static or dynamic behaviour becomes increasingly important. The design procedures emphasize the need to take into account the overall behaviour of the laminate. The elastic part is nowadays already considered for optimizing the material. The reason for this is that the main type of loading (internal pressure with restrained end closure) of the pipes induces a multiaxial stress loading in the laminate. This stress field is used to determine the best stacking sequence (from the elastic standpoint). By means of this method, the "optimum" laminate is a [+55,-55]$_n$ one. This optimization does not take into account the viscoelastic behaviour. Nevertheless, this type of material is the one most often used for transporting fluids.

Although the choice of the laminate is based on the well-known elastic relationship between stress and strain, it is obvious that the actual behaviour is more complex. In fact, the behaviour of the laminate exhibits many phenomena, such as viscoelasticity, damage and moisture absorption, which can adversely affect the proper functioning of the pipes. In order to improve the reliability and the design rules, it is necessary to take these phenomena into account, especially damage, which is the main cause of weepage.

Many studies have been carried out on the behaviour of laminates. Among these, the viscoelastic behaviour has been studied by Lou and Schapery (1971) and Xiao and Cardon (1986). Damage under static loading has also been studied, for example by Laws et al (1983) who used the self-consistent method to express the effect of damage on the stiffnesses of a unidirectionally reinforced composite. As regards Ladeveze and Ledantec (1992) based their work on modelling, also on a mesoscopic scale, for which they proposed kinetics describing the variation of damage variables. In the field of fatigue, a formulation is proposed by Reifsnider and Stinchcomb (1986) for the residual strength of a damaged composite which enables the fatigue lifetime to be determined. The behaviour of laminate pipes can be studied by using the general methods developed for laminates.

Soden et al (1989) has studied the fracture of ± 55 filament wound pipes. Hoa (1991) has concentrated on the development of design rules.

As regards fatigue behaviour, we may mention, for example, the work of Frost (1993) who studied the influence of lifetime and compared the results with static performances.

Our work concentrates mainly on modelling the effects of frequency on lifetime and on damage.

2 EXPERIMENTAL RESULTS

2.1 Materials and experimental apparatus

The material studied is a $[+55,-55]_3$ laminate. The matrix is an epoxy (Ly 556 - Hy 917 - Dy 30) from Ciba Geigy. The fibres are E-glass. The material is in the form of tubes 60 millimetres in diameter and 2.8 m in length. Test specimens 35 cm in length were cut out from these tubes. The test specimens were fitted into the testing machines by means of self-tightening jaws consisting of a set of polyamide cones and metal pieces. This system and the extensometer are described by Joseph (1995). The machines used are hydraulic triaxial or biaxial tension/torsion/internal-pressure machines from Schenck (1500 bar, 63 kN, 1000 MN) and from Instron (2000 bar, 100 kN).

2.2 Tests

The tests were carried out under controlled force and pressure. Loadings with three different $\dfrac{\sigma_{\theta\theta}}{\sigma_{zz}}$ ratios were applied ($\dfrac{\sigma_{\theta\theta}}{\sigma_{zz}}=0,2,\infty$). The effect of frequency on lifetime and on damage were studied ($0.02\leq$frequency$\leq$5Hz). All the tests were carried out with a sinusoidal loading, such that $R = \sigma^{mini} / \sigma^{maxi} = 0$. Figure 1 gives the various notations and identifiers used below.

2.3 Lifetime

In this section we present the results relating to the lifetime of the materials. Lifetime is naturally tied up with the definition of the failure phenomenon. Several definitions may be adopted. A first type of failure may be related to the loss of function of a pipe. Since the main function of a pipe system concerns the transportation of fluids, weepage constitues the functional definition of failure. A second type of failure may be related to the loss of a mechanical function, for example the variation in the stiffness of test specimens by a certain percentage. The third type is closest to the physical definition of fracture, in the sense that it

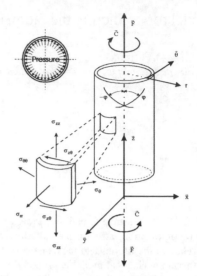

Figure 1 : Orientation of the reference axes

corresponds to separation of the test specimen into two parts.

No direct connection exists *a priori* between these various definitions. The lifetime obtained for the first two types is always less than that obtained by using the third type. Technically, it is very difficult to use weepage as a means of detecting tensile fracture. This is why we have adopted the third definition of failure in order to measure the lifetime of our test specimens, knowing that, in all cases, monitoring the damage would lead to the possibility of using this parameter and therefore of being able to use another means of detecting failure. In the following, the term failure will mean separation of the test specimen.

2.3.1 Static characteristics

In the first instance, it is useful to be able to identify static failure of the material so as to be able to define a failure criterion, that is to say a lifetime corresponding to half a cycle. Table 1 gives the results of the static tests.

Table 1 : Static failure stresses (MPa)

$\dfrac{\sigma_{\theta\theta}}{\sigma_{zz}^{static}}=0$	$\dfrac{\sigma_{\theta\theta}}{\sigma_{zz}}=2$	$\dfrac{\sigma_{\theta\theta}}{\sigma_{zz}}=\infty$
$\sigma_{\theta\theta}^{static}=75\pm2$	$\sigma_{\theta\theta}^{static}=605\pm5$	$\sigma_{\theta\theta}^{static}=550\pm10$

The effect of frequency was studied in the case of tensile loading. To do this, tests were carried out at various rates. Figure 2 gives the various results. It shows in particular that the tensile strength is independent of the rate of loading for the range of rates used in this study. It will also be noted that the tensile strength always exhibits very low scatter about the mean value.

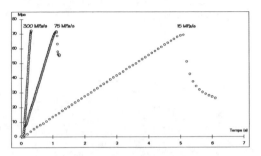

Figure 2 : Effect of rate of loading on the tensile strength

2.3.2 Morphology of static fracture surfaces

The photographs (1,2,3) show the various types of fracture. Tensile fracture is characterized by the presence of a macrocrack parallel to the direction of the fibres and over the entire length of the test specimen. The presence of delamination is noted around this defect; failure in tension is governed by the matrix. In contrast, failures in pure internal pressure and in internal pressure with restrained end closure are related to fracture of the fibres which lead to local macrocracking in the test specimens.

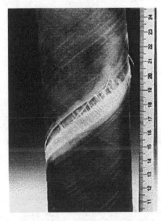

Photo 1 : Tension fracture mode

This defect stemming from failure is similar to an explosion phenomenon. The fracture surfaces in both these cases are the same as those obtained in fatigue.

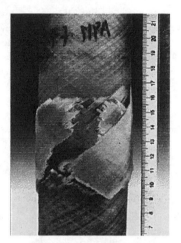

Photo 2 : PIP fracture mode

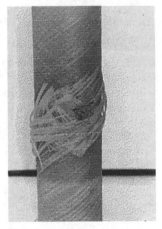

Photo 3 : PICE fracture mode

2.3.3 Fatigue lifetime

Figure 3 shows the variation in the number of cycles to failure for the three types of loading for the frequency of 0.2 Hz. In this figure, the stresses are normalized, that is to say, on the ordinate axis, we have plotted the ratio $\dfrac{\sigma^{max}}{\sigma^{static}}$. It is thus possible to compare the various types of stresses. It is observed that, for a fixed ratio value, the pure internal pressure (PIP, $\sigma_{zz} = 0$) and internal pressure with restrained end closure (IPEC, $\sigma_{zz} = \frac{\sigma_{\theta\theta}}{2}$) loadings are more detrimental to lifetime than tensile loading. Figure 4 shows these results in the form of an isonumber of cycles to failure. Analysis of these results is consistent with those already observed in the case of tension/torsion loading (Perreux et al ; 1989).

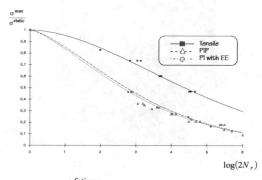

Figure 3 : $\dfrac{\sigma_{max}^{fatigue}}{\sigma_{fracture}^{static}}$ as a function of the number of cycles to failure (Nr) for a frequency of 0.2 Hz.

It is thus possible to express the variation of these curves by noticing that, if $F(\sigma_{zz}, \sigma_{\theta\theta}) = 1$ denotes the equation of the static criterion, the curves of isonumbers of cycles to failure may be written in the form:

$$F\left(\sigma_{zz}^{maxi} - U(Nr), \sigma_{\theta\theta}^{maxi} - V(Nr)\right)$$
$$= 1 - B \, Log \, (2 \, Nr) \qquad (1)$$

U(Nr) and V(Nr) are experimentally determined functions.

Equation (1) simply expresses that obtaining a curve of isonumber of cycles to failure is achieved by isotropic and kinematic hardening of the static fracture criterion.

In the specific case of a single loading direction, these results are described in the form of a Wolher curve. Equation (1) may then be simplified. For example, in the case of tension, we may obtain .

$$\sigma_{zz}^{maxi} = A - B \log 2Nr \qquad (2)$$

where:

$$A = \sigma_{zz}^{static} - U(Nr) \qquad (3)$$

In fact, $U(Nr) = U$ (for $Nr > 50$) may be considered as a constant (Joseph and Perreux ;1994), but dependent on frequency. Of course, these results are valid for $Nr > 10^6$. Beyond this, our tests do not allow us to settle on the linear form of the Wolher curves in tension.

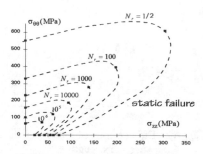

Figure 4 : Curves of the isonumbers of cycles to failure (f = 0.2Hz)

The influence of frequency on lifetime appears in the U(Nr) expression. Whatever the type of loading, the lifetime (number of cycles to failure) increases with frequency. Thus, the PIP tests show that the mean lifetime is multiplied by 3 when the frequency goes from 0.2 Hz to 1 Hz. This factor of 3 is also found in the IPEC tests. The tensile tests (Figure 5) themselves are slightly less sensitive since, for the same increase in frequency, the mean lifetime is multiplied only by 2.

The effect of frequency has already been observed by many authors, but the observations are sometimes contradictory. For Thionet and Renard (1994) for example, the frequency does not have an influence on the lifetime whereas, for Sims and Gladman (1978), it increases the lifetime, this being consistent with our own observations.

In fact, the role of frequency is relatively complex since it acts at two levels. In the first place, taking into account the viscoelastic behaviour of a glass-epoxy laminate, its increase leads to a rise in temperature of the test specimen, thus modifying the test conditions. In this case, the lifetime of the test specimen depends on its geometrical shape and therefore on its ability to dissipate the heat generated by the cycles. In other words, the lifetime obtained may not be intrinsic to

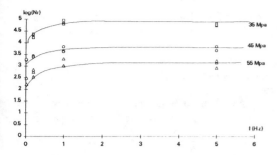

Figure 5 : Effect of frequency on Nr

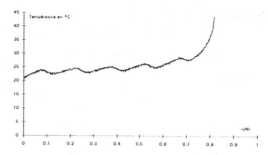

Figure 7 : Thermocouple located on the macrocrack

the material for high frequencies but is a function of the shape and size of the test specimens.

As regards the second level, this concerns the interactions between fatigue and creep. The reason for this is that, the number of cycles being identical for a lower frequency, the total loading time is greater and therefore the lowering of the frequency can have an influence on the lifetime by the simple fact of the loading time leading to the test specimen creeping to a greater or lesser extent.

In our case, we may show that it is effectively the second influence level which dominates. In fact, complementary experiments in tension were carried out (J & P ;1994). The test specimens, instrumented with thermocouples, showed (Figure 6) that the temperature of the test specimen at the highest frequency was virtually constant during the test except (Figure 7) for the thermocouples placed at the point where the macrocrack appeared, at which the temperature could rise by 40°C after the appearance of the macrocrack, that is to say at approximately 80% of the lifetime. (We will use this point again in the part concerned with damage). In addition, low-frequency tests, for which the test specimens were placed in a temperature-regulated

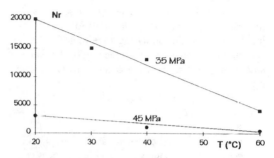

Figure 8 : Effect of the temperature, imposed on the test specimen, on the lifetime at 0.2 Hz

chamber, showed that the temperature rise decreased the lifetime (Figure 8).

These two complementary experiments show that, in our case, the thermal effect of frequency is not important. The reason for this is that the temperature of our test specimens is virtually constant and, moreover, an increase in temperature (related to the increase in frequency) would decrease the lifetime. However, this is exactly the contrary of our observations.

In conclusion, the frequency effect is related to the fatigue/creep interactions. This result will, moreover, be used in the modelling of the effect of frequency on damage.

2.4 Damage

2.4.1 Damage indicator

The anisotropy of the material leads to anisotropic damage. In addition, if it is desired to measure damage, it is necessary to determine the minimum indicator number. In fact, previous studies (Perreux and Thiebaud ; 1995) have shown that, in the case of homogeneous damage through the thickness of the material, it was possible to

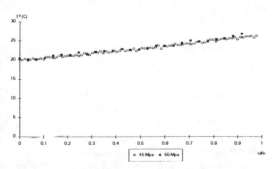

Figure 6 : Monitoring the surface temperature of a test specimen at 5 Hz

95

determine the variation in the stiffness tensor $\tilde{\underline{C}}$ of the damaged laminate by knowing the variation in the axial modulus of the test specimen $D_{ZZ} = \dfrac{\Delta E_{ZZ}}{E_{ZZ}}$ through:

$$\tilde{\underline{C}} = \underline{C} + D_{ZZ}\underline{C}_1 + D_{ZZ}^2\underline{C}_2 \qquad (4)$$

$\underline{C}$ is the virgin stiffness, $\underline{C}_1$, $\underline{C}_2$ are two tensors obtained by means of a homogenization method (P & T ; 1995, Perreux and Oytana ; 1993)].

Thanks to this result, it is only necessary to know D_{ZZ} in order to obtain the variation in all the stiffnesses of the material. The hypotheses allowing the use of Equation (4) are satisfied for the type of loading used in this study (damage must be due to fibre/matrix debonding or to cracking of the matrix, and it must have the same crack density in each ply).

2.4.2 Variation in D_{zz}

Figure 9 shows the typical variation in D_{ZZ} for the three types of loading.

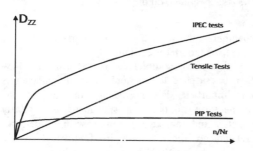

Figure 9 : Typical variation in D_{ZZ}

Given the appearance of local delamination in the vicinity of the macrocrack in the tensile tests beyond 80% Nr, the damage can only be monitored for cycles below this threshold. Above it, the measurement of D_{ZZ} is not significant for quantifying the damage.

Figure 9 shows that the variation in D_{ZZ} is different, depending on the type of loading. In tension, D_{ZZ} varies linearly. For the PIP tests, D_{ZZ} varies rapidly during the first cycles and then very quickly stabilizes. For the IPEC tests, D_{ZZ} varies rapidly in first cycles and thereafter more slowly (approximately linearly). Comparison of these variations seems to indicate that, for the IPEC tests ($\sigma_{\theta\theta} = 2\sigma_{zz}$), the damage is the sum of the damages

obtained for a PIP test ($\sigma_{zz} = 0$) and for a tensile test ($\sigma_{\theta\theta} = 0$). Moreover, this idea has been used for empirically modelling damage (J ; 1995).

2.4.3 Frequency effect

As on the lifetime, the frequency has a very great effect on the rate of damage per cycle $\left(\Delta D_{zz}(n) = \dfrac{dD_{zz}}{dn}\right)$. However, this effect is not observed with all loading types. In particular, the PIP tests are not very sensitive to the increase in frequency whereas, on the contrary, in tension $\Delta D_{zz}(n)$ varies rapidly with frequency. Given the linearity of the D_{zz} variation in the tensile tests, $\Delta D_{zz}(n)$ is independent of the cycle n in question. Figure 10 shows, for 3 levels of maximum fatigue stress in tension, the variation in ΔD_{zz} as a function of frequency.

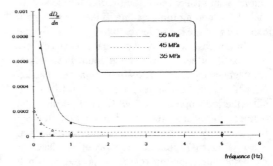

Figure 10 : ΔD_{zz} in tension as a function of frequency

Given this result, the level of damage obtained for n = 80% Nr is very sensitive to frequency. This level corresponds to a critical damage of the test specimen. These results are qualitatively valid for the IPEC tests.

3 MODELLING DAMAGE IN FATIGUE LOADING

3.1 General comment

The experimental analyses have shown that lifetime and damage are sensitive to frequency. Since thermal effects may be neglected in our tests, the only possible cause for explaining the influence of frequency resides in coupling between fatigue and creep. In addition, if it is hoped to account for damage in a fatigue test correctly, a sufficiently developed model is required which can at the same time be used to describe creep damage. The entire problem therefore resides in choosing a suitable description of the kinetics since, if the effect of a microcrack is independent of the fact that it may be created either by a creep stress or a fatigue stress, it is necessary to ascribe, to this crack, growth kinetics independent of the type of stress that creates it. This means that the number of cycles or the creep time are not variables which can be used to describe the variation in damage but only means of presenting it. The variation in damage must therefore be written in an incremental form, the rate of damage per cycle or for a creep time being obtained by integration of the incremental form.

3.2 Basic model

The basis of the model that we present may be summarized in 3 principal equations. Firstly, Equation (4) which allows us to determine $\tilde{C}$ as a function of D_{zz}. Secondly, in order to obtain the variation in D_{zz}, it is necessary to define its associated variable (Y_{zz}) with respect to the free energy density Ψ:

$$Y_{zz} = \frac{\partial \Psi}{\partial D_{zz}} = \frac{1}{2}(\underline{C}_1 + 2D_{zz}\underline{C}_2):\varepsilon_e:\varepsilon_e \quad (5)$$

and then thirdly by the choice of a dissipation potential Φ we obtain:

$$\dot{D}_{zz} = -\frac{\partial \Phi}{\partial Y_{zz}} \quad (6)$$

For a fatigue test, $\Delta D_{zz}(n)$ may be written in the form:

$$\Delta D(n) = \int_{(n-1)/f}^{n/f} \dot{D}_{zz}dt = \int_{(n-1)/f}^{n/f} -\frac{\partial \Phi}{\partial Y_{zz}}dt \quad (7)$$

Where f represents the frequency which appears here explicitly.**The entire problem is to determine Φ !**

3.3 Proposal for Φ

This analysis has been partially undertaken in the case of loading of the creep type (Thiebaud and Perreux ; 1995) and allows us to separate, in Φ, those aspects of an instantaneous type from those aspects which are time-dependent. This allows us to rewrite (6) in the form:

$$\dot{D}_{zz} = -\lambda \frac{\partial f_D}{\partial Y_{zz}} - \frac{\partial \varphi^*}{\partial Y_{zz}} \quad (8)$$

where λ_D is a Lagrangian multiplier, and f_D is a damage criterion whose form may be provided by (Perreux et al 1995):

$$f_D = <-Y_{zz}> - R_D - Y_C \le 0 \quad (9)$$

where Y_C is a threshold and R_D is an isotropic hardening variable. Figure (11) gives the form of this criterion in the stress plane $(\sigma_{zz}, \sigma_{\theta\theta})$ as well as the variation of its form due to the hardening produced by the damage described by (P&T ;1995). By using this criterion, it is possible to determine the value of D_{zz} given by the first fatigue half-cycle. After the 2nd cycle, the variation in damage is then only a function of φ^* (according to Equation (7), in which φ^* replaces Φ). In order to describe the creep stresses (T&P ; 1995), it is proposed to write φ^* in the form:

$$\varphi^* = \frac{K_D}{n_D +1}\langle\langle -Y_{zz}\rangle - Y_c + g(Rd,\overline{Dzz})\rangle^{n_D +1} \quad (10)$$

Where K_D, and n_D are constants and $g(Rd,\overline{Dzz})$ a function defined in (T&P ; 1995). It may be shown that this form is unsuitable for fatigue-type stresses (Perreux and Joseph ; 1994). In fact, the problem is related to the fact that, for a creep stress equal to the maximum fatigue stress, the rate of damage in creep is extremely low whereas it is relatively high in fatigue. In other words, the use of the difference $<-Y_{zz}>-Y_C$ as a driving variable of the damage is insufficient; an equilibrium variable, which we denote by Z, must necessarily be introduced. In order to allow correct simulation of the creep and fatigue stresses, Z kinetics and a potential φ^* in the following form may be proposed :

$$\varphi^* = \frac{K_D}{(n_D +1)(D_c - D_{zz})^2}\langle\langle -Y_{zz}\rangle(D_c - D_{zz})^2 \left.\begin{array}{l} \\ - Z - Y_c + (D_c - D_{zz})^2 g(Rd,\overline{D}_{zz}) \rangle^{n_D +1}\end{array}\right\} \quad (11)$$

and

$$\left.\begin{array}{l} \dot{Z} = 0 \text{ if } \langle -Y_{zz}\rangle(D_c - D_{zz})^2 < Y_c \\[2mm] \dot{Z} = \frac{1}{\tau_1}(\langle -Y_{zz}\rangle(D_c - D_{zz})^2 - Z - Y_c) - \frac{1}{\tau_2}Z \\[2mm] \dot{Z} = -\dot{Y}_{zz}(D_c - D_{zz})^2 \text{ if } \langle -Y_{zz}\rangle(D_c - D_{zz})^2 \\[1mm] \quad = Z + Y_c \text{ and } \langle -\dot{Y}_{zz}\rangle < 0 \end{array}\right\} \quad (12)$$

Figure (12) explains diagrammatically the choice of this kinetic description, which makes it possible to describe complex loadings ensuring consistency with the rate of damage. This apparent complexity nevertheless has the advantage of versatility.

K_D, n_D, τ_1, τ_2 are constants, D_C is the critical damage which depends on the type of stress ($D_C = 1$ in tension, $D_C = 0.8$ in IPEC, and $D_C = 0.3$ in PIP). The parameters (n_D, K_D, τ_1, τ_2) are determined by means of the tensile tests. For a non-linear regression of the Levenberg-Marquardt type.

Figures 13 and 14 give the results of simulation, obtained in tension, PIP and IPEC by using Equations (7) (8) (9) (11) (12).

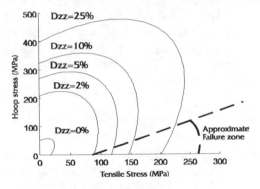

Figure 11 : Variation in the damage criterion

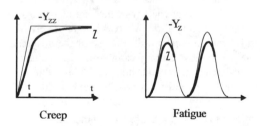

Figure 12 : Variation in Z

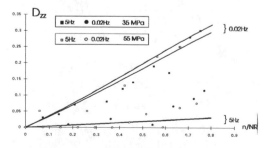

Figure 13 : Simulation in tension

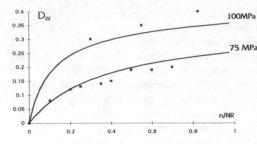

Figure 14 : Simulation in IPEC

CONCLUSION

In this study, we were concerned with modelling fatigue tests on $[+55,-55]_n$ laminates. The method may be easily generalized to another type of laminate $[+\theta,-\theta]_n$. We have examined the effect of frequency and type of stress on lifetime and on damage. By means of a micromechanical model, the damage can be described by means of a scalar, for which we have proposed variation kinetics which allow us to take into account fatigue/creep interactions which are the basis of the frequency effects.

Given the various kinds of phenomena encountered, it is clear that the thermomechanical model that we propose is relatively complex with a number of parameters which may seem large. It seems clear to us that a model of the behaviour of composite materials cannot be both a generalizing one and a simplifying one. If a model is sought which describes purely the fatigue, without the frequency effect, it is possible to simplify the equations given, but if it is desired to describe the complete phenomenon it is necessary to introduce a degree of inconvenience, such as a large number of parameters. Nevertheless the method proposed here is a stage to understand the time dependent behaviour. Indeed the équation of the Z variable must be improved to predict long term tests.

Further progress may be achieved in order to improve the performance characteristics of the model ; on the one hand, it is necessary to refine the procedure for identifying the parameters which are expensive in computing time and, on the other hand, in our case, it is necessary to take into account the essential function of our material which is intended to transport a fluid. It is therefore necessary to be able to take into account the effect of moisture on the behaviour so as to incorporate its effect into a model which can be used for designing a pipe.

ACKNOWLEDGEMENT

The authors thank EDF/DER/EMA and, most especially, Mr D. Paris for the support and interest which they have provided for this work.

REFERENCES

Frost S. R. 1995 ; *The fatigue performance of glass fibre/epoxy matrix filament wound pipes.* Proceeding ICCM9,5,684-691,Madrid.1993

Hoa S.V. 1991 ; *Analysis for design of fiber reinforced plastic vessel and piping.* Technomic publication

Ladeveze P and Ledantec E. 1992 ; *Damage modelling of the elementary ply for laminated composites,* Composites Science and Technology, 43, 257-261.

Laws N. Dvorak G.J. Hejazi M. 1983 ; *Stiffness changes in unidirectional composites caused by crack systems.* Mechanics of materials 2, North Holland, 123-137.

Le Courtois T.1995 ; *PWR composite materials use : A particular case of safety-related service water.* Proceeding of Enercomp95, Monréal,835-846..

Lou Y.C. and Shapery R. A. 1971 ; *Viscoelastic characterisation of linear fiber-reinforced plastics.* Journal of composite materials, Vol5,85-107.

Joseph E. Perreux D. 1994 ; *Fatigue behaviour of glass-fibre / epoxy-matrix filament wound pipes : Tension results,* Composites Science Technology, 52, 469-480.1994.

Joseph E. 1995 ; *Modélisation du comportement en fatigue d'un composite stratifié verre-époxy : Aspects théorique et expérimental.* PhD Thésis, Université de Franche-Comté,457.

Perreux D. Oytana C. Varchon D. 1994 ; *Static and fatigue fracture of composites in complex state stress.* Proceeding ECCM3,516-520.Borbeaux.

Perreux D. Oytana C. 1993 ;*Continuum damage mechanics for microcracked composites,* Journal of composites Engineering, 3, 115 - 122.

Perreux D. Joseph E. 1994 ;Endommagement de fatigue multiaxiale : influence de la fréquence et du niveau de contraintes. Proceeding JNC9 AMAC, 667-676.Saint-Etienne. Perreux D., Thiebaud F. *Damaged Elasto-plastic behaviour of [+φ,-φ] fibre-reinforced composite laminates in biaxial loading.* Composites Science Technology. to be published.

Perreux D. Varchon D. Lebras J. 1995 *The mechanical and hygrothermal behaviour of composite piping : Notes on the methodology of dimensionning and structural design.* Proceeding of Enercomp95, Monréal,819-826.

Reifsnider K.L. and Stinchcomb W.W. 1986 ; *A critical element-model of residual strength and life of fatigue load composite coupons.* ASTM STP 907,298-313.

Salama M.M. 1994 ; *Advanced composites for offshore industry : Applications and Challenges.* Proceeding of Composite materials in the Petroleum industry, IFP. Paris

Sims G.D. and Gladman D.G. 1978 ; *Effect of test conditions on the fatigue strength of glass fabric laminated. Part A/ frequency .* Plastics and Rubber : materials and applications.

Soden P.D. Kitching R. and Tse P.C. 1989 ; *Experimental failure stresses ±55° filament wound glass fibre reinforced plastic tubes under biaxial loads.* Composites, vol 20,2,125-134.

Thiebaud F. Perreux D. 1995 ; *Overall mechanical behaviour modelling of composite laminates.* European Journal of Mechanics / Solids, to be published.

Thionnet A. Renard J. 1994 ; *Laminated composites under fating loading : a damage development law for transverse cracking.* Composites Science and Technology, 52,173-181.

Xiao X. and Cardon A.H. 1986 ; *Comportement viscoélastique de composites à matrice thermoplastique (APC-2).* Proceeding of 6th JNC. AMAC,479-490.

Fatigue in graphite-epoxy composites: Experiment and analysis

J.P. Komorowski
Institute for Aerospace Research, National Research Council Canada, Ottawa, Ont., Canada

C. Randon, C. Roy & D. Lefebvre
Faculty of Applied Science, University of Sherbrooke, Que., Canada

Abstract: Compression dominated fatigue loading was applied to the first generation graphite/epoxy system, AS4/3501-6, and the toughened system, IM6/5245C. The fatigue data were analyzed using a three parameter Weibull distribution. The results showed a significant difference in the number of cycles to failure between the two systems. Photoelastic coatings, dye penetrant radiography and edge replicas were used to document damage growth. The difference between the two systems was attributed to high residual curing strains and matrix brittleness.

INTRODUCTION

The authors have previously published data on the effect of stacking sequence on damage growth in graphite epoxy laminates under compression-dominated fatigue loading (Komorowski et al. 1995). Differences in the fatigue lives of the two stacking sequences previously studied were close to two orders of magnitude. Two material systems have been studied by the authors: a first generation graphite/epoxy (AS4/3501-6) and a toughened system, graphite fiber in bismaleimide modified epoxy (IM6/5245C). Well defined stages of the fatigue process leading to final failure were identified. Most damage observations, however, were based on results obtained using the toughened, more fatigue resistant system, IM6/5245C. Damage always initiated at the edge and was related to the magnitude of interlaminar shear stresses. In this paper the results of damage growth observations in the AS4/3501-6 system will be presented. This is the most popular system in aircraft primary structures to date (i.e. F-18, AV8-B). Damage growth was monitored using X-rays, photoelasticity and edge replicas. The damage growth studies were undertaken to find a possible source for the significant differences in fatigue lives observed for the two material systems.

MATERIALS AND METHODS

Eighteen ply panels (610 mm x 305 mm) of IM6/5245C and AS4/3501-6 graphite/epoxy composites were fabricated using hand lay-up followed by a vacuum-bag autoclave cure. The lay-up used for this study was termed 'laminate 2':

[+45/-45/0/0/90/0/0/-45/+45]$_s$.

This lay-up results in negative normal edge stresses (σ_z) under compressive loading and displays relatively long fatigue lives.

The 2.5 mm thick panels were cut with a diamond-impregnated cutting wheel into specimens 290 mm long x 25.4 mm wide. The virgin specimens were inspected ultrasonically for manufacturing flaws. The specimens were tested using a 200 kN servo-hydraulic testing machine under stroke control for static tests and load control for cyclic loading. Anti-buckling guides supported the central portion of the specimens during static compression and compression-tension fatigue tests. A thin Teflon sheet was inserted between the specimen and the anti-buckling guide to minimize friction. The test fixtures provided good lateral support but left much of the specimen, including the edges, unconstrained and accessible for inspection and extensometer mounting (Matondag 1984). The load was transferred to the specimens through hydraulic grips which maintained a constant pressure of about 34 MPa over a 60 mm length at each extremity. The gripping areas of both the specimen and the re-usable steel tabs were lightly sand-blasted to prevent slippage during loading.

Fatigue tests were conducted using load control with a sinusoidal wave form under a load ratio R=P$_{min}$/P$_{max}$= -3.75 and a frequency of 2 to 5 Hz. This load ratio is close to the upper skin wing root area high loads. A commonly used ratio (R=-1) is more representative of typical vertical stabilizer loading than an aircraft wing and at higher load levels, is more likely to cause delamination initiation during the tensile part of loading cycle. P_{min} varied from 0.6 of P_{uc} (P_{uc} = ultimate static strength in compression) to 0.7, 0.8 and 0.9 of the P_{uc}. The P_{uc} load for each laminate was determined from a number of static tests and values are shown in Table 1 (Mean). The number of cycles to failure N recorded for the tests corresponds to complete loss of load carrying capability.

STATISTICAL TREATMENT OF EXPERIMENTAL DATA

Since 1987, test data for laminate 2 of both composite systems, AS4/3501-6 and IM6/5245C, have been generated. For static loading, tests were done in compression as well as in tension. A statistical analysis was carried out to estimate the parameters of the distributions for the data. All the tests were conducted under similar conditions. However, different batches of material were used over time which, with other factors, might have influenced the experimental results. An analysis of the variance was performed to check for the influence that time and material batch had on static data. It was found that these factors did not have a significant effect on the static strength.

Estimation of the distribution parameters for static data

Strength data are usually well described using the two parameter Weibull distribution (MIL-HDBK-17B 1988, Park 1982). The probability density function for the Weibull distribution is given by:

$$f(x) = m\beta^{-m}(x - x_0)^{m-1} \exp\{-[(x - x_0)/\beta]^m\} \quad (1)$$

where β, m and x_0 are the scale, shape and location parameters, respectively. For the two parameter Weibull distribution, x_0 is set equal to zero. There are several methods for the estimation of the parameters of the Weibull distribution, among them, the use of Weibull paper, least squares and the maximum likelihood method. In this study, the parameters of the distributions of the static data were estimated using the maximum likelihood method. The results are summarized in Table 1 along with the mean value and the coefficient of variation.

The distributions obtained in all the different cases have a shape that is close to the normal distribution. Goodness-of-fit tests for the Weibull and normal distributions (Anderson-Darling tests as described in MIL-HDBK-17B) were performed for the populations containing at least 10 data points.

The results are shown in the last two columns of Table 1, where a value higher than 0.05 indicates that the distribution fits the data. It can be seen that the Weibull distribution is suitable for static data even if for some cases, the normal distribution might be appropriate.

Estimation of the parameters for fatigue data

The fatigue data can best be described with the three parameter Weibull distribution. An assumption often used in the literature (Park 1982, Tanimoto 1989) is that the shape parameter does not depend on the load level of the fatigue test. This allows the data to be normalized by the average obtained for each load level and gathered in a unique set. The three parameters are then estimated for the overall data. The location parameter is determined graphically on Weibull paper or on regular scale after an appropriate logarithmic transformation (Hallinan 1993). If the value of the location parameter is not chosen properly, the plot of the data will be concave instead of straight. The method consists of choosing an initial value for x_0 (which is not bigger than the smallest data value) and iterating until the plot forms a straight line. Once the location parameter is found, the shape and scale parameters can be estimated by either least squares or maximum likelihood methods. The results for the two material systems are presented in Table 2.

The parameters for the two material systems are very similar and significantly different from those obtained for static data. The data are plotted in Figure 1 along with the normal, two parameter Weibull and three parameter Weibull distributions. From this figure and the goodness-of-fit values from Table 2, it can be seen that the three parameter Weibull distribution provides the best fit to the data.

Based on the estimation of the parameters, the number of cycles corresponding to 10%, 50% and 90% of the specimens failed have been calculated for each load level and composite systems. They are plotted in Figure 2.

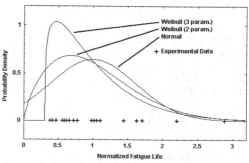

Figure 1. Distribution of the normalized fatigue data for laminate 2 of AS4/3501-6

Table 1. Two parameter Weibull distribution values, descriptive statistics and goodness-of-fit tests.

Material	Test	Number of data points	m	β MPa	Mean MPa	CV %	GoF test Weibull	GoF test normal
IM6/ 5245C	Tension	10	25.9	1283	1254	5.1	0.247	0.094
	Comp.	10	11.50	734	705	8.5	0.31	0.546
AS4/ 3501-6	Tension	11	13.21	1055	1014	9.1	0.567	0.358
	Comp.	10	16.27	750	725	8.4	0.52	0.12

Table 2. Three parameter Weibull distribution values of laminate 2 for normalized fatigue data.

Material	Number of data points	x_0	m	β	CV %	GoF test Weibull	GoF test normal
IM6/ 5245C	27	0.17	1.201	0.88	62	0.59	0.001
AS4/ 3501-6	26	0.31	1.19	0.73	74	0.77	0.004

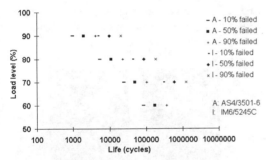

Figure 2. Fatigue results for IM6/5245C and AS4/3501-6, laminate 2

DAMAGE GROWTH STUDIES

Damage initiation

In an attempt to observe damage effect and growth, defects were introduced in some specimens. Two different methods were used. Artificial edge delaminations were introduced with the aid of a symmetrical surgical blade (2 mm wide, 0.35 mm thick) at the 8th interface (-45/+45, marked as delamination location 4 in Figure 3). This resulted in a 3.5 mm deep delamination. Two delaminations per specimen were introduced, at the same interface, in the middle of the specimen gauge length on opposite edges.

The other method used to introduce damage in specimens consisted of running 300 compression dominated load cycles at 0.9 P_{uc}, with a load ratio

of -3.75. This cycling was done at room temperature and produced delaminations at the 1st and 8th interfaces (between the 45° plies) and for some specimens, cracks in the 90° plies. This method was considered because initial tests with knife inserted damage did not seem to affect specimen life in AS4/3501-6 specimens (contrary to the experience with IM6/5245C specimens - Komorowski 1995). Since the fatigue failure mode observed at 0.9 P_{uc} was similar to that observed at other load levels it was speculated that high load cycling would shorten the pre-initiation period allowing the authors to concentrate on the damage growth period. This method of damage introduction is still under evaluation.

X-ray and photoelastic coatings comparison

Edge replication was used throughout the fatigue testing. The replicas obtained on the tape were placed between two glass slides and examined under an optical microscope. The damage observed in the replicas was correlated with that obtained from two NDI techniques used to observe damage growth in the plane of the specimens: radiography and photoelasticity.

For radiography, a zinc iodide solution was spread on the edges and left overnight to penetrate into cracks and delaminations. The excess dye was removed before the radiograph was taken. The voltage and current used were 27 kV and 2.5 mA respectively, with an exposure time of 30 s.

Photoelastic coatings were used to observe the damage progression, in real time, during testing. Since the specimens used are coupons subjected to uniform strain, any non-uniformities in the observed isochromatic patterns are related to damage accumulation in the specimen. The coatings are sensitive to the difference in the principal normal strains:

$$\gamma_{max} = \varepsilon_1 - \varepsilon_2 = kM \tag{2}$$

where γ_{max}, ε_1 and ε_2 are shear and normal strains, k is a photoelastic constant and M is the isochromatic value. The coatings used throughout this study were 0.25 mm thick, had 10% maximum elongation, an elastic modulus of 2.5 GPa and a Poisson's ratio $v = 0.38$. Experience with photoelastic coatings indicates that coating disbonds produce very distinct and easy to observe patterns.

Due to the accumulating damage, both the stiffness and the Poisson's ratio in the specimen change and thus affect the response of the coating during loading. The authors were particularly interested in recording local damage where the changes to the laminate would be different from the rest of the specimen. During the fatigue tests, at selected intervals, images were recorded with a CCD camera, using a circular polariscope and an interference filter while a static load ($0.3P_{uc}$) was applied to the specimen. Images were stored using a frame grabber installed in a PC. The PC was running software developed for automated photoelasticity (Komorowski 1991). This technique allowed damage growth to be monitored in situ as opposed to ultrasonic C-scan inspections or penetrant-enhanced radiography which generally require removal of the specimen from the test machine. The photoelastic coating technique proved to be very effective. However, it is recognized that this method may not be equally effective for thicker laminates. Figure 3 shows the relationship between the location of the damage and Poisson's ratio for the sublaminate attached to the coating for laminate 2. As can be seen the coating is less sensitive to damage (delamination) in some layers. This led to the selection of the PC based method of image recording over simple still colour photography and Tardy compensation used by the authors in previous damage growth studies.

Figure 4 shows photoelastic and radiography images of specimen 759-10. The observed damage zones with both methods are in good agreement. The photoelastic coating method while simple in principle proved to be difficult to implement in an environment were operator turnover rate was high. The method requires training in photoelasticity (abandoned by many universities) and consistent and meticulous application. The X-ray/dye penetrant method produces excellent damage definition, i.e. very small cracks in the specimen may be resolved provided that penetrant has been applied properly. For fatigue studies a question

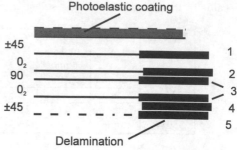

Delamination location	Sublaminate Poisson ratio v	Photoelastic effect % change (M/M^*-1)
1	0.72	- 70
2	0.65	- 42
3	0.29	18
4	0.35	8
5	0.4 (laminate 2)	0

Figure 3. Relative change in the photoelastic effect (isochromatic number) depending on the location of damage (delamination).

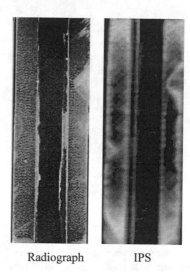

Radiograph   IPS

Figure 4. Specimen 759-10 comparison between radiograph and photoelastic image (IPS).

remains regarding the effect of the old penetrant on subsequent load cycling and new penetrant application during the next inspection period. This method requires removal of the specimen from the load frame, several hours for penetrant application and the use of X-ray equipment with associated health risks and certified personnel to operate it. These are significant impediments to the wide application of this method. The authors suggest that both methods be used, in-situ photoelasticity for quick real time assessment of damage and radiography for precise measurement and assessment of the accumulated damage. Edge replicas are needed to determine the location of damage within the laminate thickness.

DISCUSSION

Data on the life of the specimens, Figure 2, show significant differences between the two material systems (more than one order of magnitude in the number of cycles to failure). Since the static compressive strength differed by only 3%, the cause of this large difference was not obvious. It was hoped that more detailed damage growth observations would help to explain this difference. The typical fatigue damage sequence observed in IM6/5245C specimens (Komorowski 1995) was divided into three phases. Phase one, preinitiation, during which no damage was recorded, lasted for approximately half the specimen life. In phase two, damage appeared in the 8th and 9th plies ($\pm 45^0$) and progressed into the center of the specimen from one or more locations along the edges. At about 95% of the specimen life, damage was observed in the 90^0 ply which led to the final failure of the specimen (phase three).

Damage was also observed to occur in the $\pm 45^0$ and 90^0 plies for the AS4/3501-6 specimens which suggested that the fatigue and failure processes in both systems were essentially similar. However, subsequent studies involving artificial damage in the $\pm 45^0$ plies revealed that the damage did not affect the AS4/3501-6 specimen life while it did shorten the life of IM6/5245C specimens. Examination of X-ray images after only 300 load cycles and again at 20,000 cycles (approximately one tenth of the specimen life) revealed a high incidence of cracks in the 90^0 ply (Figure 5 and 6).

The early appearance of these 90^0 cracks (as compared to IM6/5245C specimens) might explain the significantly shorter fatigue lives observed in AS4/3501-6. This was unexpected since the loading was predominantly compressive. Even at 0.9 P_{uc} loading the maximum strain on the tensile cycle was $\varepsilon = 0.26$ ($\varepsilon_{90u} = 0.56$ for AS4/3501-6). It seems that the early appearance of these cracks was most likely related to the high residual tensile curing strain. The 90^0 cracks do not appear early in the IM6/5245C specimens since $\varepsilon_{90u} = 0.72$ and the matrix is tougher than the 3501-6 epoxy. Further substantiation for the curing residual strain being responsible for lowering the fatigue life of the specimens comes

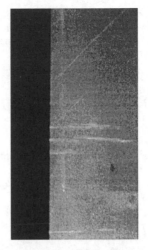

Figure 5. Radiograph showing cracks in 90^0 ply (perpendicular to specimen edge), specimen 760-13 at 20,000 cycles.

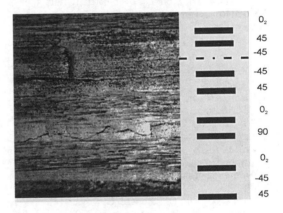

from a study being carried out in parallel with the work described here. That study involves humidity and temperature effects on fatigue of composite systems. Wet AS4/3501-6 specimens cycled at -50^0C did not contain 90^0 ply cracks up to 20,000 load cycles. The moisture effectively lowered the residual strains thus precluding early appearance of the 90^0 ply cracks.

CONCLUSIONS AND RECOMMENDATIONS

The three parameter Weibull distribution best describes the fatigue data for compression dominated loading.

An order of magnitude improvement in the number of cycles to failure was observed for the toughened graphite/epoxy system over the first generation brittle system.

It is speculated that high residual stresses significantly lowered the fatigue life of the first generation system.

It is recommended that moisture content be controlled in specimens used for fatigue studies since it modifies the residual strains and then damage growth during cycling.

Photoelastic coatings and dye penetrant radiography in conjunction with edge replication were very effective in documenting damage growth under fatigue loading.

ACKNOWLEDGEMENTS

This work has been performed under NRC Project 3G3, Aerospace Structures, Structural Dynamics and Acoustics, Sub-Project JGK-00, Fatigue and Environmental Test Techniques for Composites. The work has been partially funded by the Department of National Defence Financial Encumbrance 220794NRC08.

REFERENCES

Hallinan, A.J., Jr. 1993. A Review of the Weibull Distribution, *Journal of Quality Technology*, Vol. 25, No. 2, pp. 85-93.

Komorowski J.P., M. Foster 1991. Image Processing System with special routines for automated analysis of photoelastic images, *User Manual*, National Research Council Canada, IAR ST-613.

Komorowski, J.P., D. Lefebvre, C. Roy and C. Randon 1995. Stacking Sequence Effects and Delamination Growth in Graphite/Epoxy Laminates Under Compression-Dominated Fatigue Loading, *Composite Materials: Fatigue and Fracture, ASTM STP 1230*, R. H. Martin, Ed.

Matondag, T.H., D. Schutz 1984. The Influence of Anti-Buckling Guides on the Compression Fatigue Behavior of Carbon Fibre-Reinforced Plastic Laminates, *Composites*, Vol. 15, No. 3, (July 1984), pp.217-221.

MIL-HDBK-17B 1988. Military Handbook - Polymer Matrix Composites, Vol. 1 Guidelines, *Department of Defense, Washington, D.C.*

Park, W.J., R.Y. Kim 1982. Statistical Analysis of Composite Fatigue Life, *Progress in Science and Engineering of Composites*, ICCM-IV, Tokyo, pp. 709-716.

Tanimoto, T., Hir. Ishikawa, Hid. Ishikawa 1989. Fatigue Design and Reliability Analysis of CFRP Composites, *5th International Conference on Structural Safety and Reliability*, pp. 1571-1578.

Progress in Durability Analysis of Composite Systems, Cardon, Fukuda & Reifsnider (eds)
© 1996 Balkema, Rotterdam. ISBN 90 5410 809 6

The fatigue and temperature dependent properties of PE-fibre reinforced polymers

K. Schulte
Polymer Composites Section, Technical University Hamburg-Harburg, Germany

R. Marissen
DSM Research, Geleen & Delft University of Technology, Netherlands

H. Omloo
DSM Research, Geleen, Netherlands

K.-H. Trautmann
DLR, Institute of Materials Research, Cologne, Germany

ABSTRACT: The Dyneema SK 60, 800dTex is a low density (0,97 g/cm^3) high performance synthetic fibre based on polyethylene. It combines high tensile strength and modulus with low specific weight, with the drawback, however, that its properties only remain stable up to a maximum temperature of about 130°C. It has the potential for use in composite materials. In the past some literature has been published on polyethylene fibre reinforced polymers, with the matrix of the epoxy type (best summarized in Peijs, 1983). However, the fatigue behaviour of polyethylene fibre reinforced epoxies has not yet been studied.

1 INTRODUCTION

Cyclic loading of a polymer, either reinforced or not, leads in general to a heat up of a sample. The amount of temperature increase mainly depends on the applied load and frequency. The kind of polymer and, in the case of a reinforced polymer, the volume fraction of fibres, their thermal conductivity and the stacking sequence influence the heat generation under cyclic loading. Fatigue induced damage additionally influences the temperature increase. Stiffness reduction and temperature change give comparable results and both can lead to a similar interpretation of a damage process (Neubert et al.,1990).

High performance polyethylene fibres show a pronounced time dependent behaviour under static loading conditions. An increase in strain rate and/or decrease in temperature results in an increase in fibre modulus and strength, but a decrease in work of fracture (Peijs 1993). It is further known, that even in unidirectional PE-fibre reinforced laminates creep can be observed. How far this specific behaviour influences the fatigue behaviour is of great interest and has to be investigated, in order to find the appropriate application for PE-composites.

2 EXPERIMENTAL

2.1 Materials investigated

Prepregs with polyethylene fibres were manufactured on a drum by wet winding of corona treated Dyneema SK 60 fibre bundles. The epoxy system used was Schering Eurepox 730 with XE 279 as a curring agent. After the winding process the prepreg was pre-cured in an oven under atmospheric pressure at 90°C for 25 min.

Laminates with two different stacking sequences: [0$_3$] and [0, 90, 0,] were cured under pressure and vacuum control at 120°C for 2h. The laminates were machined into test panels with a length of 260 mm and a width of 20 mm using a water jet cutter.

2.2 Mechanical tests

Monotonic tensile tests were performed under stroke control on both, fibre bundles, which were heat treated before tensile testing and on laminate samples. The tests were performed under stroke control at a rate of s = 1mm/min.

All fatigue tests were performed at a stress ratio of R = 0,1 and at the frequencies of 3 Hz, 10 Hz and 30 Hz. During the tests the temperature development in the test coupons was continuously measured using Fe/Constantan thermocouples glued directly on the specimen surface with an epoxy adhesive. A special cooling system was fixed between the grips and the loading system to avoid a heat flux from the actuator into the specimen. Additionally the specimens were encapsulated to prevent heat convection. Elongation was measured with an extensometer.

3 RESULTS

Tensile tests were performed on each laminate. The results are summarized in Table 1. Due to the higher amount of 0°-plies the Young's modulus and the fracture stress σ_B are higher in the unidirectional than in the cross-ply laminate. However, the [0,0,0]-laminates have a higher strain to fracture than the [0, 90, 0]-laminates.

3.1 UD-Laminates

In Fig. 1 are shown the results from a fatigue test

Table 1:   Results of the tensile tests

Laminate	Initial Young-Modulus [GPa]	Fracture strain ε_B [%]	Fracture stress σ_B [MPa]
0/0/0	57,41 ± 1,36	3,61 ± 0,03	883,5 ± 45,44
0/90/0	42,09 ± 2,65	2,59 ± 0,45	710,0 ± 39,71

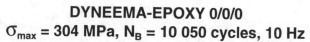

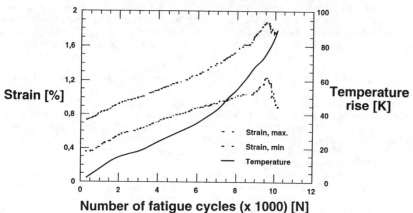

Fig.1a: Variation of maximum and minimum strain and temperature due to fatigue loading (R = 0,1), σ_{max} = 304 MPa, 10 Hz, UD-laminate

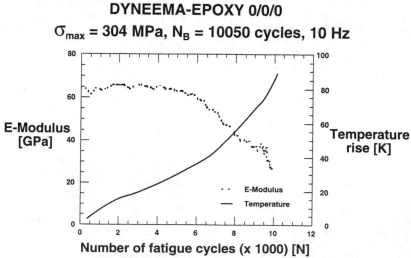

Fig.1b: Variation of Young's modulus and temperature due to fatigue loading (R = 0,1), σ_{max} = 304 MPa,10 Hz, UD-laminate

performed at a frequency of 10 Hz and a maximum stress in each consecutive load cycle which amounts to 34,4 % of the ultimate tensile strength (σ_{max}=304MPa). In Fig. 1a is plotted the variation of the maximum and minimum strain versus the number of load cycles. Incorporated into the figure is the temperature development during fatigue life. From the figure it can be seen, that during the tensile fatigue loading, a permanent creep of the composite occurs resulting in a permanent increase in maximum and minimum strain. From the strain variation the variation of the elastic modulus can be calculated. This is, together with the temperature variation summarized in Fig. 1b. Stiffness reduction and

temperature rise can be choosen as a damage analogue, in order to continuously study the fatigue dependent specimen degradation. The results show, that in contrast to composite materials reinforced with glass, aramide or carbon fibres a pronounced stiffness reduction occurs, even for unidirectional laminates. This stiffness reduction is accompanied by an increase in temperature of more 80 K. It can further be observed that for a temperature increase of more than 40 K an accelerated stiffness reduction occurs and ΔT progressively increases as well.

Tests performed at low frequencies, e.g. 3 Hz show, that no significant damage even at higher fatigue

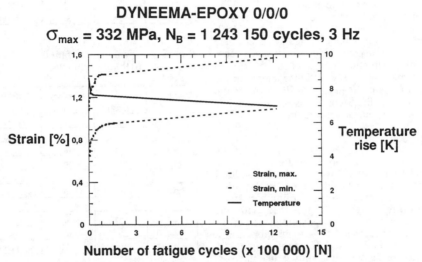

Fig.2a: Variation of maximum and minimum strain and temperature due to fatigue loading (R = 0,1), σ_{max} = 332 MPa, 3 Hz, UD-laminate

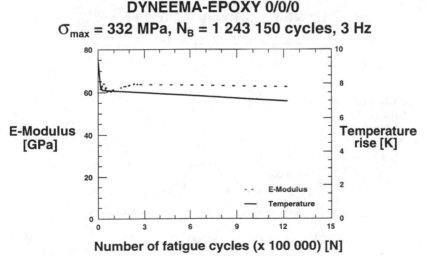

Fig.2b: Variation of Young's modulus and temperature due to fatigue loading (R=0,1), σ_{max}=332 MPa, 3Hz, UD-laminate

levels occurs (compare Fig 2). After an initial temperature rise of about 8 K it decreases first rapidly and then gradually to the value of 7 K after more than 1.2 million load cycles, when the fatigue test was stopped. However, the maximum and the minimum strain in each load cycle increases at the beginning quite rapidly (Fig. 2a). During further fatigue testing the strain variation remains constant but a continuous increase in the medium strain can be observed which gives rise to the assumption of a cyclically induced steady creep process. No significant variation in stiffness could be observed during the whole test time (Fig. 2b).

Not only the frequency, but also the maxiumum fatigue load level influences the degradation behaviour of a polyethylene composite. In Fig. 3 are summarized the results of various fatigue tests all performed at 10 Hz but with various load levels. It can be seen that when only 30% of the ultimate tensile strength (σ_{max}=270MPa) is reached in a fatigue loading situation fatigue life is close to infinity. However, with only a slight increase in the load level to about σ_{max}=304MPa a dramatic reduction in fatigue life occurs. This results also in a rapid increase in specimen temperature.

3.2 Cross-ply laminates

The behaviour of a cross-ply laminate is about the same as in UD-laminates. Only the overall load level is lower, due to the reduced amount of O°-plies. Fig. 4a shows the maximum and minimum strain variation during fatigue life together with the temperature change. Again a steady elongation of the test coupon can be observed, with in this case, a dramatic increase in the last 20% of the fatigue life. From former investigations it is known, that in cross-ply laminates with carbon, glass or aramide fibres a sudden stiffness reduction close to final failure

Table 2: Results of the fatigue tests on the [0,0,0]-laminates

Sample No.	σ_{max} [MPa]	Frequency [Hz]	Cycle No. [N]	Temperature rise [K]
2-5	270	30	9 390	95,4
2-9	270	30	7 930	98,2
2-8	270	10	(>) 720 000	14,4
2-3	270	10	(>) 720 000	11,8
2-12	304	10	10 050	87,9
2-4	332	10	8 860	93,1
2-7	332	10	7 930	90,7
2-6	332	3	(>) 1 243 150	8,6

Table 3: Results of the fatigue tests on the [0,90,0]-laminates

Sample No.	σ_{max} [MPa]	Frequency [Hz]	Cycle No. [N]	Temperature rise [K]
4-6	212	30	13 340	87,5
4-7	212	30	15 500	89,7
4-3	212	10	-	6,4
3-3	212	10	(>) 722 400	6,8
3-4	245	10	(>) 658 000	13,8
4-5	297	10	6 340	84,3
4-4	297	10	5 880	83,4
3-5	297	3	(>) 188 000	5,6

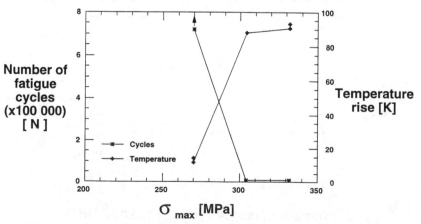

Fig.3: Total temperature and stiffness variation as a function of load level, 10 Hz, UD-laminate

occurs. This is also the case in the here investigated PE-fibre composite (Fig. 4b).

Also for a cross-ply PE/epoxid laminate a strong load dependence of the fatigue life can be observed. At a maximum load in each load cycle of about 33% of the ultimate tensile strength (σ_{max}= 245 MPa) a fatigue life of more than 650.000 load cycles can be reached, with only minor temperature increase. However, a rapid increase in specimen temperature (of more than 80K) can be observed with an increase in load to σ_{max} = 297MPa. The fatigue life is then reduced to less than 7000 cycles.

The results of all fatigue tests are summarized in Table 2 and 3, respectively.

The results show that there is independent of the laminate type, a strong dependence on the test frequency. At a frequency of 30Hz, even on the lowest maximum stress level, rupture occurs within few load cycles with a temperature rise of more than 80K, which means at more than 100°C. At a

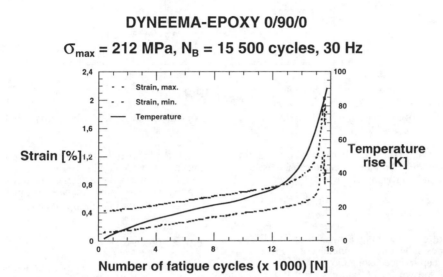

Fig.4a: Variation of maximum and minimum strain and temperature due to fatigue loading (R = 0,1), σ_{max}=212 MPa, 30Hz, cross-ply laminate

DYNEEMA-EPOXY 0/90/0
σ_{max} = 212 MPa, N_B = 15 500 cycles, 30 Hz

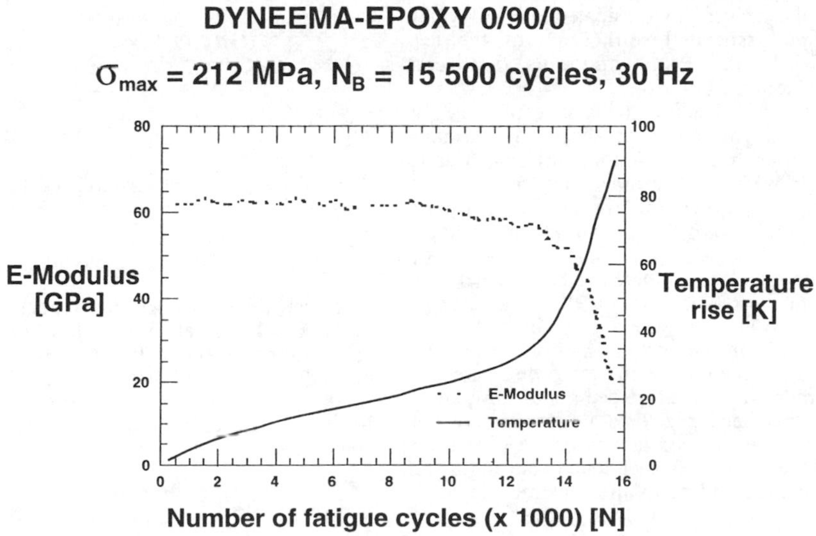

Fig.4b: Variation of Young's modulus and temperature due to fatigue loading (R=0,1), σ_{max}=212 MPa, 30Hz, cross-ply laminate

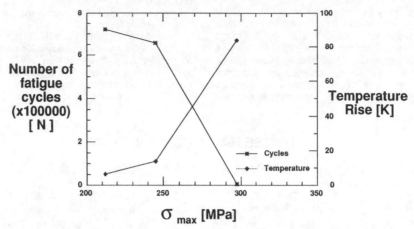

DYNEEMA-Epoxy 0/90/0, 10 Hz

Fig.5:   Total temperature and stiffness variation as a function of load level, 10Hz, cross-ply laminate

frequency of 3Hz, even at the highest maximum stress level, only a minor temperature increase can be observed and no rupture of the test coupons.

4 DISCUSSION

The results on the fatigue behaviour of PE-epoxid composites show three phenomena of interest:

1. Creep induced elongation

It is well known for PE-composites that during long term static loading a temperature dependent creep behaviour can be observed, which is similar to that of neat PE-fibres (Peijs et al. 1990).

2. Under cyclic deformation there is a continuous increase in the irreversibly dissipated energy, which results in an associated selfheating of the test coupons while the cyclic modulus decreases (stiffness reduction).

3. Due to cyclic loading, damage accumulates as matrix cracks initiate. This internal damage development has an effect on both, stiffness (due to load transfer with local strain concentration) and internal heating (due to higher local cyclic deformations which means locally higher shear stresses).

During fatigue loading there is a balance between the temperature resulting from the dissipated energy per unit volume in each cycle and the heat flux into the environment in the time t. At low frequences there is enough time for the energy to dissipate into the environment. Therefore the endurance limit under fatigue is at low frequences on a higher load level as at high frequences. In this last case, the time for a heat flux is too short. With each load cycle additional energy dissipates, leading to an accelerated internal heating with the degradation of properties.

Cyclic creep and energy dissipation can be directly made visible from the dynamic hysteresis loops. Stiffness reduction and cyclic creep result in shifting and widening of the loops.

Assuming the shape of the hysteresis loops is approximately elliptical the area of the loops, which is an equivalent to the dissipated energy ΔW, can be described by

$$\Delta W = 1,06 \, \pi a \cdot b$$

with a and b as the semiaxes of the loops (Kuksenko and Tamuzs, 1981). When testing an interval of the total energy $\Sigma \Delta W$, than ΔW has to be multiplied by the number of cycles. Kuksenko and Tamuzs gave the following expression:

$$\Sigma \Delta W = \int_{o}^{t} \frac{\psi(t) \sigma_a^2 \, \omega}{2 E_c(t)}$$

with

$\Psi = 2\Delta W / \sigma_a \cdot \varepsilon_a$ = absorption coefficient
σ_a = stress amplitude
ε_a = strain amplitude
E_c = dynamic(cyclic) elastic modulus

However, this paper will not repeat the caclculation of Kuksenko et al. We will only state that the calculations proposed can be used to describe the dependence of $\Sigma \Delta W$ on the durability of composites. In the case of creep, as it is present in the here investigated material, this has additionally to be considered.

REFERENCES

Neubert, H., Schulte, K. and Harig, H. (1990). Evaluation of the damage development in carbon fiber reinforced plastics by monitoring load-induced temperature changes. ASTM STP 1059, pp. 435.

Peijs, A.A.J.M.,(1993). High performance polyethylene fibers in structural composites? Dr. thesis, Technical University Eindhoven, The Netherlands.

Peijs, A.A.J.M., Catsman, P., Govaert, L.E. and Lemstra, P.J. (1990). Hybrid composites based on polyethylene and carbon fibres. Part 2: Influence of composition and adhesion level of polyethylene fibres and mechanical properties. Composites 21.pp 513.

Kuksenko, V.S. and Tamuzs, V.P. (1981). Fracture micromechanics of polymer materials, Martinus Nijhoff Publishers.

Schwartz, P., Netravali, A. and Sembach S. (1986). Effects of strain rate and gauge length on the failure of ultra high strength polyethylene fibres, Textile Research Journal, Aug. 19886, pp 502-508.

3 Damage

Progress in Durability Analysis of Composite Systems, Cardon, Fukuda & Reifsnider (eds)
© 1996 Balkema, Rotterdam. ISBN 90 5410 809 6

A synergistic damage mechanics approach to durability of composite material systems

Ramesh Talreja
School of Aerospace Engineering, Georgia Institute of Technology, Atlanta, Ga., USA

ABSTRACT: Micromechanics and continuum damage mechanics are reviewed in the context of damage mechanisms in composite material systems with different fiber architectures. The limitations and undesirable aspects of each approach in isolation are discussed. In search towards an approach which connects with materials aspects at one end and structural analysis at the other, a synergistic approach is proposed. This approach enters selected micromechanics results in a continuum damage framework for thermomechanical response based on internal variables. The advantage is illustrated by an example of a class of composite laminates for which prediction of stiffness reduction with transverse cracking becomes possible with input from a simple numerical analysis and a few simple coupon tests.

1 INTRODUCTION

Damage mechanics plays a central role in durability analysis of composite components. Figure 1 illustrates schematically a common procedure in conducting such analysis. The first step is usually a stress analysis to identify "hot spots" where failure in long term service conditions would be likely. The stress-strain-time histories in these spots are to be estimated using the service loading and assumed constitutive behavior of the composite. For the fiber architecture and matrix material under consideration the mechanisms of long term degradation must be assessed. This knowledge has been gathered in a generic form for several composite systems such as laminates of graphite/epoxy, and the mechanisms observed are usually categorized as cycle-dependent and time-dependent. Examples of the first category are multiple transverse cracking in off-axis plies, delamination and fiber failure (which itself may not be cycle-dependent, but can become accumulative due to cycle-dependent matrix or interface failure). In the second category are matrix viscoelasticity, physical and chemical aging, fiber oxidation, etc. The material (and therefore structural) response changes with service loading as a result of the damage mechanisms. The response characteristics such as the instantaneous stiffness and strength in selected directions can degrade to values below acceptable levels set by performance requirements. The assessment of this degradation is therefore vital to the safe and reliable long term performance of composite structures. The emerging field of damage mechanics aims at achieving this objective.

COMPONENT DURABILITY ANALYSIS

POLYMER MATRIX COMPOSITES

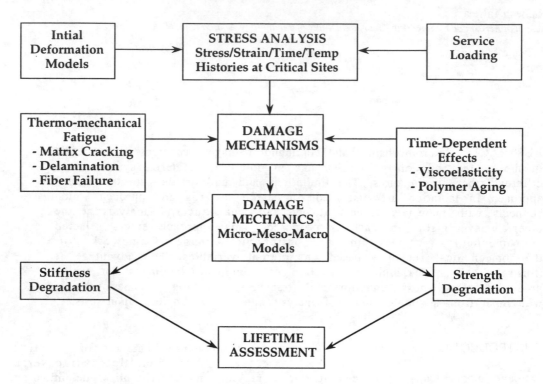

Fig. 1. A procedure for durability assessment of composite structural component

Two approaches in damage mechanics of composites have been developing separately and sometimes in a competetive spirit. One approach, commonly referred to as micromechanics, has the idealistic goal of determining the macro-level response characteristics from given properties of the constituent elements and the geometric configuration in which the constituents are placed. In situations where damage is present the entities of damage (cracks, disbonds, voids, etc.) are regarded as constituent elements and their effect on the overall (macro) response is estimated. The early works in micromechanics considered the matrix as a homogeneous, isotropic and often linearly elastic body containing inclusions of various shapes and sizes placed in regular, periodic patterns. The overall (average) properties were estimated resulting in expressions which could serve towards developing new material combinations. Later, cracks in isotropic or transversely isotropic bodies were considered and the effective properties of an equivalent homogeneous body of infinite extent were estimated. An excellent account of various results for such problems is given in Nemat-Nasser and Hori (1993). Also, a review by Kachanov (1987) is recommended.

The other approach, known as continuum damage mechanics (CDM), in its present form smears out the inhomogeneities and represents these by a field of internal variables. The inhomogeneities that do not evolve during deformation of the composite can be smeared out first and the matrix with these inhomogeneities can be replaced by a homogeneous medium of average properties and resultant symmetries. The evolving inhomogenieties (damage entities) can then be considered embedded in this homogeneous medium and be smeared out and represented by internal variables. Several general frameworks of material response based on thermodynamics with internal variables have been proposed and these provide bases for constructing CDM formulations for composites. The proposed CDM formulations for composites differ mainly in how damage is characterized, which also affects implementation of the resultant procedure for determining the response characteristics. A recent volume on damage mechanics of composite materials presents expositions on different CDM formulations (Talreja 1994).

Our purpose here is to scrutinize the two approaches for their abilities to realistically meet the needs of durability assessment of composite material systems. The two areas to examine for this purpose are stiffness-damage relationships and damage evolution. We shall mainly consider the former since the latter is immaturely developed in both approaches. In a separate paper damage evolution prediction by micromechanics has been critically assessed (Talreja and Akshantala 1995).

2 MICROMECHANICS

The range of scenarios of damage mechanisms in composites is at least as wide as the spectrum of fiber architectures. For instance the patterns of cracks resulting from a given loading mode in laminates of straight-fiber plies is different from those in woven fabric laminates. Even within laminates of straight-fiber plies the crack patterns depend on the ply orientations, the stacking sequence of plies and ply thicknesses. Also, local delaminations which are subdued at low monotonic loads become significant when time-varying loads are applied, and their intensity and distributions show high sensitivity to the fiber architecture; see for instance Schulte et al. (1987).

In micromechanics one must first identify a unit cell which repeats with certain periodicity in a representative material volume. Solutions are then sought to the stress fields within the unit cell under selected imposed traction and/or displacement conditions which are often taken to be uniformy distributed on the unit cell boundaries. A variety of methods for determining or estimating the stress fields have been devised. A survey of these methods is not the purpose here but we note that very few exact solutions have been possible. Even for the very simple case of a cross ply laminate with transverse cracks assumptions concerning variation of stresses had to be made before bounds to the overall properties could be estimated (Hashin 1985). We shall examine this problem here since it provides a good illustration of the capabilities of micromechanics.

Figure 2 shows a cross ply laminate with cracks restrained to the middle transverse plies and loaded by uniformly distributed forces normal to the crack planes. The unit cell for this case is the volume bounded by cross-sections containing planes of two adjacent cracks and the two free surfaces normal to the crack planes. One

119

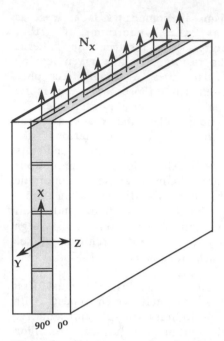

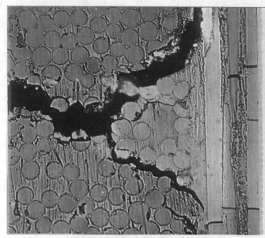

Fig. 2. A cross ply laminate with transverse cracks loaded in axial tension

Fig. 3. Examples of typical crack-up zones showing crack-tip branching (top) and crack-tip diversion into the interface (bottom). Courtesy J. Varna.

approach to determining the stress field in this unit cell would be to seek solutions in terms of the crack-tip stress intensity factors for the part of the stress field in the vicinity of crack tips. Close observations of crack-tip zones in cross ply laminates have shown that the crack tips tend to branch out before reaching the interfacial planes or divert into these planes (Varna et al 1993). Figure 3 shows examples of typical crack tip zones. Thus, singularity-based solutions to stress fields in the crack-tip zones would be questionable. Hashin (1985) forwarded different arguments for the lack of severity of the stress field in the crack-tip zones and then assumed that the axial normal stress does not vary along the crack plane. He took this stress component to also be constant along the thickness of the uncracked plies. The problem of stress field could still not be solved exactly and was estimated by invoking the principle of minimum complementary

energy. By averaging the stress field over the unit cell a lower bound to the axial Young's modulus was calculated which agreed well with the experimental data examined by Hashin (1985). However, it is found that the axial Young's modulus of cracked cross ply laminates is estimated fairly well by even crude methods such as the so-called shear-lag method which is known to give incorrect stress field in the unit cell. Hashin's analysis has been improved by Varna and Berglund (1991) who have examined the elastic moduli prediction and crack density evolution by this type of analysis. Based on careful and thorough examination of a large set of data, they concluded that

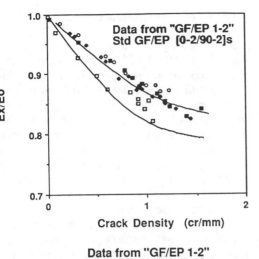

Data from "GF/EP 1-2"
Std GF/EP [0-2/90-2]s

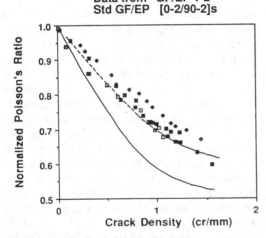

Data from "GF/EP 1-2"
Std GF/EP [0-2/90-2]s

Fig. 4. Comparison of Young's modulus and Poisson's ratio predictions with experimental data for cross ply laminates of glass-epoxy. The lower curves are given by Hashin's analysis and the upper curves are by an improved analysis of Varna and Berglund (1991). Courtesy J. Varna.

Hashin's approximate stress field results in inaccurate prediction of the axial Young's modulus and the major Poisson's ratio, the error in prediction of the latter being unacceptably large. Figure 4 reproduces their comparison of test data with Hashin's model and their improved analysis. Most other studies have only compared the Young's modulus data, which for cross ply laminates of stiff graphite fibers show only a few percent reduction of this property.

What is of interest for us here is to note the fact that accurate stress field determination in the unit cell is required for accurate prediction of the effective property reduction with damage. Analyses such as that by Hashin, referred to above, and several shear-lag type approximations have left a different impression by comparing the model predictions with the Young's modulus only. As far as cross ply laminates with transverse cracks are concerned, however, the problem of accurate stress field determination has been resolved by the works of Varna and Berglund, referred to above, and more recently by McCartney et al (1995).

When damage in general laminates with straight fibers is considered two problems emerge. The first is related to inclined cracks, i.e., cracks along fibers in plies which are inclined at other angles than 90 degrees to the longitudinal laminate symmetry axis. In such cases the problem cannot be simplified to a two-dimensional boundary value problem. Any attempts to solve this problem approximately, e.g. by considering an infinite medium with oriented cracks must be taken with caution in view of what has been illustrated above for cross ply laminates concerning the accuracy of stress solution. The other problem, even for transverse cracks in general laminates, is due to the uncracked plies having orientations different from the longitudinal. An example of this is $(0, \theta_1, \theta_2, 90)_s$ laminate with cracks only in the 90 degree plies. It is common in the literature, in particular in one-dimensional analyses, to replace the uncracked plies by an "equivalent" layer with the average axial properties. Thus, if $\theta_1=\theta$, $\theta_2=-\theta$, the sublaminate $(0, \pm\theta)$ is replaced by an equivalent 0 degree ply of reduced axial stiffness. Such approximations have only been checked against the longitudinal Young's modulus of the cracked laminate. The Poisson's ratios or the shear modulus were not checked. In a study to be reported (Saha and Talreja 1995) it was found that the major Poisson's ratio calculated for a cracked $(\pm\theta, 90_2)_s$ laminate using this assumption did not agree with experimental data. The experimental observations on this type of laminates also show that delaminations can occur more readily when the off-axis angle deviates significantly from the longitudinal direction. These effects are yet to be analyzed by micromechanics.

From the experience with cracked straight-fiber laminates it can be appreciated that for more complex fiber architectures the stress field problems could present even more demanding challenges. For instance the stress fields even in uncracked woven fabric laminates present difficulties that are yet to be satisfactorily surmounted. A range of approximate methods for estimation of the overall elastic moduli of uncracked textile composites has been devised (see Chou 1992 for a review) but each method appears to be good for some moduli and not reliable for a general case.

In view of the analytical difficulties in micromechanics one could take the numerical route. The attractiveness of this approach would then reduce drastically. One numerical study reported in the literature (Adolfsson and Gudmundson 1995) illustrates the work involved in conducting stress analysis for one crack geometry in one laminate configuration. The lack of extension of such efforts to other cases displays their futility.

3 CONTINUUM DAMAGE MECHANICS

The concept of internal variables is central to CDM. If distinct entities are embedded in a given medium such that the spatial distribution of the entities is smoothly varying (i.e., no localization), then the entities can be smeared out and represented by a set of internal variables. When the entities remain unchanged under deformation of the medium a representation with internal variables usually does not have an advantage since the medium itself can then be replaced by a homogenized medium of average properties. Examples of such cases are a matrix (e.g. metal, polymer or ceramic) reinforced with particles, whiskers, fibers, etc. When the non-evolving entities have preferred orientations the homogenized medium is rendered certain symmetry properties determined by those orientations. In such a homogenized, anisotropic medium other entities can emerge which evolve under thermomechanical loading. The evolving entities are typically voids and cracks of various geometries and orientations. These have been called damage entities in this author's works and their representation with internal variables has been treated (Talreja 1985a, 1985b, 1987, 1990).

The smearing out of distributed entities and representing these with internal variables allow elegant development of constitutive relationships for generally anisotropic media with evolving damage. Sound frameworks for this purpose were developed in the 1960s using the principle of equipresence (Truesdell and Toupin 1960) and in compliance with the laws of thermodynamics. One such framework is due to Coleman and Gurtin (1967) which was adopted by this author for development of stiffness-damage relationships for composite laminates with different damage modes

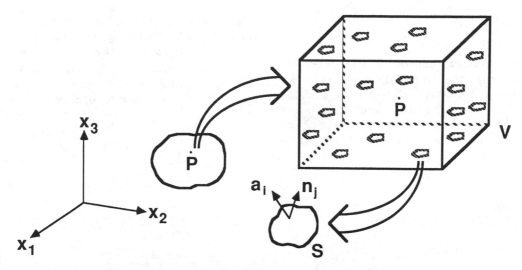

Fig. 5. A representative volume element containing damage entities.

(Talreja 1985a). A recent exposition by this author (Talreja 1994) treats characterization of damage with tensorial internal variables and provides stiffness-damage relationships for several damage modes in composites. For the purpose at hand it will be useful to recapitulate the essentials of those treatments.

Consider a heterogeneous body with damage. Let the stationary inhomogeneities be smeared out such that the resulting homogenized continuum is generally anisotropic and let the evolving damage entities be embedded in this continuum. Consider now a generic point P (see Fig. 5) with which is associated a representative volume element (RVE) of volume V. This RVE can in general contain damage entities that are separable in distinct damage modes (Talreja 1990). Consider for simplicity one damage mode consisting of damage entities which evolve by changing their surface area S. Examples of such damage entitics are voids and cracks. (Inclusions are thus not damage entities since their surface areas remain constant.) Place two vectors **n** and **a** at a point on a damage entity surface, where **n** is a unit outward normal to the surface at that point and **a** represents a preassigned quantity associated with an influence of the damage entity on the surrounding medium. For cracks embedded in a deforming medium **a** may be taken as the displacement vector.

With reference to a Cartesian coordinate system a damage entity tensor is defined by the surface integral of the diadic product of the two vectors. Thus,

$$d_{ij} = \int_S a_i n_j \, dS \qquad (1)$$

A simple volume average of the damage entities within a RVE is defined as a damage tensor

$$D_{ij} = \frac{1}{V} \sum_k (d_{ij})_k \qquad (2)$$

For cracks which evolve by undergoing opening displacement of crack surfaces the damage entity tensor becomes symmetrical, given by,

$$d_{ij} = \int_S a n_i n_j dS \qquad (3)$$

The damage tensor D_{ij} is taken to represent the internal evolving state and is entered as the internal variable field in a framework for thermomechanical response of materials. Under isothermal condition this response is given by

$$\sigma_{ij} = \sigma_{ij}(\varepsilon_{kl}, D_{mn}),$$
$$\psi = \psi(\varepsilon_{kl}, D_{mn}), \qquad (4)$$
$$dD_{ij} = dD_{ij}(\varepsilon_{kl}, D_{mn}),$$

and the restriction

$$dD_{ij} R_{ij} \geq 0 \qquad (5)$$

where σ_{ij} is the Cauchy stress tensor, ψ is the specific Helmholz free energy, ε_{kl} is the strain tensor and R_{ij} is the thermodynamic force conjugate to the damage tensor. The incremental stress-strain relations are given by

$$d\sigma_{ij} = C_{ijkl} d\varepsilon_{kl} + K_{ijmn} dD_{mn}, \qquad (6)$$

where

$$C_{ijkl} = \frac{\partial^2 \psi}{\partial \varepsilon_{ij} \partial \varepsilon_{kl}} \qquad (7)$$

$$K_{ijmn} = \frac{\partial^2 \psi}{\partial \varepsilon_{ij} \partial D_{mn}} \qquad (8)$$

Thus the stiffness coefficients C_{ijkl} at a given state of damage are obtainable from the incremental stress-strain response during unloading when $dD_{ij} = 0$.

The stress-strain-damage relationships have been developed for various modes of damage in composites in the author's previous works and reviewed in Talreja (1994). For intralaminar cracks in a laminate the RVE is illustrated in Fig. 6. When cracks are in the transverse direction, i.e. $\theta = 90$ degrees, e.g. in a $(\theta_1, \theta_2, 90)_s$ laminate, the unloading moduli for inplane response of the laminate can be written as

$$E_1 = E_1^0 + 2\frac{\kappa t_c^2}{ts}\left[c_3 + c_7\left(v_{12}^0\right)^2 - c_{13}v_{12}^0\right]$$

$$E_2 = E_2^0 + 2\frac{\kappa t_c^2}{ts}\left[c_7 + c_3\left(v_{12}^0\right)^2 - c_{13}v_{12}^0\right] \qquad (9)$$

$$v_{12} = v_{12}^0 + \frac{\kappa t_c^2}{ts}\left[\frac{1 - v_{12}^0 v_{21}^0}{E_2^0}\right]\left(c_{13} - 2c_7 v_{12}^0\right)$$

$$G_{12} = G_{12}^0 + \frac{2\kappa t_c^2}{ts}c_{11}$$

where c_3, c_7, c_{11} and c_{13} are material constants for the given laminate and κ is a constant assumed to represent the opening displacement of transverse cracks such that $a = \kappa t_c$, where t_c is the thickness of the transverse plies which is also the length of cracks in the thickness direction. Certain remarks concerning these constants are in order.

The constants c_3, c_7, c_{11} and c_{13}

It can easily be shown that these constants are independent of the deformed state (ε_{ij}) and the internal state (D_{ij}). Thus they can only depend on the material constitution, i.e. the smeared-out stationary continuum in which damage entities become embedded. It is, however, not clear whether in case of laminates the constants depend only on the fiber and matrix properties or also on the laminate configuration, i.e. fiber orientations and ply stacking sequence. A study of the $(\pm\theta, 90_2)_s$ laminates indicated that these constants did not change with the angle θ (Saha and Talreja 1995).

124

The constant κ

This constant was introduced as a coefficient of crack surface displacements for actual cracks in composites (a = κt_c). Thus for a laminate with cracked ply thickness t_c, κ accounts for the effect on the crack surface displacements of the local constraints and heterogeneities of microstructure. The dependency of κ on the angle θ of the $(\pm\theta, 90_2)_s$ laminates has been studied (Saha and Talreja 1995) and will be discussed later.

It may be noted that the products of constants, i.e. κc_3, κc_7, κc_{11} and κc_{13} appear in the linearized stiffness-damage relationships, Eq. (9). An experimental procedure to determine these products was described in Talreja (1987).

Implementation of CDM rests on determination of the type of constants discussed above. These constants may be viewed as carriers of information concerning material properties and damage entity behavior into a common framework. The generality thus generated has the cost of requiring these constants for every case. It is this author's contention that this situation can be improved by combining CDM with micromechanics in a judicial manner such that synergy is generated. This is explored in the following.

4 SYNERGISTIC DAMAGE MECHANICS

From the discussion above of micromechanics and CDM it can be appreciated that micromechanics is most useful for materials development but is less attractive for component durability assessment, while CDM has potential to form basis for the latter purpose due to the generality it offers. However, the problems of material constants evaluation provide limitations to CDM. It would seem, therefore, that if micromechanics can be used for the limited purpose of material constants evaluation, reducing thereby the need for experimental evaluation of the constants, the resulting approach could become appealing. This idea has been applied to one class of composite laminates with good results and will be described in the following.

Consider the class of laminates given by $(\pm\theta, 90_2)_s$. Under an axial tensile loading cracks will develop in the 90 degree plies (assuming the off-axis angle θ to be sufficiently small). A unit cell viewed in the longitudinal section of the laminate is shown in Fig. 7. As discussed above in the section on micromechanics a stress field solution for this unit cell has only been obtained (approximately) for the special case of θ = 0. Thus the only way to approach this problem in micromechanics today would be by numerical stress analysis, e.g. by a finite element method. Our discussion above related to cross ply laminates would suggest that accurate stress analysis would be required to estimate the stiffness properties reduced by cracking. This was confirmed by a study (Saha and Talreja, 1995) which evaluated the cases of θ = 0, 15, 25 and 40 by a finite element method which used two-dimensional (plain strain) as well as three-dimensional elements. It was found that the sensitivity of the effective Poisson's ratio to the accuracy of the stress field was high, in line with the behavior displayed by cross ply laminates.

The regular CDM approach to this problem is to apply the stiffness-damage relationships given by Eqs. 9. Note that in these equations the elastic moduli with superscript "0" refer to the virgin state values for a given θ in $(\pm\theta, 90_2)_s$ laminates. Similarly, the material

125

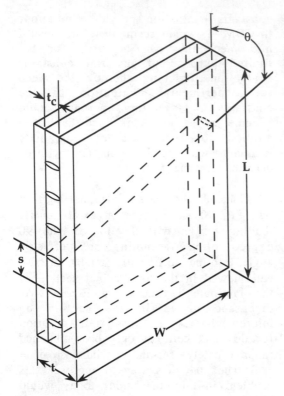

Fig. 6. A representative volume element containing intralaminar cracks.

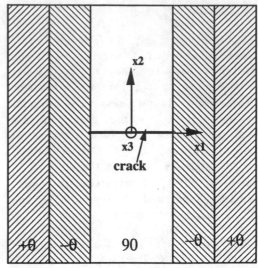

Fig. 7. A unit cell in a $(\pm\theta, 90_2)_s$ laminate with transverse cracks.

constants c_3, c_7, c_{11} and c_{13} as well as the constant κ would depend on the angle θ. Determination of all these constants for each θ value is obviously undesirable. The alternative, synergistic approach to this problem would be seek

Crack Opening Profile

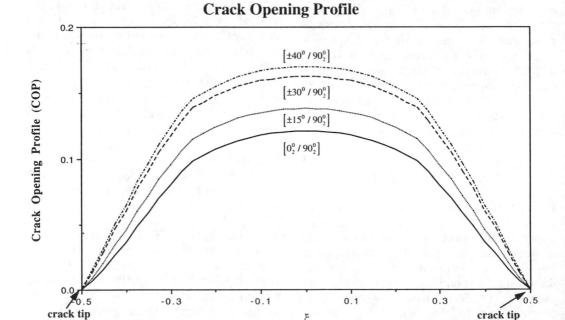

Fig. 8. Crack opening profiles of transverse cracks in $(\pm\theta, 90_2)_s$ laminates.

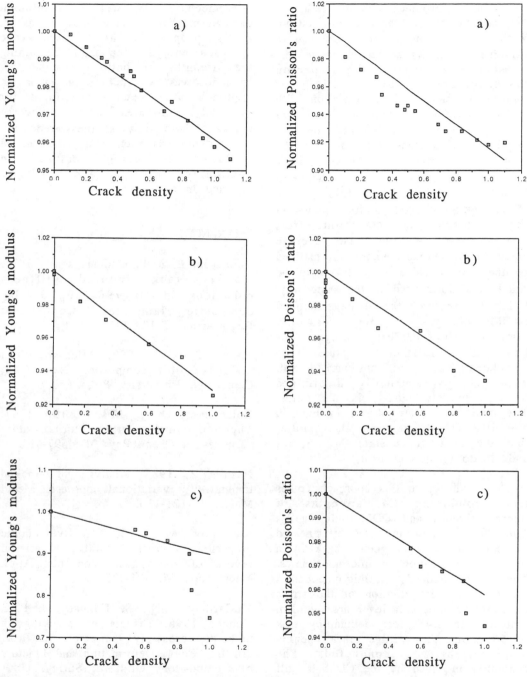

Fig. 9. Normalized Young's modulus predicted by CDM compared to experimental data for a) (±15, 90_2)$_s$, b) (±30, 90_2)$_s$, c) (±40, 90_2)$_s$ laminates.

Fig. 10. Normalized Poisson's ratio predicted by CDM compared to experimental data for a) (±15, 90_2)$_s$, b) (±30, 90_2)$_s$, c) (±40, 90_2)$_s$ laminates.

aid from micromechanics in reducing the burden of constants determination. With this in mind the constant κ was examined first. As explained above this constant represents the opening displacement characteristic of transverse cracks and its variation with θ can therefore be expected to be in some proportion to the average opening displacement of transverse cracks in the $(\pm\theta, 90_2)_s$ laminates. These crack opening displacements were calculated by a finite element method for $\theta = 0, 15, 25$ and 40. The crack profiles shown in Fig. 8 display the constraint effect induced by the $\pm\theta$ plies. The average crack opening displacements normalized by the $\theta = 0$ value were taken to be equal to κ normalized in the same way. Using this κ and the damage-induced moduli changes for the cross ply laminate the moduli changes for laminates with other θ values were calculated. The predictions agreed well with the experimentally determined Young's modulus and Poisson's ratios. The comparisons are shown in Figs. 9 and 10. This agreement also implies that the material constants c_3, c_7, c_{11} and c_{13} do not depend on θ.

The synergy to this approach comes from combining the strengths of micromechanics and CDM individually. The "loss" brought in CDM by smeared out microstructure is gained back by a judicious application of micromechanics. The severe demands on micromechanics for accurate determination of the stress field at the unit cell level are avoided. Instead, the much less demanding task of estimating the average crack opening displacement is conducted. The generality provided by the CDM is still maintained.

5 CONCLUDING REMARKS

The difficulties and limiting aspects in micromechanics as well as continuum damage mechanics have been discussed. It is proposed that a synergistic approach will develop from combining the strengths of both approaches. A specific way of achieving such an approach has been demonstrated for $(\pm\theta, 90_2)_s$ laminates with transverse cracks, where only stiffness reduction by cracking has been analyzed. Damage evolution is being analyzed in continuing research and will be reported in due time.

REFERENCES

Adolfsson, E. & P. Gudmundson 1995. Matrix crack induced stiffness reductions in $[(0_m/90_n/+\theta_p/-\theta_q)_s]_M$ composite laminates. Composites Engineering. 5:107-123.

Chou, T. -W. 1992. Microstructural design of fiber composites. New York: Cambridge University Press.

Coleman, B.D. & M.E. Gurtin 1967. Thermodynamics with internal state variables. J. Chem. Phys. 47: 597-613.

Hashin, Z. 1985. Analysis of cracked laminates: a variational approach. Mech. Mater. 4: 121-136.

Kachanov, M. 1992. Effective elastic properties of cracked solids ; critical review of some basic concepts. Appl. Mech. Rev. 45: 304-335.

McCartney, L.N., S. Hannaby & P.M. Cooper 1995. Effects of cracking in multi-ply cross-ply laminates. In Proc. 3rd Int. Cong. Deformation and Fracture of Composites., Guilford, Surrey, U.K., 27-29 March. p. 56-65. London, Institute of Materials.

Nemat-Nasser, S. & M. Hori 1993. Micromechanics: overall properties of heterogeneous materials. Amsterdam: North-Holland.

Saha, R. & R. Talreja 1995. Prediction of stiffness reduction in laminates due to transverse cracks subjected to varying constraint. In preparation.

Schulte, K., E. Reese & T.-W. Chou 1987. In F.L. Matthew et al. (eds.), Sixth Int. Conf. Comp. Mats. and Second European Conf. Comp. Mats., p. 4.89-4.99. London, Elsevier.

Talreja, R. 1985a. A continuum mechanics characterization of damage in composite materials. Proc. R. Soc. Lond. A399:195-216.

Talreja, R. 1985b. Transverse cracking and stiffness reduction in composite laminates. J. Comp. Mater. 19: 355-375.

Talreja, R. 1986. Stiffness properties of composite laminates with matrix cracking and interior delamination. Engng. Fracture Mech. 5/6: 751-762.

Talreja, R. 1990. Internal variable damage mechanics of composite materials. In J.P. Boehler (ed.), Yielding, Damage and Failure of Anisotropic Solids, p. 509-533. London, Mechanical Engineering Publications.

Talreja, R. (ed.)1994. Damage mechanics of composite materials. Amsterdam: Elsevier.

Talreja, R. 1994. Damage characterization by internal variables. In R. Talreja (ed.), Damage mechanics of composite materials. p.53-78. Amsterdam: Elsevier.

Talreja, R. & V.N.P. Akshantala 1995. Inconsistency in the analysis of damage evolution in composites using a micromechanics approach. Submitted to Int. J. Damage Mechanics.

Truesdell, C. & R.A. Toupin 1960. In S. Flugge (ed.), Encyclopedia of Physics. Vol. 3/1. p. 226-293. Berlin, Springer-Verlag.

Varna, J. & L.A. Berglund 1991. Multiple transverse cracking and stiffness reduction in cross-ply laminates. J. Comp. Tech. Res. 13: 97-106.

Varna, J., L.A. Berglund, R. Talreja & A. Jacovics 1993. A study of the opening displacement of transverse cracks in cross-ply laminates. Int. J. Damage Mech. 2: 272-289.

Progress in Durability Analysis of Composite Systems, Cardon, Fukuda & Reifsnider (eds)
© *1996 Balkema, Rotterdam. ISBN 90 5410 809 6*

A damage computational approach for composites: Basic aspects and micromechanics strength/stiffness indicators

Pierre Ladeveze
Laboratoire de Mécanique et Technologie, Cachan, France

Abstract : The basic aspects of a general damage mechanics approach for composites which is able to simulate the complete fracture phenomenon are presented. Two examples are considered : woven ceramic composites and laminate carbon/epoxy composites.

Composite design concerns both the structure and the material. A basic challenge is to compute the damage state of a composite structure subjected to complex loading at any point and at any time up until the final fracture. Damage refers to the more or less gradual development of microvoids and microcracks which lead to macrocracks and then to rupture. Brittle and progressive damage mechanisms are both present. In composite structures, damage may be the main mechanical phenomenon and is generally quite complex. There are several damage mechanisms which are highly anisotropic and present a strong unilateral feature depending on whether the microdefects are closed or open.

The present study deals with this computational challenge. We introduce a damage mechanics approach capable of predicting the behavior of any composite structure until final fracture.

The first and probably the main difficulty is to derive, at the chosen scale, a good damage model, i.e. a model compatible with all information coming from the micro, meso and macro scales. Internal variables such as inelastic strains, damage variables, hardening variables, etc, are introduced for specifying the material state. Some tools and concepts have been developed for defining such damage indicators [Ladeveze 1983 - 1993 - 1994]. They follow the line initiated by Kachanov (1958) and Rabotnov (1968). Applications to various composite materials have been carried out, in particular [Ladevèze, Le Dantec 1992], [Ladevèze, Gasser, Allix 1994] for the two material examples considered in this paper: woven ceramic composites and laminate carbon/epoxy composites. We also emphazise the interpretation of damage model constants as strength/stiffness indicators on the macro or meso scale. Other contributions to damage mechanics for composites can be found in [Highsmith, Reifsneider 1982], [Talreja 1994], [Allen 1994].

The second difficulty which appears at the structure level is now well understood. The classical damage models have serious deficiencies. For example, they do not contain classical fracture mechanics which works quite well for many cases. However, let us note that before the initiation point, such a model gives the correct response of a structure subjected to complex loadings. To remedy these shortcomings, it has been introduced recently the localization limiter concept [Bazant, Pijaudier-Cabot 1988], [Belytschko et al. 1988], [Bazant et al. 1994]. It is a regularization procedure with the additional terms built from a non-local approach or from a second gradient approach. Viscoplastic regularization can also be introduced.

Our answer for composites depends on the chosen scale. The best approach, when possible, is to use what we call a "damage mesomodel" [Ladevèze 1992-1995]. It is a semi-discrete model for which the damage state is locally uniform within the mesoconstituents . A more general approach is to introduce a damage model with a delay effect combined with a dynamic analysis. Then, a characteristic length, in fact a characteristic volume with a strong physical meaning, is introduced. Possibilities of such an approach are first given on one-dimensional bar problems [Ladevèze 1995]. To end, we shown results computed with a F.E. code under development devoted to laminate composite structures, especially with regions where large stresses are present. The code is able to compute at any time and point the "intensity" of the different damage mechanisms which are in the different layers and interfaces.

1. DAMAGE MODELLING DIFFICULTIES FOR COMPOSITES

Two materials are introduced. From experimental responses, we try to outline the main phenomena and then the primary damage modelling difficulties. A discussion on the modelling scale concludes this part of the paper.

1-1. Sic-Sic materials

The material considered is obtained by Chemical Vapor Infiltration (C.V.I.) and made at SEP (Société Européenne de Propulsion). The composite specimen is composed of 11 woven Sic-Sic plies (Fig. 1). At room temperature, the moduli of the fibers and the matrix are similar, 200 GPa and 350 GPa, respectively. The fiber volume fraction is 40 %.

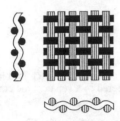

Figure 1 : A ply of a woven Sic-Sic composite

Figures 2 and 3 show combined tension/compression tests conducted at room temperature up until observed fractures for the 0° and 45° loading directions. No inelastic strains are observed. It appears that the moduli's variation is due to microcrack mechanisms which develop in tension. In compression, the microcracks are shut and thus the moduli remain at their initial values. Another modelling difficulty can be seen in Figure 3: microcracks can be orthogonal to the loading direction. So, for the 45 degree loading direction, there is both a microcrack mechanism orthogonal to the loading direction and a dual-microcrack mechanisms associated with the fiber directions. Let us note that no stiffness variations are observed in the transversal response.

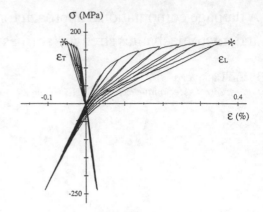

Figure 3 : 45° tension/compression test on a woven Sic-Sic

1.2. Carbon/epoxy laminates

Figure 4 shows that many damage mechanisms occur simultaneously. Apart from this new modelling difficulty, it appears that inelastic strains can be observed. They are essentially due to microcrack mechanisms.

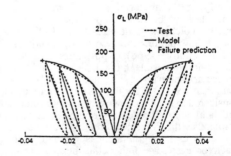

Figure 4 : Typical degradations for composites laminates

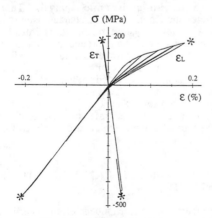

Figure 2 : 0° tension/compression test on a woven Sic-Sic

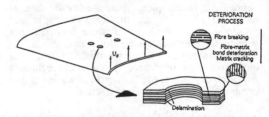

Figure 5 : Tension test on T300/914 [+45 - 45]$_{2s}$ (stress versus longitudinal and transversal strains)

1-3. The modelling scale

SiC-SiC materials:
Three scales can be distinguished:
- micro : fiber
- meso : yarn
- macro : structure

Models derived at the micro and meso scales do exist. They are probabilistic models and, unfortunately, are not able at the present time to predict what happens for complex loadings. The pragmatic damage models we have derived are deterministic macro models. We have tried to include all information emanating from the various scales [Ladevèze et al. 1994].

Composite laminates:
The different scales are :
- micro : fiber
- meso : elementary ply
- macro : structure

Here, it is possible to derive a material model at the mesoscale. The one we proposed in [Ladevèze, Le Dantec 1994] is defined by two elementary constituents :
- a single layer
- an interface which is a mechanical surface connecting two adjacent layers and depending on the relative orientation of their fibers.

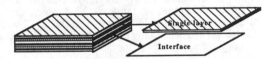

Figure 6 : Laminate modelling

The characteristic length here is the thickness of the elementary layer. To develop a mesomodel, one must add another property : the "meso" damage state is locally uniform within each elementary constituent. Thus, for laminates, the damage state is taken as uniform throughout the thickness of the elementary ply. A particular length has been introduced : the thickness of a single layer. This point plays an important role in the rupture computation, especially when one tries to simulate cracks with a damage theory. Let us note that the property assigned to the damage state can be also be specified for the other internal state variables, but this is not essential. Then, we can have only two types of macrocracks :
- delamination cracks within the interfaces;
- cracks orthogonal to the laminate, with each cracked layer being completely cracked in its thickness.

2. A DAMAGE MODELLING APPROACH FOR COMPOSITES : BASIC ASPECTS

Material models with internal variables are introduced, with the main ones being the internal damage variables. Other internal variables are the inelastic strain ε_p and hardening variables. The scale could be either meso or macro.

2.1. Damage variable concept

The idea advanced in the present paper was originally introduced by Kachanov (1958) and Rabotnov (1968) : material damage can be described by its effects on elastic coefficients. Let us start with the case where the mechanical behavior does not depend on the opening and the closure of micocracks.

The main objective is to derive, for the chosen scale - a damage kinematics - i.e. a description of the damage state which introduces a minimum number of state variables. For that, we combine information obtained at a finer scale with the geometrical description of the damage state introduced in [Ladevèze 1983, 1994] and developped in [He, Curnier 1995]. The main result is the following : the damage state is defined by two surface S_d, S_δ of the usual 3D-space, named damage surfaces. For undamaged media, they reduce to a single point. With this geometrical representation, any damage kinematics can be derived by approximating, in a geometrical sense, the damage surfaces S_d and S_δ. For example, they may be approximated by two spheres, which yield a two variable descriptions of damage that constitute the genuine isotropic damage theory. Ellipsoids can also be used.

Let us assume that damage kinematics have been chosen and denoted by d, δ, the set of scalar damage variables defining the approximations preferred for the functions d and δ.

2.2. Modelling of the microcracks' closure and opening

For many materials and, in particular, for ceramic composites, different elastic "tension" and "compression" behaviours are observed. This is due to the opening or closing of microdefects, depending on the loading. The difficulty herein is to derive models that lead, in a unilateral case, to continuous relations between stresses and strains, and that are also associated with a free energy. A general response is given in [Ladevèze 1983, 1993], wherein various possible models have been derived, some of which will be presented later. The general idea is to express this unilateral character in terms of energy rather than to distinguish between the "tension" and "compression" behaviours using stresses and strains. The latter distinctions in any case, from a general point of view, would not make much sense. We split the energy of the undamaged material into three parts : a "tension" energy, a "compression" energy and an energy identical for both tension and compression behaviors:

$$e(\varepsilon_e, 0, 0) = e_t (\varepsilon_e, 0, 0) + e_c (\varepsilon_e, 0, 0)$$
$$+ e_{tc} (\varepsilon_e, 0, 0)$$

Moreover, three sets of damage variables, (d_t, δ_t), (d_c, δ_c) and (d, δ) are now introduced to represent the damage state of the material. We then obtain:

$$e(\varepsilon_e, d_t, \delta_t, d_c, \delta_c, d, \delta) = e_t(\varepsilon_e, d_t, \delta_t)$$
$$+ e_c (\varepsilon_e, d_c, \delta_c) + e_{tc} (\varepsilon_e, d, \delta)$$

For the materials studied and, probably, for many composites, damage does not increase for compression loading, so:

$$d_c = \delta_c = 0$$

To simplify notation, we set :

$$d : (d, d_t) \quad \delta : (\delta, \delta_t)$$
$$e(\varepsilon, d, \delta) = e_t (\varepsilon, d, \delta) + e_c (\varepsilon) + e_{tc} (\varepsilon, d, \delta)$$

Sic-Sic material (see [Ladevèze et al. 1994] [Ladevèze 1995]) :
Respecting the requirement that no damage occurs for transversal responses and using approximation by "ellipsoids" of the damage surfaces, we prove that the damage state is described by a (3 x 3) symmetric matrix $\mathbb{H}$.
For plane stress, it is a (2 x 2) matrix. Therefore, our damage variable is $\mathbb{H}$. To take into account the closure and opening of microdefects, we introduce the following "tension" and "compression" energie :

$$e_c + e_{tc} = \frac{1}{2E_0} \text{Tr}[<\sigma>^- <\sigma>^-]$$

$$- \frac{v_0}{2E_0} (\text{Tr}^2 [\sigma] - \text{Tr} [\sigma\sigma])$$

$$e_t = \frac{1}{2E_0} \text{Tr}[\mathbb{H}\sigma^+ \mathbb{H}\sigma^+]$$

where σ^+ is the positive part of the stress σ in a sense that we will define later. Let us introduce the spectral decomposition of σ relative to $\mathbb{H}$, where $\underline{T}_i$, λ_i, $(i \in 1, 2, 3)$ are the eigenvectors and eigenvalues :

$$\sigma\underline{T}_i = \lambda_i \mathbb{H}^{-1} \underline{T}_i, \quad \underline{T}_i{}^t \mathbb{H}^{-1} \underline{T}_j = \delta_{ij}$$
(1 if i = j ; 0 if i ≠ j)

Taking the positive part of the eigenvalues, we get :

$$\sigma^+ = \sum_{i=1}^{3} (\mathbb{H}^{-1} \underline{T}_i) (\mathbb{H}^{-1} \underline{T}_i)^t <\lambda_i>_+$$

Then, σ^+ depends on the stress and on the damage state.

$<\sigma>^-$ denotes the classical negative part of σ.

Carbon/epoxy laminate (see [Ladevèze, Le Dantec 1992], [Ladevèze 1994] and [Allix, Ladevèze 1992]):
For the single layer, using a homogenisation calculation, we found that the only moduli which are modified are the transversal modulus E_2 and the shear modulus G_2, with "1" being the fiber direction. These properties have been confirmed by experimental observations. Three scalar damage variables, d_F, d and d', are then introduced. Considering the plane stress case, we have :

$$(e_c + e_{tc} = \frac{1}{2} [- (\frac{v_{12}^0}{E_1^0} + \frac{v_{21}^0}{E_2^0}) \sigma_{11} \sigma_{22}$$

$$+ \frac{<-\sigma_{22}>_+^2}{E_2^0} + \frac{\sigma_{11}^2}{E_1^0 (1 - d_F)} + \frac{1}{(1 - d)} \frac{\sigma_{12}^2}{G_{12}^0}$$

$$e_t = \frac{<\sigma_{22}>_+^2}{2 (1 - d') E_2^0}$$

For the interface model, we introduce three scalar damage variables, d, d_1 and d_2, related to the different delamination modes; "3" denotes the direction orthogonal to the interface. The displacement discontinuity is the kinematic quantity; its conjugate quantity is the normal stress vector $\sigma\underline{N}_3$.
We then have :

$$e_c + e_{tc} = \frac{1}{2} [\frac{<-\sigma_{33}>^2}{k_0}$$

$$+ \frac{\sigma_{13}^2}{k_0^1 (1 - d_1)} + \frac{\sigma_{23}^2}{k_0^2 (1 - d_2)}]$$

$$e_t = \frac{1}{2} \frac{<\sigma_{33}>^2}{k_0 (1 - d)}$$

More generally, for a given class of problem, let us suppose that the damage state is defined by two sets of scalar damage variables:

$$d , \delta$$

Then, we obtain:

$$e (\varepsilon_e, d , \delta)$$

2.4. Damage evolution laws

The associated damage forces to d , δ are :

$$Y_d = \frac{\partial e}{\partial\, d}\, |\, \sigma : cst$$

$$Y_\delta = \frac{\partial e}{\partial\, \delta}\, |\, \sigma : cst$$

Then, the dissipation due to damage is :

$$\omega = Y_d \circ \overset{\bullet}{d} + Y_\delta \circ \overset{\bullet}{\delta}$$

These damage forces govern the damage evolution and subsequently the rupture. For quasi-static loadings, the corresponding constitutive laws are :

$$d\,|_t = A_d\, (\, Y_d\,|_\tau,\, Y_\delta\,|_\tau,\, \tau \leq t)$$

$$\delta\,|_t = A_\delta\, (\, Y_d\,|_\tau,\, Y_\delta\,|_\tau,\, \tau \leq t)$$

which states that the values, at time t, of the damage variables depend on the previous history of the forces.

A_d and A_δ characterize the material and have to satisfy the following thermodynamic condition:

$$\omega \geq 0$$

Remark 1: Pratically, for quasi-static loadings, we consider that the value at time t of the damage variables depends on the maximum on the time of damage forces.

Remark 2: Identification of the material operators A_d and A_δ is carried out in [Ladevèze et al. 1994] for a Sic-Sic material and in [Ladevèze, Le Dantec 1992] and [Daudeville, Ladevèze 1993] for some carbon-epoxy laminates.

2.5. Inelastic strain - coupling damage - plasticity (or viscoplasticity)

The microcracks, i.e. the damage, lead to sliding with friction and thus to inelastic strains. The idea, which seems to work quite well, is to base the model on quantities called "effective". Details on the theory can be founded in [Ladevèze 1994].

2.6. Basic experiment-model comparisons

Figure 7 shows, for a Sic-Sic material, an experiment-model comparison for 0 and 45-degree tension/compression tests.

Concerning a carbon-epoxy material and $\pm\, \theta$ laminates, Figure 8 displays the experimental and computed fracture strain values. They must be compared on the right of the line $\theta = + 30°$ because on the left, delamination occurs. Delamination is not taken into account in the model used.

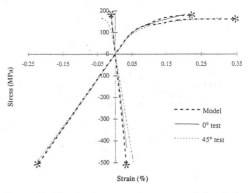

Figure 7 : Tension/compression tests and simulation for a Sic-Sic composite (0 and 45-degree)

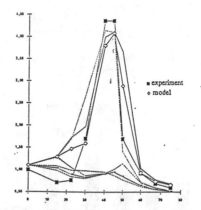

Figure 8: Tension tests and simulation for a carbon/epoxy composite $[\theta, -\theta]_{2s}$ Longitudinal strain (%) to failure versus θ

2.7. Micromechanics strength/stiffness indicators

For the Sic-Sic material studied in [Ladevèze 1995], it is proved that we get two damage mechanisms. They are defined by the two material functions h et k given figures 9 and 10. A part the initial thresholds, they depend on the slopes and a critical value, i.e. three material constants which charaterize the quality of the interphase and of the fiber and the matrix. Then, these material constants are micromechanics strength/stiffness indicators. A new question is to link these constants to classical micromechanics constants.

For carbon/epoxy laminates, it is proved at room temperatrue in [Ladevèze - Le Dantec 1992] and [Ladevèze 1994] that the two damage mechanisms which occur are essentially defined by two material critical values Y_c and Y'_c. Y_c and Y'_c can be considered as micromechanics strength indicators ; Y_c is related to the matrix and Y'_c to the matrix-fiber interphase. That can be seen on the figure 11.

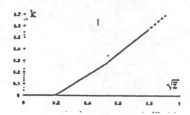

Figure 9 : Damage matrix mechanism.
Material function h

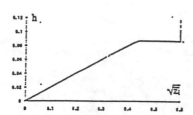

Figure 10 : Fiber-yarn mechanism.
Material function k

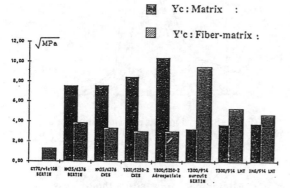

Figure 11 : Critical damage force values for different
materials

extends a code dedicated to the delamination analysis of laminate structures with an initially-circular hole [Allix 1987] and [Allix et al. 1989]. It includes our damage mesomodel based on two constituents : the elementary layer and the interface. Damage model with delay effect are also introduced. We conducted a quasi-static analysis except in the damage zones where a dynamic analysis is introduced. This code has been developed by Gornet (1995).

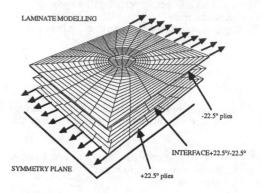

Figure 12: A structure computation example

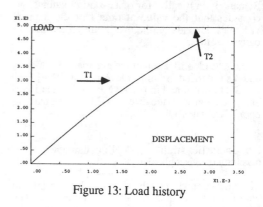

Figure 13: Load history

3. A FINITE ELEMENT CODE UNDER DEVELOPMENT

In order to obtain a consistent computational damage approach, we have introduced, in combination with a dynamic analysis, damage models with delay effect [Ladevèze 1989]. A dynamic analysis introduces a characteristic speed and a damage model with delay effect, a characteristic time. Then, a characteristic volume can be obtained. Even for our laminate mesomodel, we have to introduce such a damage model because, in this case, a mesomodel introduces only a characteristic length and not a characteristic volume.

A Finite Element code is under development for laminate composite structures, especially for structural parts where large stresses are present. It

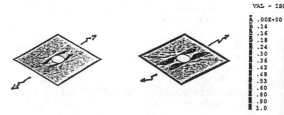

Figure 14: Damage variable d of the interface at times T_1 and T_2

Let us consider the structural computation example defined in Figure 12. It is a plate [+ 22.5° , -22.5°]$_s$ with a hole submitted to

136

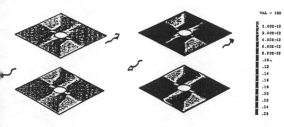

Figure 15: Damage variable d of the layers at times T_1 and T_2

tension. The load history is shown in Figure 13. At any point and time, the code is able to give the "intensity" of the different damage mechanisms up until the ultimate fracture. The main damage mechanism herein is delamination, i.e. the deterioration of the interface [22.5°, -22.5°]. Figures 14 gives the value of the damage variable d at times T_1 and T_2. The increase of the delaminated area is very significant. The layer damage mechanisms are weakly excited (Figure 14).

CONCLUSION

The proposed damage computational approach introduces damage models with delay effects which have a strong mechanical significiance. A dynamic analysis is conducted on the heavily damaged structural parts. This approach contains classical fracture mechanics and is thus a consistent one. To this end, we would emphasize the need to link, what we call micromechanics strength/stiffness indicators, to caracteristics of the matrix, the fiber and the interphase. The combined material/structure design cannot be carried out without such relationships.

REFERENCES

ALLEN D.H., 1994
Damage evolution in laminates
Damage Mechanics of Composite Materials, Talreja R. ed, Elsevier, 79-114.

ALLIX O., 1987
Delaminage - Approche par la Mécanique de l'Endommagement
Calcul des Structures et Intelligence Artificielle, Fouet J.M., Ladevèze P., Ohayon R. eds, Pluralis, 39-53.

ALLIX O., LADEVEZE P., 1989
Damage analysis for laminate delamination
Proceedings of the 5th Int. Symp. on Num. Meth. in Engng., Springer, 347-354.

ALLIX O., LADEVEZE P., 1992
Interlaminar interface modelling for the prediction of laminate delamination
J. Composite Structures, 22, 235-242.

ALLIX O., LADEVEZE P. AND CORIGLIANO A., 1995
Damage analysis of interlaminar fracture specimen
J. Composite Structures, 31(1), 61-74.

BAZANT Z.P. AND PIJAUDIER CABOT G., 1988
Non local damage : continuum model and localisation instability
J. of Appl. Mech., ASME, vol 55, 287-294.

BAZANT Z.P., BITTNAR Z. AND JIRASEK M., 1994
Fracture and Damage in Quasibrittle Structures (eds)

BELYTSCHKO T. AND LASRY D., 1988
Localisation limiters and numerical strategies for strain-softening materials
Cracking and Damage, J. Mazars, Z.P. Bazant eds, Elsevier, 349-362.

DAUDEVILLE L., LADEVEZE P., 1993
A damage mechanics tool for laminate delamination
J. Composite Structures, 25, 547-555.

DUMONT J.P., LADEVEZE P., POSS M. AND REMOND Y., 1987
Damage mechanics for 3D composites
Composite Structure 8, 119-141.

GORNET L., 1995
Simulation des endommagements et de la rupture dans les composites stratifiés
Thesis to appear

HE Q. AND CURNIER A., 1995
A more fundamental approach to damaged elastic stress-strain relations
Int. J. Solids Structures,32 (101), 1433-1457.

HIGHSMITH A.L. AND REIFSNEIDER K.L., 1982
Stiffness reduction mechanism in composite materials
Damage in Composite Materials - ASTM-STP 775, 103-117.

KACHANOV L.M., 1958
Time of the rupture process creep conditions
Izv Akad Nauk S.S.R. Otd Tech Nauk, 8, 26-31.

LADEVEZE P., 1983
Sur une théorie de l'endommagement anisotrope
Rapport Interne n°34, Laboratoire de Mécanique et Technologie, Cachan

LADEVEZE P., 1989
About a damage mechanics approach
Mechanics and Mechanisms of Damage in Composite and Multimaterials, Baptiste D. ed, MEP, 119-142.

LADEVEZE P., 1992
A damage computational method for composite structures
J. Computer and Structure, 44(1/2), 79-87.

LADEVEZE P., 1993
On an anisotropic damage theory
Failure criteria of Structured Media, Boehler J.P. ed, Balkema, 355-363.

LADEVEZE P., 1994
Inelastic strains and damage
Damage Mechanics of Composite Materials, Talreja R. ed, Elsevier, 117-136.

LADEVEZE P., 1995
Modeling and simulation of the mechanical behavior of CMC$_s$
High-temperature ceramic-matrix composite, Evans A. - Naslain R. eds, Ceramic Transactions, 53-64.

LADEVEZE P., 1995
A damage computational approach for composites : basic aspects and micromechanical relations
(to appear)

LADEVEZE P., GASSER A. AND ALLIX O., 1994
Damage mechanisms modelling for ceramic composites
J. Eng. Mat. Tech., 116, 331-336.

LADEVEZE P., LE DANTEC E., 1992
Damage modelling of the elementary ply for laminated composites
Comp. Sc. and Tech. 43-3, 257-268.

RABOTNOV Y.N., 1968
Creep rupture
Proc. XII, Int. Cong. Appl., Mech., Stanford-Springer

TALREJA R., 1994
Damage characterization by internal variables
Damage Mechanics of Composite Materials, Talreja R. ed., Elsevier, 53-78.

Progress in Durability Analysis of Composite Systems, Cardon, Fukuda & Reifsnider (eds)
© 1996 Balkema, Rotterdam. ISBN 90 5410 809 6

Low velocity impact damage of carbon fiber reinforced thermoplastics

H. Fukuda, H. Katoh & J. Yasuda
Department of Materials Science and Technology, Science University of Tokyo, Yamazaki, Noda, Chiba, Japan

ABSTRACT : This paper deals with a degradation of CFRTP laminates at low velocity impact. It is necessary to know the timing of replacement in the recycling of composite materials, and this is the main subject of this paper. Carbon fabric/Nylon 6 coupons were subjected to repeating 3-point impact bending and the change of bending modulus was monitored. The correlation between the reduction of modulus and the number of impact cycles has been made clear. An empirical equation to predict the modulus has been derived. A so-called craze was observed on the impact surface and it has been made clear that each craze corresponds to each kink of a fiber bundle. This kink was also successfully correlated with the modulus reduction using a simple series-parallel model.

1. INTRODUCTION

Composite materials are nowadays used in various fields of advanced technology because of their high strength-to-density or modulus-to-density ratios. The rustless property of polymer matrix composites is also another advantage.

On the other hand, due to the above rustproof property, the subject of waste disposal becomes one of the toughest issues for composite materials and this is getting a world-wide social problem from a view point of keeping Earth environment. It is more desirable if the material recycle is possible prior to disposal. For thermoplastic composites, four steps of recycle from continuous composites to dry distillation are expected (Kemmochi, et. al., 1993) in our research project.

In order to rwalize a material recycle, it becomes necessaray to know the timing when the material should be replaced. For this purpose, we need to evaluate the durability or damage accumulation of composite materials.

The objective of this paper is to make clear a relation between the damage accumulation and the reduction of macroscopic properties such as elastic modulus or residual strength. A thermoplastic composite was selected here from a view point of possible recycling.

2. EXPERIMENT

2.1 Specimens, apparatus, and test method

Test panels supplied were laminated plates made of Torayca carbon fabric/Nylon 6 prepregs. Nine plies of prepregs were hot-pressed and the size of the test panels were 400mm × 400mm × 2mm, which were supplied from Toray Industries. From this panel, test coupons of 100mm length and 15mm width were cut out with a diamond saw. To examine the effect of moisture, specimens dried in a vacuum desiccator (denoted Dry) , specimens left in an air (Leave) , and specimens dipped in the room-temperature water (Wet) were prepared.

Some specimens were tested under quasi-static 3-point bending using an Instron-tyoe testing machine. This test is for obtaining reference values. The span length was 80mm throughout the test.

A drop-weight type impact machine was used to generate damages in a coupon under three-point impact bending. Figure 1 briefly shows the diagram of the test.

Before starting the impact test, the condition that the specimen just broke at one impact was first looked for. Let the height of the drop weight at this condition be denoted by H_B.

Then from the height of either $0.9H_B$ or $0.8H_B$, the drop-weight was fallen on the specimen repeatedly.

Under this condition the specimen did not fail at one time of impact. After each impact, the bending modulus was measured using an Instron type testing machine. The surface of the specimen was also inspected with a naked eye. Some specimens were cut to observe the inside with a microscope.

2.2 Test result

Typical load-deflection curves of undamaged coupons under quasi-static 3-point bending are shown in Fig.2 and Table 1 summarize the static test results. The Moisture content of Dry specimen is unknown and the value 0.1% was assigned tentatively. As is seen in Fig.2, the maximum load or bending strength is strongly affected by the specimen condition, i.e., Dry, Leave or Wet. This degradation due to water absorption is probably caused by the degradation of the matrix as was reported, for example, by Judd (1975) .

The impact 3-point bending was then conducted where the mass of the drop weight was 500g and the radius of the drop-weight tip was 5mm. The span was the same as the static case. Figure 3 demonstrates typical load-deflection curves of quasi-static bending and low velocity impact ; no serious difference was

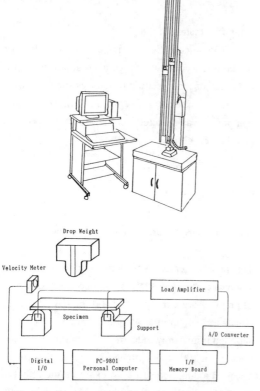

Fig.1 Schematic diagram of the test procedure.

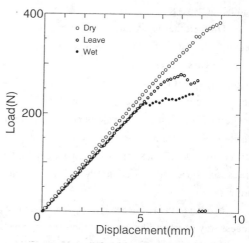

Fig.2 Example of load-deflection curves.
(static 3-point bending)

Table 1 Static test results

Specimen	Mointure contents(%)	Size(mm)	Max. Load(N)	Young's Modulus(GPa)	Bending Strength(MPa)
Dry	0.1	100×15×2	313.2	50.3	626.5
Leave	0.4	100×15×2	280.3	47.2	560.7
Wet	1.0	100×15×2	243.1	44.6	486.0

observed between the two. Figure 4 shows some examples between the relation of the modulus reduction and the number of impact cycles, where circles and squares are the experimental data. Solid lines are empirical values which will be discussed later. The bending modulus reduced rapidly at the early stage of the impact cycle and afterwards the modulus satulated to a certain value.

After the impact, small cracks or crazes were observed on the impact surface. The size of the craze corresponded to the size (width) of a fiber bundle of the woven fabric. By careful observation of the side or the inside of the specimen with a microscope, each

kink of a fiber bundle was recognized just beneath each craze. Figure 5 is a microscope picture of a cut side where a kink is recognized ; this kink is a typical compressive failure pattern of a composite.

To correlate this kink or craze with the modulus reduction, the following data reduction was tried. First, the modulus retention was plotted against the number of impact cycle. Figure 6 shows each one example of Dry, Leave and Wet specimen, although the data are different from Fig.4. In the present case, the drop-weight height was H/HB=0.9. In the experiment, the number of crazes was counted and it was plotted against the number of impact cycle, as shown in Fig.7. The location of the crazes was also measured in the experiment. From these figures, data were converted to the relation between the modulus retention and the number of crazes. Figures 8-10 are the cases.

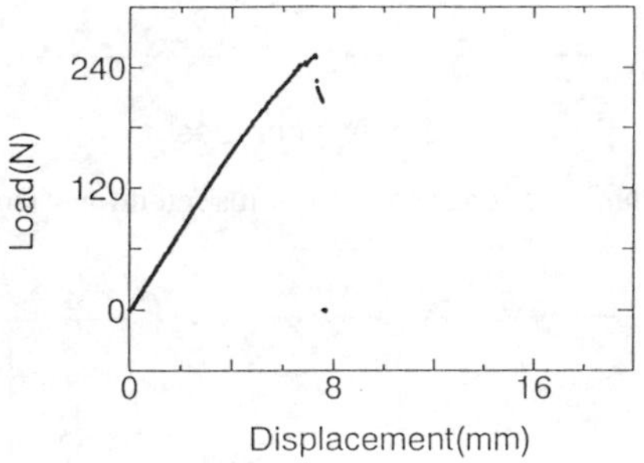

Fig.3 Comparison of load-deflection curves between quasi-static bending and impact bending.

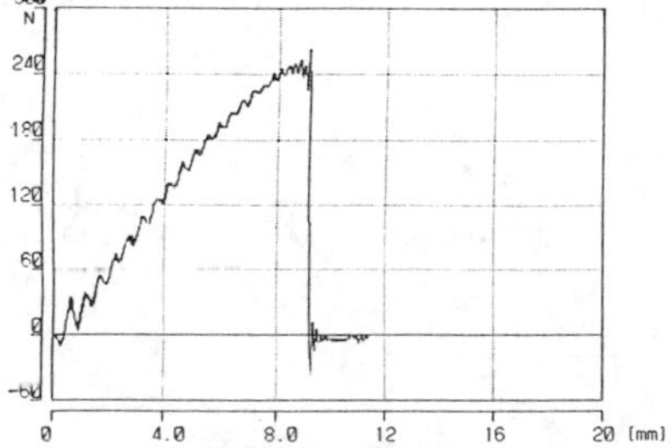

(a) quasi-static bending

(b) impact bending

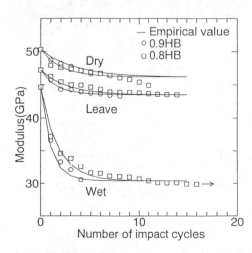

Fig.4 Number of impact cycle vs. bending modulus.

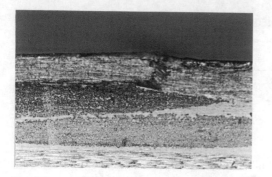

Fig.5 An example of a kink generated by an impact.

3. DISCUSSION

3.1 Modulus vs. number of impact cycle

Figure 4 indicates that the bending modulus is the function of moisture content, drop-weight height, and the number of impact cycles. To deal with these experimental values, we assumed the following empirical equation, although its physical meaning is not necessarily clear :

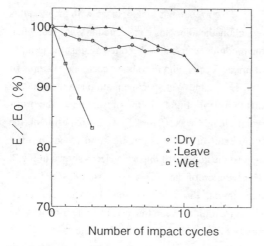

Fig.6 Number of impact cycle vs. bending modulus.

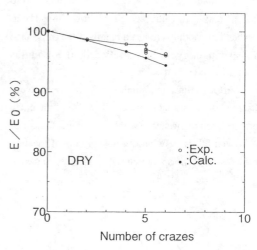

Fig.8 Number of crazes vs. modulus retention (Dry).

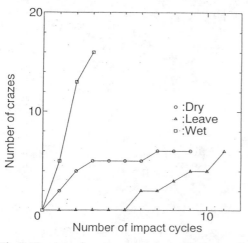

Fig.7 Number of crazes vs. number of impact cycle.

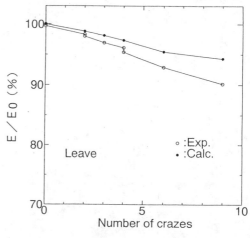

Fig.9 Number of crazes vs. modulus retention (Leave).

$$E = (E_0 - E^*) \exp(-kN) + E^* \quad (1)$$

In the above equation, E is the bending modulus after impact, E_0 is the initial bending modulus with no impact damage, E^* is the lower limit of the bending modulus, N is the number of impact cycles, and k is a factor which includes the moisture content, the drop-weight height, and the material fracture properties.

Since the load-deflection curves of the quasi-static bending is nearly the same as that of impact as shown in Fig.3, we supposed that the secant modulus

at just before fracture under static bending would represent the lower limit of modulus, E^*.

From the characteristics of eq. (1), the value k will increase with increasing the moisture content and the impact energy (drop weight height), and will decrease with increasing the fracture toughness. Assuming a linear relation among them, the value k was taken here as

$$k = H (m - m^*) n / H_B \quad (2)$$

where m is the apparent moisture content, m^* is a shift

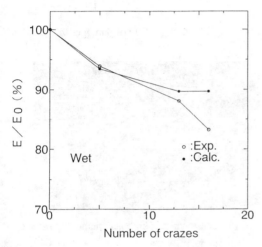

Fig.10 Number of crazes vs. modulus retention (Wet).

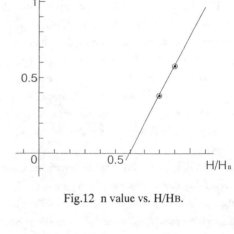

Fig.12 n value vs. H/H_B.

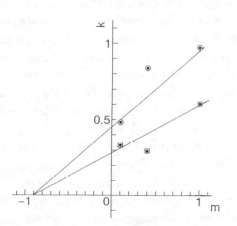

Fig.11 k value vs. apparent moisture content, m.

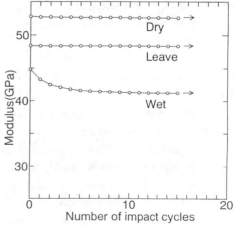

Fig.13 Modulus retention at repeating impact from the drop-weight height of H/H_B.

factor to an absolute 0% moisture content, n is an empirical coefficient. The value HB was adopted here to represent the fracture toughness.

Each group of the experimental data of Fig.4 was curve-fitted to eq. (1) by means of a least square method and the value k in eq. (2) was calculated. Figure 11 is the relation between k and the apparent moisture content. From this, $m^*=-0.9\%$ was obtained which means the Dry specimens still contain 0.9% of moisture. The value n was 0.576 for $H/HB=0.9$ and 0.382 for $H/HB=0.8$. Figure 12 is the n value against H/HB and

$$n=1.19H/HB-1.17 \qquad (3)$$

is an empirical value, although the number of data is too few, only two.

The value n should be positive from the characteristics of eq. (2), hence Fig.12 suggests that the modulus reduction would not take place at the region of $H/HB<0.6$. To confirm this, the repeated impact from the height of $H/HB=0.5$ was tried ant the results were shown in Fig.13. Except for Wet specimens, the modulus reduction was very small which suggests indirectly the validity of the present assumption.

The solid lines of Fig.4 are the predicted results which is fairly close to the experimental values. Thus, we may estimate the modulus reduction using eq. (1). Although the number of impact cycles to failure could not be determined, we may judge, by monitoring the modulus, whether or not the material is towards the end of use.

3.2 Modulus vs. number of crazes

When a kink is generated, the fiber bundle will no more carry the load near the kink. The damaged region was assumed here as in Fig.14. That is, assuming the symmetric damage with respect to the longitudinal direction, the area within the outermost crazes was supposed to be damaged. The damaged region was also assumed here to be limited in the top layer among nine layers. Then, using a simple series-parallel model (Jones,1975), the flexural rigidity of the damaged part

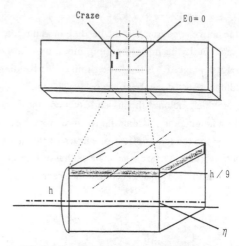

Fig.14 Modelization of damaged zone.

Fig.15 A case where a kink is generated in the second layer.

was calculated and finally the overall bending modulus was estimated. Solid marks of Figs.8-10 are the results.

In Fig.8, the calculated value was lower than the experimental result whereas Figs.9 and 10 exhibit opposite tendency. The model of Fig.14 will underestimate the modulus because the region of zero modulus was taken too wide. On the other hand, the kink of fiber bundle was sometimes observed in the second layer as shown in Fig.15. Since we assume the kink of the first layer only, our calculation may overestimate the modulus. The complex results of Figs.8-10 will be due to the combination of the above two opposite effects. Anyway, the predicted values are close to the experimental data and therefore, it may be concluded that the modulus reduction is due to the kink formation.

4. CONCLUSIONS

In the present paper, the damage accumulation of CFRTP coupons under low velocity impact was examined especially paying attention to the modulus reduction.

First, we examined the modulus reduction as a function of the number of impact cycle. An empirical equation was derived which agreed fairly well with the experimental data.

Next, somewhat microscopic discussion was done taking into account of the craze or the kink. The present conclusion is that the modulus reduction is due to the kink formation. Since this kink corresponds to a craze which is observed on the impact surface, we may get useful information for the timing of replacement of CFRTP material by monitoring the crazes which appears on the surface of the Specimen ; this will be a kind of nondestructive evaluation.

ACKNOWLEDGEMENTS

This work is supported by the Special Coordination Funds of the Science and Technology Agency of the Japanese Government. Test panels were supplied from Toray Industries. We also thank Mr. M. Kawano for his assistance in experiments.

REFERENCES

Jones, R.M., 1975, Mechanics of Composite Materials, Scripta, 90.

Judd, N.C.W., 1975, 30th Ann. Tech. Conf. SPI, 18-A.

Kemmochi, K., et. al., 1993, Intl. Workshop on Environmentally Compatible Materials and Recycling Technology, Tsukuba, 139.

Progress in Durability Analysis of Composite Systems, Cardon, Fukuda & Reifsnider (eds)
© 1996 Balkema, Rotterdam. ISBN 90 5410 809 6

The effect of overload on residual durability of laminated GRP

V.Tamuzs, I.Gruseckis & J.Andersons
Institute of Polymer Mechanics, Riga, Latvia

ABSTRACT: Composite structures in most applications apart from normal service loads may occasionally undergo overloads with magnitude close to their design limit. Such overloads are caused either by malfunction of the system they are part of, or by external loads of unforeseen magnitude. Usually their duration is very small compared to the design life of composite member. Even if overloads do not lead to immediate failure, they can affect residual material properties and damage accumulation rate.

The effect of short duration high amplitude cycling on residual lifetime of GRP at a lower load level is studied. Autogenous heating temperature of specimens was employed as a damage parameter. The applicability of Miner's rule was evaluated.

1 INTRODUCTION

Fatigue life is one of the basic wind turbine blade design parameters. Wingblades are subjected to spectrum of cyclic load levels during service, including short duration overloads due to storm etc. Therefore it is important to evaluate the effect of overloads of magnitude comparable to static strength on both residual durability of composite and damage accumulation rate.

A number of models have been developed for describing composite material response to cyclic loading, see reviews Oldirev et al. 1989, Sendecky 1992, Andersons 1993. Two approaches to phenomenological modeling of fatigue can be discerned: empirical description of damage accumulation relating the latter to effective material parameters (Oldirev et al. 1989), and residual strength and stiffness based fatigue modeling (Sendecky 1992). Both of them are applicable to variable amplitude loading. We will consider fatigue models which employ as damage metrics residual strength (Yang 1978, 1981) and specimen temperature increase (Oldirev et al. 1977, Parfeev et al. 1979) caused by energy dissipation during loading.

Linear damage accumulation model is routinely applied in blade fatigue design. It has been shown elsewhere that fatigue life of GRP used as blade material can be predicted with reasonable accuracy by Miner's rule for standardized wind load spectrum. Therefore evaluation of the applicability of Miner's rule in the case of overloads is important.

2 FATIGUE MODELS

2.1 Residual strength

Residual strength based models introduce rate equation for σ_R. We shall use the form (Yang 1978):

$$\frac{d\sigma_R}{dn} = -\frac{f(\sigma, r)}{c\sigma_R^{c-1}(n)} \quad (1)$$

where σ_R, σ, r are residual strength, peak stress, and stress ratio correspondingly. Integrating (1) and using $\sigma_R = \sigma$ at n = N as fatigue failure criterion, the following form of function f is obtained:

$$f(\sigma, r) = [\sigma_0^c - \sigma^c]/N \quad (2)$$

Weibull distribution is used for static strength:

$$F(\sigma_0) = 1 - \exp\left[-\left(\frac{\sigma_0}{\beta}\right)^{\alpha}\right] \qquad (3)$$

Static strength scatter leads to fatigue life scatter at a given applied stress σ. High cycle fatigue lifetime distribution is accordingly (Yang 1978):

$$F_N(n) = P[N \le n] = P\left[\left(\sigma_0^c - \sigma^c\right)/f(\sigma, r) \le n\right] =$$
$$P\left[\sigma_0 \le \left(n \cdot f(\sigma, r) + \sigma^c\right)^{1/c}\right]$$

Using (3) and taking into account $\left(\dfrac{\sigma}{\beta}\right)^c \ll 1$:

$$F_N(n) = 1 - \exp\left[-\left(\frac{n}{\beta^c/f(\sigma,r)}\right)^{\frac{\alpha}{c}}\right] \qquad (4)$$

It follows from (4) that $\beta^c/f(\sigma,r)$ is characteristic lifetime as a function of stress level. Equaling it to constant amplitude fatigue diagram in the form:

$$K\sigma^b N = 1 \qquad (5)$$

one obtains $f(\sigma,r) = \beta^c K \sigma^b$. Since parameters K and b depend on stress ratio r, so does function f.

Finally, integrating (1) and substituting the expression for f derived, the following relation for residual strength under constant amplitude loading is obtained (Yang 1978):

$$\sigma_R^c(n) = \sigma_0^c - \beta^c K \sigma^b n \qquad (6)$$

There are five model parameters to be determined: α, β - form static strength distribution, K, b - from fatigue diagram, and c - from residual strength data. Knowing these material parameters one can predict residual strength for spectrum loading.

2.2 Damage monitoring

The second approach allowing reasonably simple estimation of overload effect on damage accumulation in specimen is based on damage-related specimen property monitoring during fatigue loading. Several mechanical characteristics of GRP composite (e.g. modulus, mechanical loss factor, as well as autogenous heating temperature related to the parameters mentioned) possess a generic pattern of variation during loading which reflects damage level and accumulation rate (Oldirev et al. 1989). For instance, rapid growth of specimen temperature takes place initially, followed by slow, almost linear increase over most of lifetime, and rapid temperature increase is observed again shortly before failure. Damage monitoring method was applied to accelerated fatigue testing, determination of temperature dependence of fatigue strength, rest effect evaluation (Oldirev et al. 1977, 1989, Parfeev et al 1979).

3 MATERIAL CHARACTERIZATION

Static strength, fatigue life, and residual strength under constant amplitude loading were determined and residual strength model parameters calculated for glass fiber/polyester laminated composite.

The reinforcement was E-glass with the following characteristics: $E \approx 73000$ MPa, $\nu = 0.15$, $\rho = 2610$ kg/m^3, matrix - polyester resin ($E \approx 3600$ MPa, $\nu = 0.38$, $\rho = 1200$ kg/m^3). Specimens comprised six layers, outer ones were CSM, while inner - fabric reinforced with fiber ratio in longitudinal (x) and transverse (y) direction 9:1. Elastic properties of fabric-reinforced laminae: $h = 1.13$ mm, $E_x = 29000$ MPa, $E_y = 8500$ MPa, $\nu_{xy} = 0.29$, $G_{xy} = 2150$ MPa, CSM laminae: $h = 0.63$ mm, $E_x = E_y = 11500$ MPa, $\nu_{xy} = 0.42$, $G_{xy} = 403$ MPa. Average fiber mass fraction $\mu_m = 59\%$ and volume fraction $\mu_v = 39\%$.

Static tests were carried out in compliance with ASTM standard D 638 on MTS test setup at temperature $T = 10 \div 12°C$ on dog-bone shape specimens. Strain was measured by two extensometers with 25 mm gauge length. Outer plies tended to fracture and delaminate during loading which hampered limit strain measurement at static loading and modulus variation monitoring during fatigue loading. Fatigue tests were performed on ZDM PV-10 setup with load control at 17 Hz frequency and $T = 10 \div 12°C$ temperature. Stress ratio was r=0.1. Self-heating temperature of the specimen was monitored during loading.

Tensile strength data of 18 specimens are as follows: σ_0 =639; 602; 714; 601; 665; 646; 595;596; 625; 723; 730; 710; 713; 544; 688; 668; 574; 591 MPa.

Mean strength is $\langle\sigma_0\rangle$ = 646 MPa, variance ρ = 58 MPa, coefficient of variation ν = 9%. Strength distribution function (3) parameters are α = 13.5, β = 671 MPa. Experimental data and Weibull distribution function (3) of normalized static strength $x = \sigma / \beta$ are plotted in Fig. 1.

Fatigue data of 24 specimens tested at r=.1 are shown in Fig. 2. Characteristic fatigue curve (5) parameters are K = 0.79 · 10^{-33}, b = 11.5.

Residual strength degradation was measured at σ =

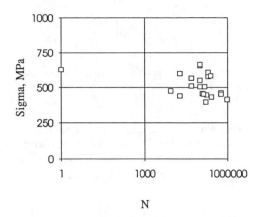

N

Fig. 3. Residual strength degradation at σ = 264.5 MPa.

264.5 MPa and data obtained are shown in Fig. 3. Large scatter of residual strength values is observed. The data are used to determine model parameter c as a value minimizing deviation between static strength distribution and equivalent static strength σ_{0i} of specimens with residual strength σ_{ri}:

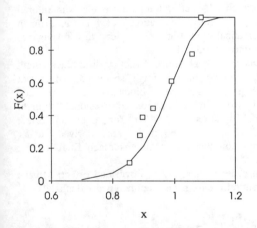

Fig. 1. Static tensile strength distribution.

$$\sigma_{0i} = [\sigma_{Ri}^c + \beta^c K \sigma^b n_i]^{\frac{1}{c}} = [\sigma_{Ri}^c + \beta^c \frac{n_i}{\overline{N}}]^{\frac{1}{c}}$$

i = 1,...,k where k is the number of specimens tested for residual strength, n_i - number of load cycles applied to i-th specimen. The parameter c was found to be c≈14. Having determined material parameters α, β, c, K, and b, one can predict fatigue life distribution at a given load level:

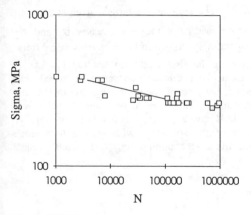

N

Fig. 2. S-N diagram for r=.1

$$F_N(n) = 1 - \exp\left[-\left(\frac{n + \sigma^c / \beta^c K \sigma^b}{1 / K \sigma^b}\right)^{\alpha / c}\right] =$$

$$1 - \exp\left[-\left(\frac{n}{\overline{N}} + \left(\frac{\sigma}{\beta}\right)^c\right)^{\alpha / c}\right]$$

Because $\left(\dfrac{\sigma}{\beta}\right)^c \approx 2 \cdot 10^{-6} \ll \dfrac{n}{\overline{N}}$ for load level σ = 264.5 MPa , this term can be neglected, and life distribution corresponding to $\overline{N}$ = 200000 is:

149

$$F_N(n) = 1 - \exp\left[-\left(\frac{n}{200000}\right)^{.96}\right]$$

Experimental and theoretical distributions are plotted in Fig. 4. Reasonable correspondence is observed.

Residual strength model allows one to predict life distribution under block loading. Let us consider the effect of short duration n_2 overload σ_2 imposed on service loading σ_1. Then residual strength is:

$$\sigma_R^c\left(n_1 + n_2\right) = \sigma_0^c - \beta^c\frac{n_1}{\overline{N}_1} - \beta^c\frac{n_2}{\overline{N}_2} \qquad (7)$$

where

$$\overline{N}_1 = 1/K\sigma_1^b, \qquad \overline{N}_2 = 1/K\sigma_2^b$$

The specimen is cycled till failure at service load level σ_1 after overload, therefore its life distribution is:

$$F_N(n) = 1 - \exp\left(-z^\alpha\right)$$
$$n_1 = \overline{N}_1 \cdot \left(z^c - \sigma_1^c\right) - n_2\frac{\overline{N}_1}{\overline{N}_2} \qquad (8)$$

4 FATIGUE OVERLOADS

The effect of the overload depends not only on its magnitude and duration, but also on the specimen lifetime fraction at which overload is introduced. One would expect most detrimental results for overloads occurring close to design life of the specimen when its residual strength is already degraded.

Three-block loading was performed with the second block representing overload. Peak stress level of the first and last blocks was $\sigma_1 = 264.5$ MPa, and first block duration $N_1 = 200000$ cycles was chosen close to average lifetime. The second (overload) block duration was $n_2 = 500$ cycles for the following overload levels $\sigma_2 = 392; 372; 353;$ and 301 MPa; $n_2 = 1000$ cycles for $\sigma_2 = 372$ MPa. The range of overload stress values was chosen based on residual strength degradation at σ_1 so that failure during overload is avoided. Combination of given loading parameters produced partial Miner's sum values D_0 of overload in the range of 0.01...0.24. Specimen self-heating temperature T was monitored during loading, variation of T is shown schematically in Fig. 5. Typical test results are presented in Table 1.

Due to the high value of first loading block length only stronger than average specimens underwent subsequent loading blocks. It precluded straightforward application of residual strength model relations (7), (8) for overload effect prediction, because they are derived under assumption that all specimens reach final loading block.

In order to quantify the overload effect, Miner's sum at failure for each specimen was calculated:

$$D = \frac{n_1}{\overline{N}_1} + \frac{n_2}{\overline{N}_2}$$

where $\overline{N}_1$, $\overline{N}_2$ are average fatigue lives at constant amplitude loading with peak stresses σ_1 and σ_2 correspondingly, n1 - number of cycles at σ_1 (total duration of the first and third loading blocks). D values calculated this way consistently exceed 1, as expected, because of the censoring due to the first loading block. Therefore Miner's sum D value was normalized by a factor accounting for relative position of the specimen with respect to average fatigue life of the population. Normalized damage parameter D^*:

$$D^* = \frac{D}{N_f^* / \overline{N}_1}$$

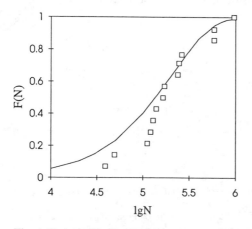

Fig. 4. Fatigue life distribution at $\sigma = 264.5$ MPa

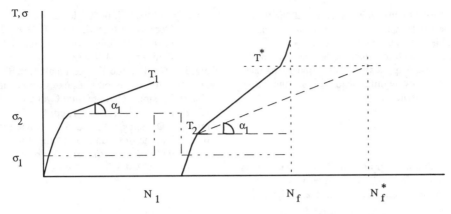

Fig. 5. Specimen temperature variation during block loading

where N_f^* is the fatigue life of a given specimen at σ_1. There is obviously no direct way available of measuring N_f^* - the same specimen cannot be led to failure twice. N_f^* value for each block-loaded specimen was estimated based on damage accumulation kinetics manifested by surface temperature increase during loading. It was assumed that critical damage level at failure (and self-heating temperature T^*) depends only on the stress level immediately preceding failure (Oldirev et al. 1989). Having determined damage accumulation rate at σ_1 (slope α_1 of the linear part of temperature diagram in the first loading block) and critical temperature level T^*, one can estimate fatigue life given specimen would have at constant amplitude loading as shown in Fig. 5:

$$N_f^* = n_1 + \frac{(T^* - T_2)}{ctg\,\alpha_1}$$

Calculated N_f^* values for 7 specimens are shown in Table 1. To test the accuracy of the proposed method, average estimated fatigue life lgN_f^* was compared to the average life at constant peak stress

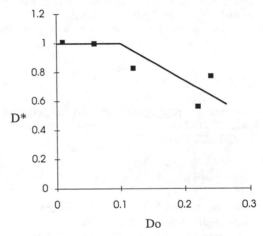

Fig. 6. Miner's sum at failure vs. overload severity.

Table 1

N	σ_2 MPa	n_2	D_o	N_f	$ctg\,\alpha_1$ $\cdot 10^3$	$ctg\,\alpha_2$ $\cdot 10^3$	N_f	D^*
1	392	500	0.22	323000	0.025	0.110	652000	0.56
2	372	500	0.12	285000	0.055	0.143	373000	0.83
3	353	500	0.06	267000	0.055	0.060	280000	1.
4	304	500	0.01	217000	0.107	-	217000	1.01
5	372	1000	0.24	282000	0.038	0.175	476000	0.69
6	372	1000	0.24	277000	0.041	0.210	405000	0.8
7	372	1000	0.24	256000	0.071	0.310	368000	0.82

σ_1 of those specimens which survived 200000 load cycles. The difference of the corresponding values 5.6 and 5.7 is statistically insignificant.

D* decreases with increasing overload magnitude. It means that application of linear damage accumulation law would lead to unconsevative residual life estimate.

CONCLUSIONS

Residual strength degradation model parameters were determined for GRP. The model was shown to correctly predict constant amplitude fatigue life distribution.

A method of individual lifetime estimation of specimens having been subjected to overload is proposed. It is based on autogenous heating monitoring, and is applicable if prolonged linear temperature increase periods both before and after overload application are observed.

The effect of short duration overload with partial Miner's sum D_0 less than .1 can be predicted by Miner's rule. It yields unconservative estimates of residual life after fatigue overload with $D_0 = 0.1...0.24$, and discrepancy increases with growing overload severity.

ACKNOWLEDGMENTS

This work was partially supported by CEC contract JOU 2-CT92-0085, supplementary agreement ERBCIPDCT930323.

REFERENCES

Andersons, J. 1993. Theoretical Methods of Fatigue Life Prediction of Composite Laminates (An Overview). *Mechanics of Composite Materials*. Vol. 29, N 6, pp. 741-754.

Oldirev, P.P., V.M. Parfeev & V.I. Komar 1977. Refined Method of Fatigue Life Determination for Polymeric Materials Based on Self-Heating Temperature. *Mechanics of Polymers*. N 5, pp. 906-913.

Oldirev, P.P. & V.P. Tamuzs 1989. High-Cycle Fatigue of Composite Materials. *J. of Mendeleev's Sov. Chem. Soc.* Vol. 24, N. 5, pp. 545-552.

Parfeev, V.M., P.P. Oldirev & V.P. Tamuzs 1979. Damage Summation under Non-Stationary Cyclic Loading of GRP. *Mechanics of Composite Materials*. N. 1, pp. 65-72.

Sendecky, G. 1992. Life Prediction for Resin-Matrix Composite Materials. *Comp. Mat. Ser. Vol. 4. Fatigue of Composite Materials*, ed. by K.L.Reifsnider, pp. 431-483

Yang, J. 1978. Fatigue and Residual Strength Degradation for Graphite/Epoxy Composite Under Tension - Compression Cyclic Loading. *J. of Comp. Mat.*, Vol. 12, N 1, pp. 19-39.

Yang, J. & D. Jones 1981. Load Sequence Effects on the Fatigue of Unnotched Composite Materials. *Fatigue of Fibrous Composite Materials, ASTM.STP 723*, pp. 213-232.

Progress in Durability Analysis of Composite Systems, Cardon, Fukuda & Reifsnider (eds)
© *1996 Balkema, Rotterdam. ISBN 90 5410 809 6*

Damage and failure of composite materials: Application to woven materials

M. Chafra, T. Vinh & Y. Chevalier
ISMCM, LISMMA (Groupe Rhéologie et Structures), Saint Ouen, France

A. Baltov
Institut de Mécanique, Sofia, Bulgaria

ABSTRACT

In this work, a tridimensional modeling is used to study advanced fiber and woven composite materials taking into account unilateral aspects. First, the frame-work of linear or non-linear elasticity is adopted, the second step concerns plasticity and eventually damage and fracture. We define a damage oriented parameter which is dependent on the level of plastic strain.

Our theory takes into account the features of loading (type and direction) and of plastic or nonlinear elastic behavior of a composite. The incremental procedure for stresses and strains permits the step-by-step computations, with any type of loading which is also decomposed into increments.

1. INTRODUCTION :

Fibre orientation and reinforced composite structure create an anisotropy which is to be taken into account in mechanical behavior modeling. This anisotropy permits for an equal weight, and equal volume of material to obtain a better strength than for classical materials. From the theoretical point of view , this anisotropy (often very marked) gives rise to some difficulties either in the description of reversible process (elasticity) or in the irreversible one (plasticity, damage, fracture). It is in the nature of composites to create a sensitivity with respect to a loading mode. This effect is accounted for in our paper . Our attempt is to present our research in a more general framework including a variety of reinforcements (unidirectional fibers with various orientations in layers, woven composites with different elementary models : taffeta, satin , serge with one or many layers). Woven composites are extensively used in industry in their large variety (glass/resin, graphite/resin, glass/polyester, graphite/graphite, SiC/SiC) (Léné 84).

We adopt the isothermal and static loading , and study the composite in short time intervals with small strains. Elastic, plastic and non-linear elastic properties are modeled with anisotropy and influence of loading orientation (off axes tests). Experimental studies on composites are widely developped but often with preferential proportional loadings applied on samples which are cut at various angles with respect to fiber orientation (Vinh 81), (Fuju 92) and (Baste 91).

A variety of models which takes into account anisotropy (Ladevèze1 93) is available. In some of them, sensitivity with respect to loading modes in tension and compression is examined (Ladevèze2 93). Ladeveze's approach is particularly interesting (Ladevèze3 93), it closely follows different material orientations in a composite. The aim of our study is to propose a modeling of the mechanical behavior of composite materials in more general framework including a large range of geometrical variables used in industry.

2. THE ELASTIC BEHAVIOR

To take into consideration all main effects in a composite material we introduce two 6-dimension vectorial spaces.

(i) *stress space* Σ : presented with element of the base of vectors $\{\Sigma_\alpha\} \in \Sigma$ defined by :

$$\{\Sigma_\alpha\} = \{\Sigma_1 = \sigma_{11}, \Sigma_2 = \sigma_{22}, \Sigma_3 = \sqrt{2}\sigma_{12}, \Sigma_4 = \sigma_{33}, \Sigma_5 = \sqrt{2}\sigma_{13}, \Sigma_6 = \sqrt{2}\sigma_{23}\}^T$$

$(\alpha = 1, 2, \ldots 6)$

where σ_{ij} (i, j = 1, 2, 3) are the components of the stress tensor

(ii) *strain space* E: presented with element of the base of vectors $\{E_\alpha\} \in E$ defined by :

$$\{E_\alpha\} = \{E_1 = \varepsilon_{11}, E_2 = \varepsilon_{22}, E_3 = \sqrt{2}\varepsilon_{12}, E_4 = \varepsilon_{33}, E_5 = \sqrt{2}\varepsilon_{13}, E_6 = \sqrt{2}\varepsilon_{23}\}^T$$

$(\alpha = 1, 2, \ldots 6)$

Where ε_{ij} (i, j = 1, 2, 3) are the components of the strain tensor

According to the definition of the two vectors $\{E_\alpha\}$ and $\{\Sigma_\alpha\}$ the energy density of proportional loading can be written in the following manner :

$$W = \frac{1}{2}\sigma_{ij}\varepsilon_{ij} = \frac{1}{2}\Sigma_\alpha E_\alpha \qquad (1)$$

(i, j = 1,2,3); (α = 1, 2,....6)

We define two tensorial invariants, the stress intensity $\Sigma_o = \sqrt{\Sigma_\alpha \Sigma_\alpha}$ and the strain intensity $E_o = \sqrt{E_\alpha E_\alpha}$, deduced from the second order strain (or stress) tensor .

To explain the unilateral effect we propose to characterize a loading in direction and type :

- *Loading direction* : defined by a unit vector $\{\sigma_\alpha\}$ $\in \Sigma$ with $\sqrt{\sigma_\alpha \sigma_\alpha} = 1$ and $\Sigma_\alpha = \Sigma_o \sigma_\alpha$.

In the case of a proportional loading $\{\sigma_\alpha\}$ = constant

- *Type of loading* : is defined as a function with various possibilities associated with the sign of the stress tensor, hence the stress space is decomposed into zones characterized by the sign attributed to each component of stress.

We define a parameter (χ) which corresponds to the number of lines of the matrix $[I_{\chi\alpha}]$ and is equal to the line:{sign(σ_1), sign(σ_2),.....sign (σ_6)}

$$\text{sign}(\sigma_\alpha) = \begin{cases} 1 \text{ if } \sigma_\alpha \geq 0 \\ -1 \text{ if } \sigma_\alpha < 0 \end{cases}$$

We consider all the combination of signs in the stress tensor, where the matrix $[I_{\chi\alpha}]$ is composed of 2^6 rows : (α=1, 2,....6; χ = 1, 2, ...2^6)

$$I_{\chi\alpha} = \begin{bmatrix} 1 & 1 & 1 & 1 & 1 & 1 \\ 1 & 1 & 1 & 1 & 1 & -1 \\ 1 & 1 & 1 & 1 & -1 & -1 \\ - & - & - & - & - & - \\ - & - & - & - & - & - \\ - & - & - & - & - & - \end{bmatrix}$$

In the one dimensional case, the parameter (χ) is equal to 1 or 2 , we have two spaces with differences in behavior, and recognize the theory of Ladeveze (Ladevèze3 93). This differences in behavior is due to cases of closing (in compression) and opening (in traction) of micro-cracks.

When α, (α = 1, 2, ...6) is the dimension of the subspace of stresses (χ = 1, 2,....,2^α) we have 2^α zones of loading with different behavior.

A sign convention is taken according to the following rules : a positive sign is taken for the tensile stress and a negative sign for the compressive stress. For the shearing stress, with regard to the main diagonal: if stress diverges we affect a positive

sign (Fig 1a), if stress converges, we affect a negative sign (Fig 1 b)

During a proportional loading $\{\sigma_\alpha\}$ = constant , we define a generalized modulus which relates a stress intensity to a strain intensity.

$$\Sigma_o = E^{(\chi)} E_0 \text{ or } E_o = H^{(\chi)} \Sigma_0 \qquad (2)$$

We associate to each type of loading (2^6) , a Hook's operator,

$$\Sigma_\alpha = E_{\alpha\beta}^{(\chi)} E_\beta \text{ or } E_\alpha = H_{\alpha\beta}^{(\chi)} \Sigma_\beta \qquad (3)$$

and in the case of non-proportional loading, the constitutive law is defined as an incremental elastic linear law:

$$\Delta\Sigma_\alpha = E_{\alpha\beta}^{(\chi)} \Delta E_\beta \text{ or } \Delta E_\alpha = H_{\alpha\beta}^{(\chi)} \Delta\Sigma_\beta \qquad (4)$$

To describe the variation of elastic compliance in linear domain which depend on the type and the direction of loading , we construct an isostrain quadratic surface in the space of the stress Σ:

$$\phi_E^{(\chi)} \equiv \mathfrak{R}_{\alpha\beta}^{(\chi)} \Sigma_\alpha \Sigma_\beta - E_o^2 = 0 \qquad (5)$$

E_0 = constant , (α, β = 1,....6)

$\mathfrak{R}_{\alpha\beta}^{(\chi)}$ is related to the components of compliance matrix
$$\mathfrak{R}_{\alpha\beta}^{(\chi)} = H_{\alpha\gamma}^{(\chi)} H_{\gamma\beta}^{(\chi)} \qquad (6)$$

and is related to the generalized modulus $E^{(\chi)}$ (2) :

$$E^{(\chi)} = \frac{1}{\sqrt{\mathfrak{R}_{\alpha\beta}^{(\chi)} \sigma_\alpha \sigma_\beta}} \qquad (\{\sigma_\alpha\} = \text{constant}) \quad (7)$$

In the case of biaxial loading (Σ_1, Σ_2), the matrix $[I_{\chi\alpha}]$ is defined by :

$$I_{\chi\alpha} = \begin{bmatrix} 1 & 1 \\ -1 & 1 \\ -1 & -1 \\ 1 & -1 \end{bmatrix} \quad (\alpha = 1, 2; \chi = 1, 2, 3, 4) \quad (8)$$

The constitutive law :

$$\Delta E_1 = H_{11}^{(\chi)} \Delta\Sigma_1 + H_{12}^{(\chi)} \Delta\Sigma_2$$
$$\Delta E_2 = H_{21}^{(\chi)} \Delta\Sigma_1 + H_{22}^{(\chi)} \Delta\Sigma_2 \qquad (9)$$

and the isostrain criteria is wreitten as follows :

$$\mathfrak{R}_{11}^{(\chi)}\Sigma_1^2 + 2\mathfrak{R}_{12}^{(\chi)}\Sigma_1\Sigma_2 + \mathfrak{R}_{22}^{(\chi)}\Sigma_2^2 - E_o^2 = 0 \quad (10)$$

(E_0 = constant , χ = 1, 2, 3, 4)

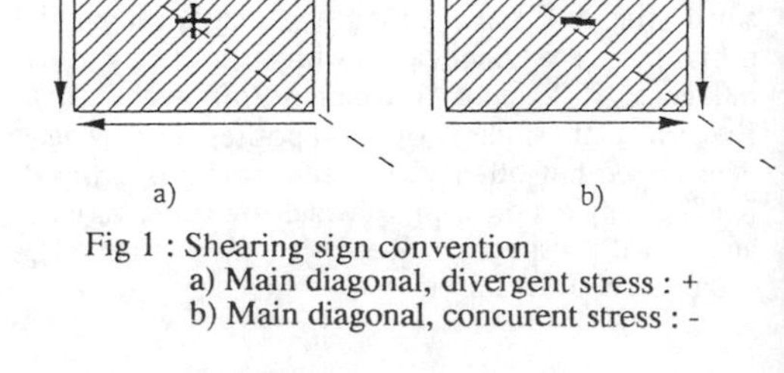

Fig 1 : Shearing sign convention
a) Main diagonal, divergent stress : +
b) Main diagonal, concurrent stress : -

154

3. IDENTIFICATION OF THE PARAMETERS IN THE CASE OF BIAXIAL LOADING

In the case of biaxial loading we came to the following relation between Σ_1 and Σ_2 (Fig 2)
$$\Sigma_2 = K_a \Sigma_1 \;,\; K_a = tg\alpha \text{ or } \Sigma_1 = C_b \Sigma_2 \;,\; C_b = tg\beta \quad (11)$$

$$\sigma_1 = \frac{1}{\sqrt{1+K_a^2}} = \frac{C_b}{\sqrt{1+C_b^2}} \text{ and } \sigma_2 = \frac{K_a}{\sqrt{1+K_a^2}} = \frac{1}{\sqrt{1+C_b^2}}$$

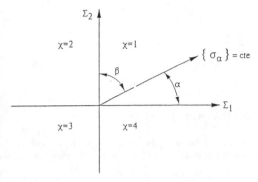

Fig 2 : Various load in biaxial loading case

Using the relations (10) and (11) , we obtain

$$\mathfrak{N}_{11}^{(\chi)} \frac{1}{1+K_a^2} + 2\mathfrak{N}_{12}^{(\chi)} \frac{K_a}{1+K_a^2} + \mathfrak{N}_{22}^{(\chi)} \frac{K_a^2}{1+K_a^2} = (H'^{(\chi)}(\alpha))^2 \quad (12)$$

where $H'^{(\chi)} = \dfrac{E_0(\alpha)}{\Sigma_0(\alpha)}$, $\overline{\Sigma}_0 = \sqrt{\Sigma_1^2 + \Sigma_2^2}$

The criterion is in the stress plane. We have three parameters and we should have three stiffnesses in various directions : 0° , 90°, and 45°. Hence the parameters of the surface (11) are :

$$\begin{cases} \mathfrak{N}_{11}^{(\chi)} = (H'^{(\chi)}(0°))^2 \\ \mathfrak{N}_{22}^{(\chi)} = (H'^{(\chi)}(90°))^2 \\ \mathfrak{N}_{12}^{(\chi)} = (H'^{(\chi)}(45°))^2 - \frac{1}{2}[\; H'^{(\chi)}(0°))^2 + H'^{(\chi)}(90°))^2 \;] \end{cases} \quad (13)$$

We often use an experimental method using samples out off axes for various directions. The results provided by a test is the longitudinal modulus, while the results used in the models is the generalized modulus. By means of the transformation method, we establish the following relation between Σ_1, Σ_2 and σ_L

$$\begin{cases} \Sigma_1 = \sigma_L \cos^2 \theta \\ \Sigma_2 = \sigma_L \sin^2 \theta \\ \frac{1}{\sqrt{2}} \Sigma_3 = -\sigma_L \sin^2 \theta \cos^2 \theta \end{cases} \quad (14)$$

$$tg(\alpha) = tg^2(\theta)$$

we define the longitudinal modulus and Poisson ratio: $E_L^{(\chi)}(\theta) = \dfrac{\sigma_L}{\varepsilon_L}$, $\nu_{LT}^{(\chi)}(\theta) = -\dfrac{\varepsilon_T}{\varepsilon_L}$ (15)

and then we calculate
$$\Sigma_0 = \sqrt{\Sigma_1^2 + \Sigma_2^2 + \Sigma_3^2} = \sigma_L \;,\; \overline{\Sigma}_0 = \frac{\Sigma_0}{\sqrt{\cos^4\theta + \sin^4\theta}} \quad (16)$$
$$E_0 = \sqrt{\varepsilon_L^2 + \varepsilon_T^2} = \varepsilon_L \sqrt{1 + (\nu_{LT}^{(\chi)})^2}$$

using the relation (14), (15), (16), we establish the elastic moduli
$$E'^{(\chi)}(\alpha) = \frac{\overline{\Sigma}_0}{E_0} = \frac{E_L^{(\chi)}(\theta)\sqrt{\cos^4\theta + \sin^4\theta}}{\sqrt{1 + (\nu_{LT}^{(\chi)}(\theta))^2}} \quad (17)$$

With three values of $H'^{(\chi)}(\alpha)$ for α = 0°, 90° and 45° , we identifiy the parameters of surface (10) by the relation (13), and following for these parameters we express the elastic characteristic for the constitutive law (9) :

$$H_{11}^{(\chi)} = \sqrt{\frac{\mathfrak{N}_{11}^{(\chi)}}{1 + (\nu_{LT}(90°))^2}} \;,\; H_{22}^{(\chi)} = \sqrt{\frac{\mathfrak{N}_{22}^{(\chi)}}{1 + (\nu_{LT}(0°))^2}} \quad (18)$$
$$H_{12}^{(\chi)} = \sqrt{\frac{\mathfrak{N}_{11}^{(\chi)} + 2\mathfrak{N}_{12}^{(\chi)} + \mathfrak{N}_{22}^{(\chi)}}{1 + (\nu_{LT}(45°))^2}} - \sqrt{\frac{\mathfrak{N}_{11}^{(\chi)}}{1 + (\nu_{LT}(90°))^2}}$$

The model is identified by the finite elements method. We present an application with woven carbon/epoxy composite materials studied by Ishikawa (ISH 77), with the following mechanical characteristic:
for fibers : E_L = 137,3 GPa ; E_T = 10,79 GPa ; G_{LT} = 5,394 GPa et ν_{LT} = 0,26
for matrix : E_m = 4,511 GPa et ν_m = 0,38

By means of identification method we obtain the parameters of surface (10) (table1) .
$$\mathfrak{N}_{11}^{(1)} = 5,43 \, 10^{-4} (GPa)^{-2}; \; \mathfrak{N}_{22}^{(1)} = \mathfrak{N}_{11}^{(1)}; \; \mathfrak{N}_{12}^{(1)} = 25,04 \, 10^{-4} (GPa)^{-2}$$

Tab 1 The values of elastic compliance in various directions

α	θ	$E'^{(cal)}$ (GPa)	$H' = (E')^{-1}$
0°	0°	42.9	0.0233
4.2°	15°	32.4	0.0309
18.2°	30°	21.4	0.0467
45°	45°	18.1	0.0552

we calculate $\Sigma_0(\alpha)$ for α = 0°, 90° , 45° and E_0= 0.1 % , E_0 = 0.5 %, the results are given in figure 3.

Experimental results which are in agreement with the quadratic surface permit to describe the variation of elastic compliance in linear domain with the type and the direction of a loading.

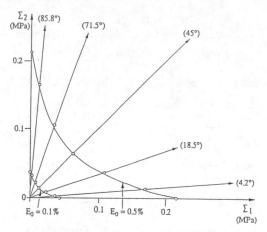

Fig 3 Isostrain surfaces for a woven carbon/epoxy composite materials

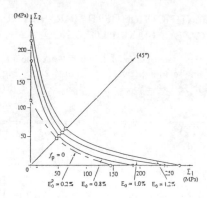

Fig 4 : Isostrain surfaces in non linear range

4 MODELING OF A NONLINEAR BEHAVIOR

We introduce a condition of non linearity similar to the Hill condition with various types of loading dependent parameters.

$$f_p^{(\chi)} \equiv f_{\alpha\beta}^{(\chi)} \Sigma_\alpha \Sigma_\beta = 0 \qquad (19)$$

$f_{\alpha\beta}^{(\chi)}$ is related to the elastic limit wich is direction dependent

$$\Sigma_0^P = \frac{1}{\sqrt{f_{\alpha\beta}^{(\chi)} \sigma_\alpha \sigma_\beta}} \quad , \{\sigma_\alpha\} = \text{constant} \qquad (20)$$

In order to describe a nonlinear behavior after the yielding point, we decompose the process of loading into increments, successively evaluating tangent modulus (superscript T) $E^{T(\chi)}(i) = \dfrac{\Delta\Sigma_0(i)}{\Delta E_0(i)}$, and

secant modulus (superscript S) $E^{S(\chi)}(i) = \dfrac{\Sigma_0(i)}{\Delta E_0(i)}$ for

each level (i) of loading (i = 1, ...N), and in the same manner of the linear case we construct two quadratic isostrain surfaces

$$\phi^{S(\chi)} \equiv \mathfrak{R}_{\alpha\beta}^{S(\chi)}(i)\Sigma_\alpha \Sigma_\beta - (E_0(i))^2 = 0 \qquad (21)$$

$$\phi^{T(\chi)} \equiv \mathfrak{R}_{\alpha\beta}^{T(\chi)}(i)\Delta\Sigma_\alpha \Delta\Sigma_\beta - (\Delta E_0(i))^2 = 0 \qquad (22)$$

$E_0(i)$ and $\Delta E_0(i)$ being constant in (21) and (22) .

Tangent moduli are used in incremental constitutive equation :

$$\Delta E_\alpha(i) = H_{\alpha\beta}^{T(\chi)}(i)\Delta\Sigma_\beta(i) \text{ or } \Delta\Sigma_\alpha(i) = E_{\alpha\beta}^{T(\chi)}(i)\Delta E_\beta(i) \quad (23)$$

$$(\alpha,\beta=1,....6 \; ; \; i = 1,....N)$$

ΔE_α , $\Delta\Sigma_\beta$ are strain and stress increments.

Experimental results concerning this formulation are given in figure 4, with woven glass/epoxy composite materials studied by Oytana and Varchon (92).

5. DAMAGE MODELING

A directional damage which depends on the type and the orientation of loading and the residual strain is examined by using a parameter $D^{(\chi)}$:

$$D^{(\chi)}(E_{OM}^P) = \frac{(E^{(\chi)} - E^{*(\chi)}(E_{OM}^P))}{E^{(\chi)}} \qquad (24)$$

where $E^{(\chi)}$ is an initial elastic modulus , $E^{*(\chi)}$ is a modified elastic modulus , E_{OM}^P (residual elastic deformation) .

The model of a damaged material is a linear model with incremental aspects and isostrain criteria.

$$\phi_E^{*(\chi)} \equiv \mathfrak{R}_{\alpha\beta}^{*(\chi)}(i)\Sigma_\alpha \Sigma_\beta - (E_0')^2 = 0 \qquad (25)$$

$E_0' = \sqrt{E_\alpha E_\alpha'} = \text{constant}$ and E_{OM}^P known $(E_0' = E_0 - E_{OM}^P)$

Σ_α and E_α are components of the stress and strain tensor. The parameters $\mathfrak{R}_{\alpha\beta}^{*(\chi)}$ depend on the elastic moduli of the damaged material :

$$E^{*(\chi)} = \frac{1}{\sqrt{\mathfrak{R}_{\alpha\beta}^{*(\chi)} \sigma_\alpha \sigma_\beta}} \qquad (26)$$

6. FAILURE MODELING

is expressed by Hill criteria :

$$f_F^{(\chi)} \equiv F_{\alpha\beta}^{(\chi)} \Sigma_\alpha \Sigma_\beta = 0 \qquad (27)$$

in which coefficient are dependent on the type of loading , and are related to the failure limit in each

direction $\Sigma_F(\sigma_\alpha) = \dfrac{1}{\sqrt{F_{\alpha\beta}^{(\chi)} \sigma_\alpha \sigma_\beta}} \qquad (28)$

Experimental results concerning this formulation are given in figure 5 , with carbon/epoxy T300/5200 composite materials studied by Chevalier (90)

156

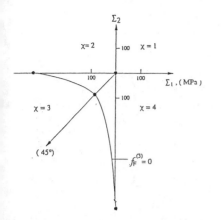

Fig 5 Failure surface of a T300/5208 carbon epoxy
composite materials in a loading zone ($\chi=3$)

7. CONCLUSION

In this work an isostrain criterion is used to describe the variation of elastic compliance with the type and direction of loading.

A new description of degradation in composites is proposed, taking into account a structural anisotropy and the type of loading.

The Hill criterion is adopted with some type of loading dependent parameters.

It would be interesting to achieve some tests with non proportional loading and to verify if the suggested model is always valid.

8. REFERENCES

S. BASTE - Comportement non linéaire des composites à matrice fragile, théorie et mesure de leur endommagement - *Habilitation, Univ.Bordeaux* I, n°21, 1991

M.CHAFRA- Comportement mécanique des matériaux composites fibres et tissés, Modélisation tridimensionnelle, endommagement et rupture *Thèse CNAM* Paris 1994

Y.CHEVALIER - Matériaux composites = critères de rupture - *Document, ISMCM*, 1990

T.FUJU, S.AMUJIMA, FAN LIN - Study on strength and non linear stress - strain response of plan woven glass fiber laminates under brassical loding -*J.Comp.Mat.vol.26, n°17, 1992*, p.2493-2510

T.ISHIKAWA, K.KOYAMA, S.KOBAYASHI- Elastic moduli of carbon-epoxy composite and carbon fibers - *J.Comp.Mat., vol.11, 1977*, p.332-344

P.LADEVEZE- Damage in composites Materials, studies in *Appl. Mechanics - N°34*, Elsevier, London, 1993

P.LADEVEZE, O.ALLIX, C.CLUZEL - Damage modelling at the macro and meso-scales for 3D composites- Damage in comp.Mat., Ed, *GZ.Voyiadjis, 1993*, p.195-215

P.LADEVEZE - Mécanique de l'endommagement des composites, une nouvelle approche des composites par la mécanique de l'endommagement-Ed,O.ALLIX,P.LADEVEZE, *Cachan, 1993*, p.97-127

J.LEMAITRE, J.L CHABOCHE. -Mécanique des matériaux solides - *Dunod, Paris, 1985*.

F.LENE - Contribution à l'étude des matériaux composites et de leur endommagement -*Thèse-Paris VI* - 1984

O.OYTANA, M.D.VARCHON.- Greco - communication, *CMCF - 10*, 1992

VINH. T - Mesures ultrasonores des constantes élastiques des matériaux composites, *Sci. et techniques de l'armement, 1981*, vol.54, p.265-289.

Progress in Durability Analysis of Composite Systems, Cardon, Fukuda & Reifsnider (eds)
© *1996 Balkema, Rotterdam. ISBN 90 5410 809 6*

Fatigue behaviour of aged glass-epoxy composites

E. Vauthier & A. Chateauminois
Ecole Centrale de Lyon, Département MMP, URA CNRS, Ecully, France

T. Bailliez
Direction des Etudes Renault, Rueil-Malmaison, France

ABSTRACT: The effects of hygrothermal aging on the fatigue properties of an unidirectional glass/epoxy composite have been investigated. Damage nucleation was monitored in terms of fibre breakage by in-situ microscopic observation of the tensile face of the specimens during flexural tests. The decrease in the endurance properties of the composite after water aging has been found to be related to a drop in the in-situ fibre strength. This fibre weakening may be associated to an increase in the pH of the aging media due to the leaching of unreacted hardener during water sorption.

INTRODUCTION

Predicting service life of structural components depends primarily on the overall fatigue behaviour of the material considered. Since polymer based composites have proven to be highly moisture sensitive, understanding both the aging mechanisms and the influence of moisture intake on the overall fatigue life of the material has been the scope of this study.

During water exposure of glass-epoxy composites, three major kinds of aging processes may occur :
- reversible plasticisation of the epoxy network resulting from water diffusion (Mc Kague 1978, Mikols 1982, Chateauminois 1994),
- macroscopical damage such as matrix cracking and/or fibre/matrix debonding which could arise from irreversible chemical degradation or swelling stresses (Jones and Mulheron 1983),
- loss of bearing properties of the glass reinforcement due to chemical attack in moist environments.This effect is generally attributed to flaw creation on the glass surface due to a thermally activated ion exchange mechanism (Charles 1958, Metcalfe 1972). Such a process has been also found to be pH sensitive. It is therefore of importance to evaluate the strength of the glass fibres in an physico-chemical environment similar to this encountered in the composite matrix.

Fatigue behaviour must be considered in terms of tolerance to first nucleated defects. The latter, such as fibre breakage, initiate during the first loading cycles. It has previously been shown that the overall fatigue behaviour was dependant on these first nucleated defects (Vincent 1988).

As hygrothermal aging enhances damage, the purpose of our study was hence to focus on the coupling effects between fatigue behaviour and environmental aging.

1 MATERIALS AND METHODS

1.1 Materials

The material used in this work was an E-glass-epoxy unidirectional laminate with a DGEBA/DICY type matrix. Plates were obtained by compression moulding of prepregs using a specific cure cycle of 5 minutes at 165°C followed by a 2 h postcure at 140°C. This procedure resulted in a fully crosslinked network with a Tα transition at 150°C as measured by DMTA at 1 Hz.

1.2 Environmental exposure

The material was aged by immersion in distiled water at four different temperatures, namely 50°C, 60°C, 70°C and 90°C. The effect of environment on flexural fatigue is presented here only on samples immersed in distiled water at 60°C and on samples aged at 50°C and 50% relative humidity. Gravimetric analysis of moisture uptake was monitored by weighing the samples to within 0.1 mg using an analytical balance.

Prior to water exposure, the as-cast specimens were dried at 50°C in vacuum until they reached constant weight. The latter was used as a reference for the calculation the relative weight gain Mt (%).

During water aging, UV spectrophotometry, performed between 190 nm and 380 nm, was used to investigate the occurrence of some leaching of low molecular weight products such as the unreacted

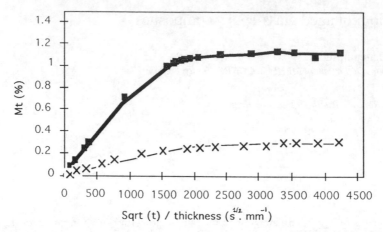

Figure 1. Sorption behaviour of the composite immersed in distiled water at 60°C (■) and aged at 50°C and 50% RH (×)

hardener. This UV analysis was also associated with pH measurements of the aging solutions.

1.3 Monotonic tests

In situ microscopic observation of the tensile face of the specimens was also performed during monotonic tests. The latter were performed under three-point bending conditions using a span-to-depth ratio of 20 to minimise shear stress.

An optical device linked to a camera and a Quantitative Analysis Software allowed the detection and location of broken fibres on an area of 5 x 2 mm^2 located beneath the loading nose. The following incremental loading procedure was used to monitor fibre breakage: The samples were loaded at 2 mm/min up to successive strain levels. At each strain level, the tensile face of the specimen was scanned using the microscopical device which was moved by a micropositioning unit. This took approximately 10 minutes. The tests were stopped when macroscopic cracks appeared and it was thus no longer possible to identify all fibre breaks.

1.4 Dynamic fatigue tests

Three-point bending flexural fatigue behaviour of the composite was investigated at imposed deflection. Test frequency was chosen as 25 Hz to avoid extra heating of the sample tested at room temperature and 50% R.H. The ratio of minimum strain ε_{min} to maximum strain ε_{max} was taken as R = 0.1. Continuous recording of stiffness loss during the fatigue tests provided damage monitoring. Final breakage of the specimen into two parts is generally not observed during bending fatigue as opposed to during tensile fatigue. A 10% stiffness loss was therefore selected as a conventional failure criterion.

The conclusions of this study will, however, remain the same whatever the criterion chosen.

2 SORPTION BEHAVIOUR

The resulting sorption curves exhibited a Fickian behaviour (Fig.1). This observation is only valid for both time and temperatures considered.

When aging the samples by immersion in distiled water, pH measurements of the aging medium showed that it gradually became basic (see Fig. 2). UV analysis of the aging media revealed the progressive appearance of an absorption peak located at 215 nm. This peak proved to be characteristic of DICY hardener in aqueous solution. It was concluded that the unreacted hardener left in the epoxy network was actually leached out in the water and hydrolysed (Jones 1985), which accounted for pH increase. Such solutions with a pH > 9 have deleterious effects on the fibre ultimate properties (Gaur and Miller 1990). Local basic solutions may however be trapped at the fibre /matrix interface after hygrothermally induced debonding occurred and lead to enhanced crack nucleation on the fibre surface.

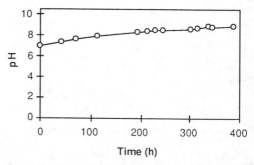

Figure 2. pH increase of the aging medium at 60°C.

Table 1. Ultimate bending properties of the composite before and after aging

Aging conditions	Mt (%)	Strain to failure ε_R (%)
Non aged		3.6
50°C and 50% RH	0.25	3.5
Immersion at 60°C	1.55	1.7

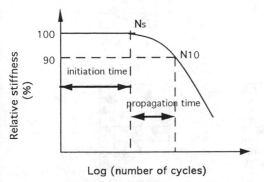

Figure 3: Typical stiffness loss curve.

Figure 4. View of broken fibres with interfacial debonding at fibre tips (non aged composite; magnification x10).

3 MECHANICAL MONITORING

After the samples were soaked in distiled water at 60°C or aged at 50°C and 50% RH, their residual monotonic properties were measured under 3-point bending conditions. Aging in immersion induced a severe decrease in strain and stress to failure as shown in Table 2. From these values, a range of applied strain levels was chosen to perform the dynamic 3-point bending fatigue tests.

3.1 Fatigue properties before aging

Analysis of the stiffness loss curves recorded during fatigue tests reveals that the fatigue life can be divided into two steps (Fig.3):

i) During the first step, no significant stiffness loss is recorded and first defects such as fibre breaks nucleate on the tensile face of the specimens. Due to delayed fibre fracture and stress redistribution, the density of these non interactive defects will progressively increase with time.

ii) Above a critical density of broken fibres, the nucleated defects coalesce and give way to progressive macroscopic crack propagation. As a result, a continuous stiffness loss is recorded during this second step.

According to these remarks, we have chosen to scan the tensile face of unaged specimens after increasing numbers of fatigue cycles. The samples were subsequently tested in fatigue for a fixed number of cycles chosen as a fraction of N_S. They were then loaded under monotonic conditions at 2 mm/min up to the average loading strain used for the fatigue tests. The samples were hence scanned for broken fibres appearing as on Figure 4. Figure 5 gives the various times after which scanning was performed.

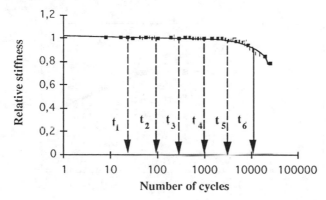

Figure 5: Fatigue stiffness loss curve of unaged material tested at 70% of εr - Sampling periods for scanning of tensile face.

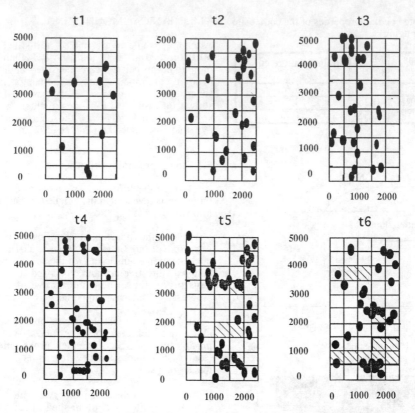

Figure 6: Mapping of tensile face of unaged material at sampling periods t1 to t6.
(Shaded areas correspond to macroscopic damage)

At lower number of fatigue cycles, broken fibres were randomly distributed on the tensile side of the specimens and were non-interactive (Fig. 6). Closer to N_S, i.e. the beginning of loss of macroscopic stiffness, these defects coalesced and gave way to macroscopic cracks which propagated through the specimen thickness.

3.2 Fatigue properties after aging

A drastic drop in the endurance properties may be observed after aging (Fig.7). This phenomenon may firstly be related to shorter macroscopic damage initiation times (N_S) - as shown on Fig. 8 - due to

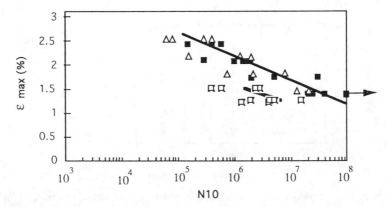

Figure 7. S-N plots with Wöhler curves for a N_{10} criterion (■ non aged, □ aged in immersion at 60°C, △ aged at 50°C and 50% RH)

162

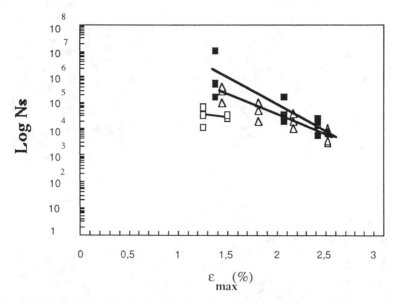

Figure 8: Damage initiation times N_S as a function of applied strain (■ non aged, ◻ aged in immersion at 60°C, aged at 50°C and 50% RH ▲).

reduced in-situ strength properties of the glass fibres. Secondly, crack propagation times (N_{10} - N_S) are higher (Fig. 9). This could be explained by the weakening of the glass fibres and by an increase in the probability to find a hygrothermally induced defects on the glass fibres located at the crack tip.

As static and dynamic fatigue tests show the same type of damage mechanisms (Fournier 1991), it seemed interesting to monitor the increase in the density of the early nucleated defects during monotonic loading.

The number of broken fibres detected on the tensile side of the specimen was mapped for increasing strain levels (Fig. 10). In the case of the unaged

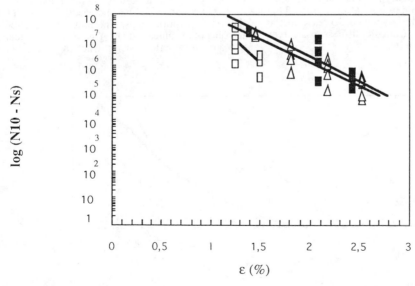

Figure 9: Damage propagation times N_{10}-N_S as a function of applied strain (■ non aged, ◻ aged in immersion at 60°C, ▲ aged at 50°C and 50% RH)

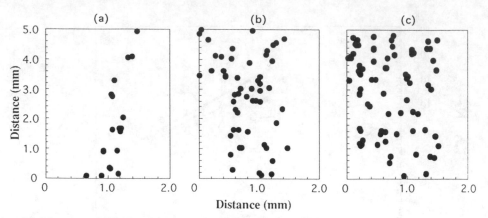

Figure 10. Mapping of fibre breaks on the tensile face of an non aged specimen as a function of the applied strain ((a) 50 % ε_R; (b) 60 % ε_R; (c) 80 % ε_R)

specimens, cracks extended into the matrix near the broken fibre ends. After aging, the broken ends of the fibres were located in a columnar shaped crack which appeared to contain all the cracks, yielding, debonding and/or fibre pull-out at the interfacial region. This suggested that hygrothermal debonding occurred.

An average concentration of broken fibres was derived from this mapping. The results plotted on Fig. 11 show that, due to the statistical distribution of fibre strength, fibre breakage initiates at strain levels well below the ultimate properties of the material. Aging in immersion resulted in a shift in the statistical distribution of the ultimate fibre properties towards lower values. Furthermore, fibre breakage initiates for a lower threshold value after aging. This observation may be linked to the decrease in the initiation times derived from the stiffness loss curves of the dynamic fatigue tests.

4 CONCLUSIONS

This study has shown the importance of the fibres'contribution to the overall mechanical performances of aged composites. Hygrothermal aging of the composite studied has been found to induce substantial drop in the endurance properties during fatigue loading. It has been related to a strong decrease in fibre strength which may be measured by in-situ mapping of fibre breakage during monotonic loading. As a result, higher densities of early nucleated defects are achieved during the first fatigue loading. This affected both the time for the onset of crack propagation and the subsequent crack propagation rates. The detrimental effect of moisture on the load bearing ability of the glass fibres is probably enhanced by pH increase as noted during the water sorption steps. The latter was related to leaching of the unreacted DICY hardener remaining in the composite after curing. Such a process is of

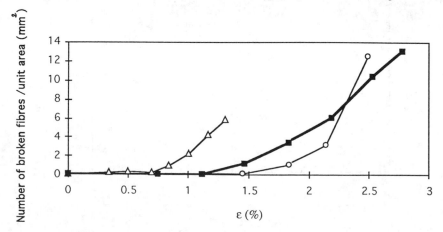

Figure 11: Density of broken fibres versus applied strain (Monotonic loading, ■ unaged, ∆ aged in immersion at 60°C, o aged at 50°C and 50% RH)

164

importance when considering the local pH increase which can occur at water clusters or hygrothermally induced microvoids at the fibre/matrix interface.
Future scope for this work is to derive damage maps where the evolution of the various mechanisms presented here could be reported as a function of the aging conditions.

ACKOWLEDGMENTS

The authors wish to thank RENAULT for sponsoring the work presented herein.

REFERENCES

Mikols, W.J., Seferis, J.C., Apicella , A., Nicolais, L. 1982. Evaluation of Structural Changes in Epoxy Systems by Moisture Sorption-Desorption and Dynamical Mechanical Studies, *Pol. Comp.,* 3 (3): 118-124.

Mc Kague,L. 1978. Environmental Synergism and Simulation in Resin Matrix composites, *ASTM STP 658* :93-204

Chateauminois, A., Chabert, B., Vincent L. and Soulier, J.P. 1994. *Polymer Composites* : in press.

Jones, F.R. and Mulheron M. 1983. The Effect of Moisture on the Thermal Expansion Behaviour and Thermal Strains in GRP, *Composites,* 14 (3): 281-287.

Charles, R.J. 1958. Static Fatigue of Glass, *J. Appl. Physics,* 19 (11): 1549-1553.

Metcalfe, A.G. and Schmitz, G.K. 1972. Mechanism of stress corrosion in E-glass fibres, Glass Technology, 13 (1): 5

Vincent, L., Fiore, L., Puget, P. and Gourdin, C. 1988. Effect of microstructure on fatigue breaking of glass-epoxy. *ECF7, Failure Analysis - Theory and Practice* : Budapest.

Jones, F.R., Shah, M.A. and Bader, M.G. 1985. The Effect of Curing Conditions on the Equilibrium Moisture Levels and Residual DICY in GRP Laminates, *Mech. Prop. Comp.* :87-93

Gaur, U. and B. Miller, B. 1990. Effects of Environmental Exposure on Fiber/Epoxy Interfacial Shear Strength, *Polymer Composites,* 11 (4): 217.

Fournier, P., Vincent, L. Elmendorp, J.J. and Van Veelen, A. 1991.Flexural fatigue behaviour of unidirectional glass fibre reinforced epoxy resin based composites, *1st international conference on deformation and fracture of composites*: 25-27. Manchester, UK: The Plastics and Rubber Institute.

4 Accelerated testing

Progress in Durability Analysis of Composite Systems, Cardon, Fukuda & Reifsnider (eds)
© *1996 Balkema, Rotterdam. ISBN 90 5410 809 6*

A framework for long-term durability predictions of polymeric composites

Mark E.Tuttle
University of Washington, Seattle, Wash., USA

ABSTRACT: The long-term durability of polymeric composites is associated with a gradual change in physical properties, which occurs with the passage of time and loading. The many mechanisms which impact composite long-term durability is grouped into three categories: nonlinear constitutive behavior (i.e., viscoelastic or viscoplastic behavior), mechanical degradation (i.e., matrix cracking, fiber failures, delaminations, etc), and "aging" effects; the latter is further divided into "physical" and "chemical" aging phenomenon. Analytical and/or empirical models have been developed which account for each of these factors seperately. The challenge before the composites community is to develop predictive methodologies which can account for all factors acting simultaneously.

1 INTRODUCTION

Although polymeric matrix composites have emerged as an important new class of structural materials, concerns remain regarding the "long-term durability" of composites. These concerns arise because of the observation that a composite structure which safely supports some specified thermomechanical loading at short times (on the order of weeks or months, say) may nevertheless fail due to the same thermomechanical loading at long times (on the order of years or tens of years, say). Virtually all physical properties of polymeric composites may change with the passage of time, including (for example) structural stiffness, strength, density, deflections due to loading and/or environment, or damage tolerance. These changes in durability are caused by a wide array of different mechanisms, and are brought about by the integrated effects of all of these mechanisms acting simultaneously. In general, changes occur more rapidly (i.e., are "accelerated") when the composite structure is exposed to any combination of high temperatures, high stress levels, high levels of electromagnetic radiation, and/or high levels of low-molecular weight plasticizing agents such as moisture, jet fuels, lubricants, or other hydrocarbons.

Practical methods of predicting the long-term durability of composite structures to within engineering accuracy's must be developed before polymeric composites can be confidently used in critical load-bearing structures. In this context the adjective "practical" has several fundamental implications. First, the overall goal is to predict the durability of composites after the passage of years or tens of years. It is impractical to obtain real-time measurements over these lengthy time periods. Furthermore, new and (presumably) improved composite material systems are continually brought to the marketplace, and consequently in practice there often is virtually no information regarding the long-term behavior of the composite material system of interest. These two factors imply that predictions of long-term durability must be based on measurements obtained during short-term tests, with a duration on the order of hours, days, or months, say. Second, during its' service life a composite structure will generally be subjected to changing biaxial (or triaxial) stress levels as well as changing environmental conditions. The desired method of predicting long-term durability must be able to account for these changing service conditions, or at least be useful in identifying the most severe combination of factors, such that reasonably accurate estimates of durability can be obtained. Third, changes in composite durability occur due to the combined effects of a variety of factors, as already mentioned. The mechanisms involved are usually coupled and are all highly nonlinear. Analytical and/or empirical models have been developed which can account for many of the factors involved acting seperately. The challenge before the composites community is to develop predictive methodologies which can account for all factors acting simultaneously. Despite these complexities, the "practical" requirement implies that all theories used to model long-term behavior (as well as any associated experimental measurements) must be relatively simple to implement in practice.

Although advances have been made in recent years, methods of predicting long-term durability which satisfy all the above requirements are still well beyond the state of the art. The primary objective of this paper is to suggest a framework in which the complex problem of composite durability predictions can be discussed. Methods currently used by the author and his colleagues to predict long-term durability, which only account for a portion of the mechanisms involved, will also be discussed in passing.

2 FACTORS WHICH IMPACT LONG-TERM DURABILITY

The principal source of changes in composite durability are mechanisms which impact the polymeric matrix, although for composites in which an organic fiber is used (e.g., aramid fibers), the fiber itself may also be a contributing factor (Wang et al 1990). To begin a discussion of how to model these effects it is convenient to classify the mechanisms involved into three broad categories, as briefly discussed below:

1. Nonlinear Viscoelastic-Viscoplastic (VE-VP) Behavior: All polymers are viscoelastic to some extent, and many polymers (especially thermoplastics) are also viscoplastic. The VE-VP response of a polymeric composite typically becomes more pronounced and nonlinear under conditions of high temperature, high moisture content, long loading times, and/or high stress levels.

2. Aging Effects: Both "physical aging" and "chemical aging" of the polymeric matrix are included in this category. Physical aging is associated with a loss in "free volume", and occurs because at temperatures below the glass-transition temperature (T_g) a polymer is generally not in thermodynamic equilibrium. Consequently over time a gradual change may occur in mechanical properties measured at the macroscale.

Chemical aging is associated with changes in the chemical nature and/or structure of the polymer,

such as a increase or decrease in molecular weight, changing crosslink densities, oxidation, etc. Chemical aging is typically irreversible, and may be caused by many factors, including moisture sorption-desorption, outgassing, or electromagnetic radiation.

3. Mechanical Degradation: Included in this category are physical defects at the macroscale which develop with time. These can include matrix cracking, delaminations, fiber-matrix interfacial bond failures, fiber microbuckling, fiber failures, etc. Although mechanical degradation is usually caused by an externally-applied stress, it can also be caused by internal stresses induced by environmental factors, e.g. cyclic changes in temperature or moisture, or exposure to electromagnetic radiation.

Note that these categories are not identified with any specific form of service conditions. That is, all three categories must be considered during design of a composite structure which in service will be subjected to fatigue loading, constant (creep) loading, spectrum loading, or any combination thereof. Although these factors have been presented in separate categories, they are in reality all highly coupled, as shown schematically in Figure 1. Hence, for example, mechanical degradation in the form of matrix cracks caused by fatigue loading can lead to an increased VE-VP response at the macroscale, and/or lead to increased levels of moisture content, impacting aging effects. A prediction of the long-term durability of composites must therefore be based on the coupled effects of all of these nonlinear mechanisms.

3 NONLINEAR VISCOELASTIC-VISCOPLASTIC BEHAVIOR

For a unidirectional composite ply based on inorganic fibers (e.g., graphite fibers), no VE-VP response is observed in the fiber direction. That is, the fiber-dominated modulii E_{11} and v_{12} are considered time-independent but temperature-dependent elastic constants. In contrast, the matrix-dominated modulii E_{22} and G_{12} are observed to be time- and temperature-dependent, and hence the time and temperature variation in these properties must be modeled using a suitable VE-VP theory.

Linear VE-VP theory is fairly well-developed and understood, as typified by the well-known mechanical-analogies, Boltzman hereditary integral or the time-temperature-superposition principal. In contrast, nonlinear VE-VP has received significant attention relatively recently and is less-well developed and understood. Many researchers have reported that polymeric composites exhibit nonlinear VE-VP behavior, especially in shear (see, for example, Brinson 1982, Crossman 1978, Gates 1992, Tuttle 1995).

At the University of Washington nonlinear VE behavior is modeled at the macroscopic (ply) level using the Schapery nonlinear hereditary integral (Schapery 1969), while nonlinear VP behavior is modeled using a viscoplastic functional used by Zapas and Crissman (Zapas 1984). For example, the VE component of the transverse strain (ε_{22}^{VE}) induced by a specified thermomechanical loading is given by:

$$\varepsilon_{22}^{VE}(t) = g_0^t A_{22} \sigma_{22}^t$$
$$+ g_1^t \int_0^t \Delta A_{22}(\psi - \psi') \frac{\partial(g_2^\tau \sigma_{22}^\tau)}{\partial \tau} \partial \tau \qquad (1)$$

where:

$$A_{22} = \frac{1}{E_{22}} = \text{transverse elastic compliance}$$

$\Delta A_{22}(t) = $ transverse linear creep compliance

$$\psi = \int_0^t \frac{\partial t}{a_{\sigma T}} \qquad \psi' = \int_0^\tau \frac{\partial t}{a_{\sigma T}}$$

g_0, g_1, g_2, $a_{\sigma T}$ = stress- and temperature-

dependent nonlinearizing parameters
The transverse linear creep compliance is modeled using an exponential (Prony) series of the form:

$$\Delta A_{22}(t) = \sum_{r=1}^{k} A_r[1 - e^{-\lambda_r \psi}] \qquad (2)$$

An exponential series is used because it leads to a computationally efficient recursive algorithm, such that nonlinear viscoelastic behavior can be integrated within numerical schemes based on classical lamination theory (Tuttle 1995, Pasricha 1995a) or the finite element method (Henrickson 1984, Ha 1989).

The VP component of the transverse strain (ε_{22}^{VP}) induced by a specified thermomechanical loading is given by:

$$\varepsilon_{22}^{VP}(t) = \left\{ \int_0^t C(\sigma_{22})^N \right\}^n \qquad (3)$$

where C, N, and n are considered temperature-dependent material constants. Note that the integral embedded in Eq (3) is not a hereditary integral. That is, the viscoplastic strain at time t is modeled as a function of stress history but not of time. The total transverse strain induced by a specified thermomechanical loading is given by the sum of the VE and VP strains, i.e., $\varepsilon_{22} = \varepsilon_{22}^{VE} + \varepsilon_{22}^{VP}$. An analogous approach is used to model shear strains, i.e., ply shear strains are calculated as the sum of VE and VP components, $\gamma_{12} = \gamma_{12}^{VE} + \gamma_{12}^{VP}$.

The nonlinear VE-VP constitutive model represented by Eqs (1) through (3) has been

combined with classical lamination theory to allow prediction of the nonlinear VE-VP behavior of general multi-angle laminates. This involved developing a recursive algorithm, which can be used to integrate Eqs (1) and (3) for a specified thermomechanical loading in a computationally-efficient manner (Tuttle 1995, Pasricha 1995a). The approach has been used to predict the response of various graphite-bismaleimide laminates (IM7/5260) for times up to 6 months (Pasricha 1995a). Briefly, all material properties associated with Eqs (1) through (3) were deduced based on 10-hr creep/creep recovery tests. The properties measured during these 10-hr tests were then used to predict the response of various multi-angle laminates for times up to 4032 hrs (6 months). A typical comparison between measurement and prediction for a $[\pm 45]_{4s}$ laminate subjected to cyclic thermomechanical loading is shown in Figure 2. In this test a single thermomechanical loading cycle had a duration of 7 hrs, and consisted of 4 hrs at an axial creep stress and temperature level of 62 MPa and 121°C, respectively, followed by 3 hrs at no load and a temperature of 79°C. This creep stress level is roughly 35% of the ultimate strength of a $[\pm 45]_{4s}$ laminate, and the T_g of IM7/5260 is roughly 215°C, far higher than the maximum test temperature of 121°C. Consequently neither mechanical degradation effects nor aging effects were expected to be significant during the 6-month test. The 7-hr thermomechanical loading cycle was repeated continuously for a total of 576 cycles. As indicated in Figure 2, the first loading cycle was predicted almost exactly, but a discrepancy of 6.9% between measured vs. predicted strain developed over the 6-month test time. No obvious signs of mechanical degradation (in the form of matrix cracks, edge delaminations, etc) were present in the specimens following the 6-month test. There was evidence that modest aging effects occurred, since as the number of loading cycles increased the shape of the measured creep curves typically became "flatter," as compared to the predicted curves.

4 PHYSICAL AND CHEMICAL AGING EFFECTS

The long-term durability of composites is affected by "aging" of the polymeric matrix. Two "types" of aging have been defined in the literature, physical and chemical aging. With a few notable exceptions, factors which result in aging of the polymer generally (a) increase room temperature polymer stiffness and strength, and (b) decrease room temperature polymer ductility or toughness (i.e., increase brittleness), with a corresponding decrease in the VE-VP response. On the other hand, aging effects can either increase or decrease the T_g of a polymer, and consequently at

171

elevated temperatures aging may either increase or decrease these properties, depending on circumstance.

4.1 Physical Aging

When a polymer at a temperature above the T_g is cooled to a temperature below the T_g, the polymer molecular structure is no longer in thermodynamic equilibrium. That is, due to the drastic reduction in molecular mobility which occurs once the temperature is decreased to below the T_g, the specific volume of the polymer is generally at a non-equilibrium level. In this condition the polymer is said to contain excess "free volume," where free volume can be thought of as the unoccupied portion of the specific volume (Sullivan 1990). Although molecular mobility is low it is still finite, so that with the passage of time free volume may decrease, resulting in a significant change in mechanical properties measured at the macroscale. For example, the level of VE-VP ("creep") behavior generally decreases with increased physical aging.

Significant physical aging effects are observed at temperatures ranging from (roughly) the first secondary transition temperature T_β to about (T_g - 10°C). Significant physical aging is not observed at temperatures below the T_β because at these temperatures molecular mobility is reduced to such low levels that no physical aging can occur over practical time periods. Conversely, physical aging probably does occur at temperatures approaching the T_g (i.e., at temperatures above T_g - 10°C, but still below the T_g), but is not observed experimentally because thermodynamic equilibrium is attained quickly and hence the aging process is completed quickly.

The effects of physical aging on the creep behavior of polymers at linear stress levels has been studied extensively by Struik (Struik 1978, 1987a, 1987b, 1989), who developed a method of accounting for the effects of physical aging called "effective time" theory. Struik found that plots of log(creep compliance) vs log(time), where compliance data are obtained at different aging times, can be shifted horizontally by an amount log(a) to form a continuous "momentary master curve." The magnitude of log(a) depends on both temperature and the aging time t_a, where the aging time is defined as the elapsed time since the polymer temperature was decreased to below the T_g. The aging "shift rate", μ, is a function of temperature and is defined by:

$$\mu = -\frac{d(\log a)}{d(\log t_a)} \qquad (4)$$

Based on the preceding discussion, it would be expected that μ would have a relatively high value at temperatures approaching (T_g - 10°C), but would tend towards zero once this temperature is exceeded

since the physical aging process is completed over short times at these temperatures. Similarly, one would expect that μ decreases towards zero at temperatures approaching T_β, since the physical aging process effectively ceases at T_β and below. This general behavior has been confirmed for a wide variety of polymers (Struik 1978).

Equation (4) can be rearranged to give the shift factor a in terms of the aging shift rate and aging time:

$$a(t) = \left(\frac{t_a}{t_a + t}\right)^\mu \qquad (5)$$

During a creep test the time interval between $t = 0$ and current time t is equivalent to an "effective" time λ:

$$\lambda = \int_0^t a(\xi)d\xi \qquad (6)$$

Substituting Eq (5) in (6) and performing the indicated integration, there results (Struik 1978):

$$\lambda = \frac{t_a}{(1-\mu)}\left[\left(1+\frac{t}{t_a}\right)^{1-\mu} - 1\right] \qquad \text{(for } \mu < 1) \qquad (7)$$

Effective time is used to relate the long-term creep compliance, $\overline{A}(t)$, to compliances measured under constant aging times, $A(t_a)$:

$$\overline{A}(t,t_a) = A(\lambda,t_a) \qquad (8)$$

Struik's effective time theory has been successfully used to model the creep behavior of several polymeric composite systems at linear stress levels (Sullivan 1990, Gates 1993, Hastie 1991). A method of including the effects of physical aging in Schapery's nonlinear VE theory has recently been proposed (Pasricha 1995b). If it is assumed that the only effects of high stress on physical aging is a reduction in the shift rate, then the effects of physical aging can be incorporated into Schapery's theory by computing strains in the effective time scale:

$$\varepsilon_{22}^{VE}(\lambda) = g_0^\lambda A_{22}\sigma_{22}^\lambda$$
$$+ g_1^\lambda \int_0^\lambda \Delta A_{22}(\psi^\lambda - \psi^\xi)\frac{\partial(g_2^\xi \sigma_{22}^\xi)}{\partial\xi}\partial\xi \qquad (9)$$

where ξ is a dummy integration variable on the λ scale. These VE strains are then mapped to the test time scale, t. An analogous approach can be used to model the shear response at the ply level. A recursive algorithm which can be used to integrate Eq (9) for a specified thermomechanical loading in a computationally-efficient manner has been developed (Pasricha 1995b), but has not yet been implemented in the numerical code used at the UW.

To the author's knowledge the effect of physical aging on VP behavior has not been discussed in the literature. It can be argued that physical aging

172

will play a role in VP behavior, since all molecular motions, including those that result in time-dependent irrecoverable deformations, are likely to be impacted by the physical aging effect. On the other hand, thus far physical aging has been associated strictly with reversible effects, whereas VP behavior is by definition irreversible. A need exists to experimentally investigate whether physical aging does play a role in VP behavior. Note that if it is found that physical aging does play a role in VP behavior, then physical aging can be included in the Zapas-Crissman functional by changing the limits of integration in Eq (3):

$$\varepsilon_{22}^{VP}(\lambda) = \left\{ \int_0^\lambda C(\sigma_{22})^N \right\}^n \qquad (10)$$

The VP strains are then mapped to the test time scale, t, as is done with VE strains.

4.2 Chemical Aging

Chemical aging is associated with irreversible changes in the chemical nature and/or structure of the polymer, such as a increase or decrease in molecular weight, changing crosslink densities, oxidation, etc. From the literature it appears that the effects of chemical aging on polymeric composites has received most attention as it applies to the use of composites in space vehicles or space structures (e.g., Milkovich 1985, Milkovich 1986, Yancey 1994). In the studies performed by Milkovich et al samples of T300/934 graphite-epoxy were subjected to high levels of electron irradiation while in a vacuum, conditions which simulate the space environment. Among other things, it was found that electron irradiation caused molecular chain scission and the formation of low molecular-weight products within the epoxy network, resulting in a decrease in the T_g of about 50°C. At the macroscopic (ply) level these effects caused an increase in stiffness and strength at room temperatures or below, but a decrease in stiffness and strength at elevated temperatures. Yancey and his colleagues studied the effects of electron irradiation on the creep/creep recovery response of T300/934. It was found that irradiation caused a decrease in creep behavior at room temperatures, but a marked increase in creep behavior at elevated temperatures. These measurements are consistent with the decrease in T_g as measured by Milkovich et al.

Although the effects of chemical aging on the mechanical properties of polymeric composites have been documented, to the author's knowledge no method of predicting the long-term effects of chemical aging based on short-term data have yet been proposed. However, Valanis and Peng have developed a set of constitutive equations for "chemically unstable" polymers, which can potentially

be used to predict the creep response under conditions of changing temperatures and stress levels (Valanis 1983, Peng 1985). They reason that at the molecular level the primary effect of chemical aging is either additional crosslinking of the polymer network or, conversely, the scission of polymer chains. Consequently in their approach chemical aging is associated with the crosslink density. A set of coupled differential equations is developed which may be solved numerically to determine the coupled effects of temperature and creep stress on chemical aging (Peng 1985). To the author's knowledge this model has only been applied to unreinforced polymers, but it is likely that a similar approach may be applicable to polymeric composites as well.

5 MECHANICAL DEGRADATION

Mechanical degradation refers to physical defects which develop with thermomechanical loading and/or the passage of time. These can include matrix cracking, delaminations, fiber-matrix interfacial bond failures, fiber microbuckling, fiber failures, etc. The effects of mechanical degradation on composite stiffness and strength has been studied extensively, and many (perhaps hundreds) of failure theories have been proposed.

Failure theories may be classified according to the physical scale at which they are posed; micromechanical ("mechanistic") theories being posed at a physical scale comparable to the fiber diameter, versus macromechanical ("phenomenological") theories being posed at physical scale comparable to the ply thickness or greater. The fundamental difficulty in predicting mechanical degradation and/or failure of a multi-angle composite laminate is that, in general, all of the damage modes listed above occur simultaneously, and none are solely responsible for ultimate composite failure. As has been discussed by others (e.g., Hashin 1980), the micromechanical approach does not lead to a practical method of predicting mechanical degradation for general structural analyses, due to "intractable" mathematical complexities. This is not to say that micromechanical theories serve no purpose, since fundamental insights into the microprocesses which ultimately lead to composite failure can only be obtained by studying the composite at the microscale, and the development of new and improved material systems require such modeling. However, since the current objective is to develop a practical method of predicting long-term durability of composites at the structural level, only macromechanical failure theories are considered viable at the present time.

Even if only macromechanistic failure criteria are considered, there is still no criterion which is generally agreed to be the "best" approach. No failure

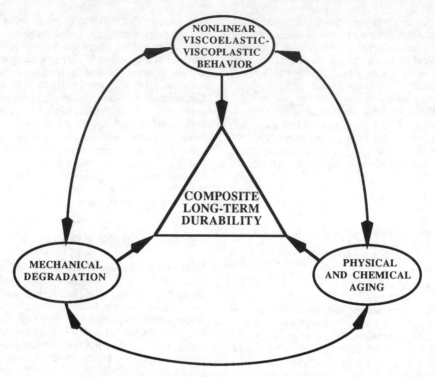

Figure 1: Schematic Classification of Factors Impacting Composite Long-Term Durability

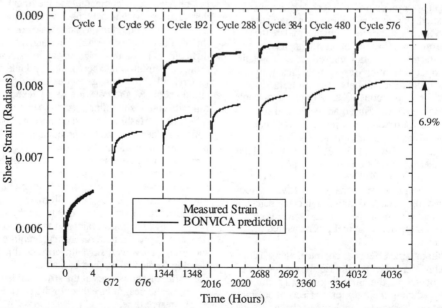

Figure 2: Measured vs Predicted Viscoelastic-Viscoplastic Response of a $[\pm45]_{4s}$ Graphite-
Bismaleimide Laminate Subjected to a Cyclic Thermomechanical Loading for 6 months

criterion is applicable for all conditions which may be encountered. Recalling the overall need for a "practical" procedure, then selection of a suitable criterion must be based in part on engineering judgment and in part on simplicity of use. At the UW mechanical degradation is modeled at the ply level using a modified form of the Tsai-Hill failure criterion in combination with Robinson's life-fraction rule and the ply-discount scheme. This approach is similar to that developed by Brinson and his colleagues (Brinson, 1982). Ply failure is predicted if the following criterion is satisfied:

$$\frac{(\sigma_{11})^2}{X(T)^2} - \frac{(\sigma_{11})(\sigma_{22})}{X(T)^2} +$$
$$\frac{(\sigma_{22})^2}{[Y(t,T)]^2} + \frac{(\tau_{12})^2}{[S(t,T)]^2} \geq 1 \qquad (11)$$

where:
$X(T)$ = temperature-dependent ply strength in the fiber direction
$Y(t,T)$ = time- and temperature-dependent ply strength transverse to the fibers
$S(t,T)$ = time- and temperature- dependent ply shear strength
Also, the transverse and shear strength are assumed to be of the following form:
$$Y(t,T) = A(T) - B(T)\log(t)$$
$$S(t,T) = C(T) - D(T)\log(t) \qquad (12)$$
Temperature-dependent material constants A, B, C, and D are obtained through a series of creep-to-rupture tests.

Equation (11) does not account for "cumulative damage" which occurs if ply stresses change with time. Ply stresses will obviously change if the externally-applied loading changes. However, even if the applied loading remains constant (as in a creep test, say) ply stresses still vary with time due to the VE-VP response. Generally speaking, VE-VP effects result in stress redistributions in which the ply fiber stresses increase with time and the matrix stresses decrease with time. During analyses performed at UW, changes in ply stresses are modeled as discrete steps in stress; i.e., ply stresses are assumed to remain constant over discrete increments in time. During each time increment, t_{inc}^i, the creep-to-rupture time, t_{rup}^i, associated with current ply stresses is calculated according to Eq (11). Ply failure due to cumulative damage is then predicted to occur according to Robinson's life-fraction rule; ply failure is predicted when:

$$\sum_{i=1}^{n} \frac{t_{inc}^i}{t_{rup}^i} = 1 \qquad (13)$$

where:
n = number of time increments

t = current total time = $\sum_{i=1}^{n} t_{inc}^i$

Once a ply is predicted to have failed its' stiffnesses are reduced according to the standard ply discount scheme (Tsai 1987).

6 SUMMARY AND CONCLUDING REMARKS

Prediction of the long-term durability of polymeric composites is a multidisciplinary problem. The many mechanisms which impact long-term durability are coupled and highly nonlinear. For the most part these mechanisms have been studied individually. The challenge before the composites community is to develop a modeling procedure, based on short-term tests results, which can be used to predict the long-term effects of all of these mechanisms, acting simultaneously. It is suggested that the problem is simplified (at least conceptually) by grouping the many factors into three categories: nonlinear viscoelastic-viscoplastic effects, aging effects, and effects due to mechanical degradation.

Past efforts at the UW have focused mainly on predicting long-term viscoelastic-viscoplastic (VE-VP) behavior. Predictions for test times up to 6-months in duration have compared favorably with measurements, for composites tested under conditions which do not involve mechanical degradation or aging effects. A method of accounting for physical aging effects on the VE-VP response has recently been developed, but predictions which include physical aging not yet been compared with experimental measurements. Also, a method of accounting for mechanical degradation is included in the UW approach, although once again predictions have not been compared with measurement.

The long-term effects of chemical aging are not accounted for in the UW approach. In fact, the author is not aware of any proposed methods which can account for the long-term effects of chemical aging based on short-term test results. Although there is a general need for further research involving the long-term durability of composites, it would seem that the investigation of the effects of chemical aging are in particular need of increase attention.

7 ACKNOWLEDGMENTS

Financial support of composite durability studies at the UW has been provided primarily by The Boeing Company and the NASA-Langley Research Center. Also, the author gratefully acknowledges the innumerable contributions made by his colleagues and former students, in particular Prof. Ashley F. Emery, Dr. Arun Pasricha, and Mr. Andrew Delaney.

175

8 REFERENCES

Brinson, H., and Dillard, D., The Prediction of Long Term Viscoelastic Properties of Fiber Reinforced Plastics, *Proc 4th Int'l Conf Comp Mat'ls*, Vol 1, Tokyo (1982).

Crossman, F.W., Mauri, R.E., and Warren, W.J., "Moisture-Altered Viscoelastic Response of Graphite/Epoxy Composites", ASTM STP 658, Philadelphia, PA (1978)

Gates, T.S.,, "Experimental Characterization of Nonlinear Rate-Dependent Behavior in Advanced Polymer Composites," *Experimental Mechanics*, Vol 32 (1), (1992).

Gates, T., and Feldman M., "Time-Dependent Behavior of a Graphite/Thermoplastic Composite and the Effects of Stress and Physical Aging," NASA TM 109047 (1993).

Ha, S.K., and Springer, G.S., "Time Dependent Behavior of Laminated Composites at Elevated Temperatures", *Jrnl Composite Materials*, Vol 23, (1989).

Hastie, R.L., Morris, D.H., "The Effect of Physical Aging on the Creep Response of a Thermoplastic Composite," CCMS Report 91-17, VPI&SU, Blacksburg, VA (1991).

Hashin, Z., "Failure Criteria for Unidirectional Fiber Composites," *Jrnl of Applied Mechanics*, Vol 47, (1980).

Henriksen, M., "Nonlinear Viscoelastic Stress Analysis - A Finite Element Approach," *Computers and Structures*, Vol 18(1), (1984)

Milkovich, S.M., Sykes, G.F., and Herakovich, C.T., "Fracture Surface of Irradiated Composites", *Fractography of Modern Engineering Materials:*, ASTM STP 948, (1985)

Milkovich, S.M., Herakovich, C.T., and Sykes, G.F., "Space Radiation Effects on the Thermo-Mechanical Behavior of Graphite-Epoxy Composites," *Jrnl Composite Materials*, Vol 20, (1986).

Pasricha, A., Tuttle, M., and Emery, A., "Time Dependent Response of IM7/5260 Composites Subjected to Cyclic Thermo-Mechanical Loading", to appear, *Comp Sci and Tech* (1995a).

Pasricha, A., Parvatareddy, H., Dillard, D., and Tuttle, M., "Physical Aging Effects on Polymeric Composites Subjected to Variable Stress History," *Proceedings*, 1995 SEM Spr Conf (1995b)

Peng, S., Constitutive Equations of Aging Polymeric Materials, *Jrnl Material Science*, Vol 20 (1985).

Schapery, R.A., "Further Developments of a Thermodynamic Constitutive Theory: Stress Formulation", Purdue Univ Report, AA&ES 69-2, (1969).

Struik, L.C.E., *Physical Aging in Amorphous Polymers and Other Materials*, Elsevier Pub Co., Amsterdam, 1978.

Struik, L.C.E., "The Mechanical Behavior and Physical Aging of Semicrystalline Polymers: 1", *Polymer*, Vol 28, (1987a).

Struik, L.C.E., "The Mechanical Behavior and Physical Aging of Semicrystalline Polymers: 2", *Polymer*, Vol 28, (1987b).

Struik, L.C.E., "The Mechanical Behavior and Physical Aging of Semicrystalline Polymers: 3. Prediction of Long Term Creep from Short Term Tests", *Polymer*, Vol 30, (1989).

Sullivan, J.L., "Creep and Physical Aging of Composites," *Composite Science and Technology*, Vol 39, (1990).

Tsai, S.W., *Composites Design*, Third Edition, Think Composites, Dayton, OH (1987)

Tuttle, M., Pasricha, A., and Emery, A., "The Nonlinear Viscoelastic-Viscoplastic Behavior of IM7/5260 Composites Subjected to Cyclic Loading", to appear, *J. Comp Matls* (1995).

Valanis, K.C., and Peng, S.T.J.,"Deformation Kinetics of ageing materials", *Polymer*, Vol 24, (1983).

Wang, J.Z., Wolcott, M.P., Teague, H., Kamke, F.A., and Dillard, D.A., "Experimental Techniques to Measure Fiber and Composite Response to Transient Moisture Exposure,"*Proceedings*, 1990 SEM Spring Conference, Albuquerque, NM (1990)

Yancey, R.N., Pindera, M-J, Slemp, W., and Funk, J., "Radiation and Thermal Effects on the Time-Dependent Response of T300/934 Graphite/Epoxy," *Proceedings*, 34th SAMPE Symposium, (1989).

Zapas, L., and Crissman, J., "Creep and Recovery Behavior of Ultra-High Molecular Weight Polyethylene in the Region of Small Uniaxial Deformations," *Polymer*, Vol 25, (1984)

Progress in Durability Analysis of Composite Systems, Cardon, Fukuda & Reifsnider (eds)
© *1996 Balkema, Rotterdam. ISBN 90 5410 809 6*

Long term prediction method for static, creep and fatigue strengths of CFRP composites

Yasushi Miyano
Materials System Research Laboratory, Kanazawa Institute of Technology, Ohgigaoka Nonoichi Ishikawa, Japan

ABSTRACT : A long term prediction method for static, creep and fatigue strengths of CFRP composites at an arbitrary frequency, temperature and stress ratio is proposed using the modified reciprocation law of time and temperature and the linear cumulative damage law based on the fact that the fracture mode is consistent for all loading types.

A satin woven CFRP laminates was chosen as specimens. The validity of this proposed prediction method is confirmed by the flexural static, creep and fatigue tests for various stress ratios which were performed at various temperatures below the glass transition temperature of the matrix resin.

1 INTRODUCTION

It is well known that the mechanical behavior of polymer resins exhibits time and temperature dependence, called viscoelastic behavior, not only above the glass-transition temperature Tg but also below Tg. Thus, it can be presumed that the mechanical behavior of FRP (Fiber Reinforced Plastic) using polymer resins as matrices also depend on time and temperature even below Tg which is with-in the normal operating temperature range. These examples are shown by Aboudi et al. (1989), Sullivan (1990) and Gates (1992), Miyano (1986, 1988) and Nakada et al. (1993).

It was previously reported by Miyano (1994), Mohri (1991) and McMurray (1993) that the flexural static, creep and fatigue strengths of a satin woven CFRP laminates were dependent on time and temperature and these strengths were different for the three types of loading for the same conditions of time and temperature. However, the fracture mode were found to be the same over a wide range of time and temperature for the three types of loading (Enyama 1993). Similar results were also reported by Karayaka (1994) for room temperature testing.

In this paper a long term prediction method for static, creep and fatigue strengths of CFRP composites at an arbitrary frequency, temperature and stress ratio is proposed using the modified reciprocation law of time and temperature and the linear cumulative damage law based on the fact that the fracture mode is consistent for all loading types.

2 PROCEDURE OF PREDICTION

The long term prediction method for static, creep and fatigue strengths of CFRP composites is explained by the procedure shown in Figure 1.

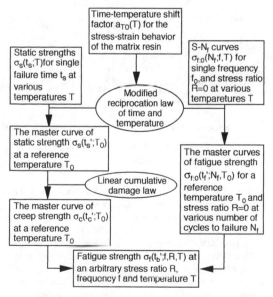

Figure 1   The procedure of long term prediction method for static, creep and fatigue strengths of CFRP composites

2.1 Modified reciprocation law of time and temperature

For the matrix resins used for CFRP composites, the time-dependent mechanical responses at a temperature T can be mapped onto the responses at the reference temperature To based on the modified reciprocation law of time and temperature by using the time-temperature shift factor $a_{T_0}(T)$ and the temperature shift factor $b_{T_0}(T)$ defined by the following equations,

$$a_{T_0}(T) = \frac{t}{t'} \quad, \quad b_{T_0}(T) = \frac{E'(t,T)}{E'(t',T_0)} \qquad (1)$$

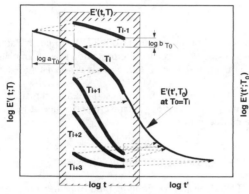

Storage moduli E'(t,T) at various temperatures T and this master curve E'(t',T₀) at a reference temperature T₀

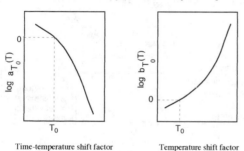

Time-temperature shift factor   Temperature shift factor

Figure 2   Modified reciprocation law of time and temperature

One means of determining these shift factors for the viscoelastic behavior of the matrix resin is by conventional deflection controlled dynamic test. As shown in Figure 2, the storage moduli E' versus time t at various temperatures T can be measured from this test and then can be used to create a master curve of storage modulus versus reduced time t' by shifting horizontally and vertically these moduli. If a smooth curve is produced then the modified reciprocation law of time and temperature holds for the viscoelastic behavior of this matrix resin.

It is assumed in this study that the same reciprocation law holds for the static, creep and fatigue strengths of

CFRP composite with this matrix resin, that is, the time-temperature shift factors of the viscoelastic behavior of matrix resin and these strengths of CFRP composites agree with each other.

2.2 Master curve of static strength of CFRP composites

Static testing is performed over a wide range of temperature, usually the same temperature range as was used with the matrix resin. As shown in Figure 3, the static strength σ_s is shifted by using the time-temperature shift factors from the viscoelastic behavior of the matrix resin to produce a master curve of static strength σ_s versus reduced time t'.

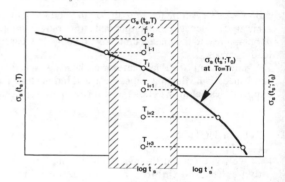

Figure 3   Static strengths $\sigma_s(t_s;T)$ for single failure time t_s at various temperatures T and this master curve $\sigma_s(t_s';T_0)$ at a reference temperature T_0

2.3 Prediction of creep strengths by linear cumulative damage law

We propose here the prediction method of creep strength σ_c based on the master curve of static strength and the linear cumulative damage law. Let $t_s(\sigma)$ and $t_c(\sigma)$ be the static and creep failure time for the stress σ. Furthermore, suppose that the material experiences a stress history $\sigma(t)$ for $0 < t < t^*$ where t^* is the failure time under the stress history. According to the linear cumulative damage law, we have

$$\int_0^{t^*} \frac{dt}{t_c[\sigma(t)]} = 1 \qquad (2)$$

Our objective is to find the creep failure time $t_c(\sigma)$ from the static failure time $t_s(\sigma)$ and the linear cumulative damage law as shown in Figure 4.

Choose an equally-spaced increasing sequence of stress, σ_i (i = 1, 2, 3, ...), and denote the associated static and creep failure time by $t_s^{(i)}$ and $t_c^{(i)}$, respectively.

In the static test, the deflection rate is kept constant and the force-deflection curves are found to be linear up to just before failure. We regard, therefore, the stress increases linearly during the static test. Further, we approximate the linear stress history by a staircase function with steps $\sigma_1, \sigma_3, \sigma_5, \ldots$. Thus, the linear stress history up to the stress level σ_4 is replaced by σ_1 for $0 < \sigma < \sigma_2$, and σ_3 for $\sigma_2 < \sigma < \sigma_4$. By the aid of the linear cumulative damage law, the creep failure time $t_c^{(2n-1)}$ ($n = 1, 2, 3, \ldots$) is expressed successively by $t_s^{(2n)}$ as follows:

$$t_c^{(1)} = t_s^{(2)}$$

$$t_c^{(2n-1)} = \frac{t_s^{(2n)}\, t_s^{(2n-2)}}{n t_s^{(2n-2)} - (n-1) t_s^{(2n)}} \quad (n = 2, 3, 4, \cdots) \tag{3}$$

For determination of t_c from Equation (3) over the entire stress range, we must have the master curve of static strength over the entire stress range. However, the portion of the curve below a certain stress can not be determined due to the experimental limitations. We, therefore, extrapolate that portion of the master curve by a decaying exponential curve with the same slope at the lowest available stress level.

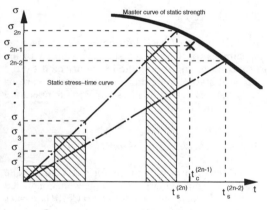

$t_s^{(i)}$: Failure time when static strength is σ_i

$t_c^{(i)}$: Failure time when creep strength is σ_i

Figure 4   Explanation of creep prediction method

2.4 Master curve of fatigue strength for zero stress ratio

The objective of this step is to find a method to estimate the fatigue strength under arbitrary frequency f, temperature T, and stress ratio R from some quick experiments and a limited number of fatigue tests. This method rests on the modified reciprocation law of time

and temperature for the fatigue strength of CFRP composites. In this step, we treat the static strength, i.e., the strength under constant deflection rate, creep strength and fatigue strength on the same grounds as much as possible, based on the fact the fracture mode is the same for all three loading types.

So, we regard the fatigue strength either as a function of the number of cycles to failure N_f or a function of the time to fatigue failure $t_f = N_f/f$ as shown in Figure 5. Furthermore, stress ratio R is defined as the ratio of minimum stress to maximum stress. To compare fatigue and creep tests results, we regard the creep test as a special case of fatigue test when R=1 as schematically explained in this figure. We consider that the fatigue strength σ_f at $N_f = 1/2$ may be represented by the static strength σ_s with the time to $t_s = 0.5/f$ when the stress ratio R is zero; this interpretations is motivated by the approximation of the sine wave by t/π over the interval $0 < t < \pi$.

Curves of fatigue strength σ_f versus the number of cycles to failure N_f, S-N_f curves, for single frequency f_0 and stress ratio R=0 at various temperatures T are then drawn as Figure 6.

The master curve of fatigue strength σ_f versus reduced time to failure t_f' can now be drawn as Figure 7, using the following relation based on the modified reciprocation law of time and temperature.

$$t_f' = \frac{N_f/f}{a_{T0}(T)} \tag{4}$$

where $a_{T0}(T)$ is the shift factor for the viscoelastic

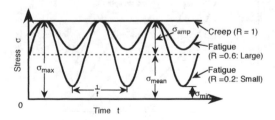

Stress Ratio: $R = \dfrac{\sigma_{min}}{\sigma_{max}}$

Mean Stress: $\sigma_{mean} = \dfrac{\sigma_{max} + \sigma_{min}}{2}$

Stress Amplitude: $\sigma_{amp} = \dfrac{\sigma_{max} - \sigma_{min}}{2}$

Time to failure : t_f(Fatigue), t_s(Static), t_c(Creep)

$t_f = \dfrac{N_f}{f}$ where N_f: Number of cycles to failre
f: Frequency

Creep test as fatigue : R=1, $t_c = t_f$

Static test as fatigue : R=0, $N_f = 1/2$, $t_s = t_f = 1/2f$

Figure 5   Interpretation of fatigue test

behavior of matrix resin. The above of this figure shows the fatigue strength σ_f versus the reduced time to fatigue failure failure t_f' at various temperature T. By connecting fatigue test points with the same number of cycles to failure N_f for the different testing conditions the master curve for the corresponding N_f can be drawn as the below of this figure.

2.5 Prediction of fatigue strength at arbitrary stress ratio

Now, we denote the fatigue strength with the time to failure t_b for an arbitrary stress ratio R and frequency f under temperature T by $\sigma_f(t_b; R,f,T)$, the fatigue

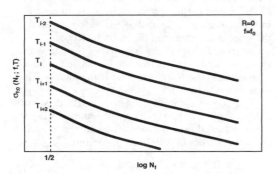

Figure 6   S-N_f curve $\sigma_{f:0}(N_f; f,T)$ for stress ratio R=0 and single frequency f_0 at various temperatures T

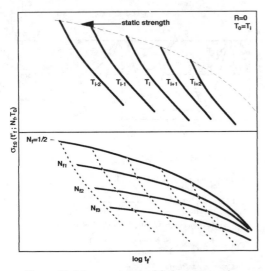

Figure 7   Master curves of fatigue strength $\sigma_{f:0}(t'_f; N_f,T_0)$ for stress ratio R=0 and a reference temperature T_0 at various number of cycles to failure N_f

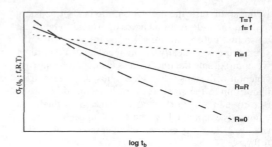

Figure 8   Fatigue strength $\sigma_f(t_b; f,R,T)$ for an arbitrary frequency f, stress ratio R and temperature T

strength for zero stress ratio by $\sigma_{f:0}(t_b;f,T)$ and the creep strength by $\sigma_c(t_b;T)$. We propose that they are related by the following equation.

$$\sigma_f(t_b;R,f,T) = \sigma_c(t_b;T)\cdot R + \sigma_{f:0}(t_b;f,T)(1-R) \quad (5)$$

We can predict σ_f-t_b relation for given R, f and T from Equation (5) if $\sigma_c(t_b,T)$ and $\sigma_{f:0}(t_b;f,T)$ is known as shown in Figure 8.

3 TRIAL TESTS

A satin woven CFRP laminates was chosen as specimens for these trial tests. The validity of the proposed prediction method is confirmed by the flexural static, creep and fatigue tests at various temperatures below the glass transition temperature of the matrix resin.

3.1 Specimen and testing method

The three-point bending tests for static (monotone loading), creep and fatigue loadings were conducted at several temperatures on satin woven CFRP laminates referred to as T400/3601.

Table 1.   Specifications and mechanical properties of fiber T400.

Diameter of mono filament (μm)	7.0
Filament number in 1 yarn	3000
Elastic modulus (GPa)	250
Tensile strength (GPa)	4.41
Ultimate strain (%)	1.8
Specific gravity	1.8

The satin woven CFRP laminates is made from carbon fibers T400 of which specifications and mechanical properties are shown in Table 1 and a matrix resin 3601 with a high glass transition temperature of T_g=236°C.

Eight prepreg sheets were stacked symmetrically about the mid-plane as shown in Figure 9. The CFRP laminates T400/3601 were formed by hot pressing these prepreg sheets into 2.7 mm thick plates. The volume fraction of the fibers in the composite is approximately 65.5%.

The set-up of three-point bending tests for static, creep and fatigue for T400/3601 is shown in Figure 10 where the span, width and the thickness used in the tests are L = 50 mm, b = 15 mm and h = 2.7 mm, respectively. To identify the states of stress and deformation of the specimen, we adopt $\sigma=3PL/2bh^2$ as the nominal maximum stress due to the load P and $\varepsilon=6\delta h/L^2$ as the nominal maximum strain when the deflection at the load point is δ.

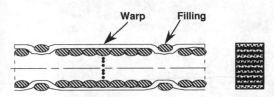

Warp   **Filling**

Figure 9   Laminate constitution of a satin woven CFRP laminates T400/3601

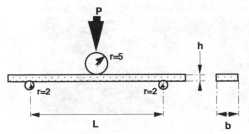

Figure 10   Three-point bending test
(L=50,b=15,h=2.7;unit mm)

Table 2   Testing conditions for satin woven CFRP

Loading type	Deflection rate [mm/min]	Frequency [Hz]	Stress ratio $\sigma_{min}/\sigma_{max}$	Temperature [°C]
Static	0.2 2 20 200	—	—	50, 75, 100, 125, 150, 175, 200, 225, 235, 250
Creep	—	—	1	50, 150, 230
Fatigue I	—	0.02 2	0.05	50, 100, 150, 200, 230
Fatigue II	—	2	0.5 0.8	50, 150, 230

The static tests were carried out at four deflection rates and ten uniform temperatures with an Instron type testing machine, as shown in Table 2. A creep testing machine with a constant temperature chamber was used to perform the creep tests at three uniform temperatures. The fatigue tests were performed by using an electro-hydrolic servo testing machine with a constant temperature chamber. The fatigue tests were also performed at various frequencies, temperatures, and stress ratios, as shown in Table 2.

3.2 Failure modes in static, creep and fatigue tests

We review here the failure modes of the CFRP laminates T400/3601 observed in the static, creep and fatigue tests. In the static test for the test temperature of 200°C or below, faint sounds were heard, without any visible sign of failure, preceeding the warp fiber buckling on the compressive side of the specimen near the loading point. The crack initiated in the matrix, at the site of the warp fiber buckling, propagated into the thickness of the specimen with some angle up to a plane slightly in the tensile side and delamination accompanying a loud sound occurred. The failure mode and process are quite similar to that reported by Karakaya and Kurath (1994) for satin woven CFRP laminates tested at room temperature. For the test temperature below 200°C, the entire failure process took a very short period of time. For the test temperature above 200°C, no sounds were heard during the failure process and the crack propagated more slowly with the increasing temperature though the same fracture process as for the test temperature below 200°C was observed.

Figure 11 shows the side views of the failed specimens under static, creep and fatigue loading at the test temperatures, T=50, 150, 230°C where V is the deflection rate in static test, f is the frequency of the fatigue loading and R stands for the stress ratio.

All these failed specimens in Figure 11 and those observed in the tests are similar regardless to loading patterns and the test temperatures. We consider, therefore, that the failure mechanisms are the same for static, creep and fatigue tests at all temperature tested. Furthermore, it can be presumed that the same reciprocation law of time and temperature for the viscoelastic behavior of matrix resin hold for these strengths of T400/3601, because the failure is triggered by the microbuckling of fibers on the compressive surface of specimen and the time-temperature dependencies of this microbuckling are controlled by the viscoelasticity of matrix resin.

3.3 Time-temperature shift factor of matrix resin

The conventional deflection-controlled dynamic tests

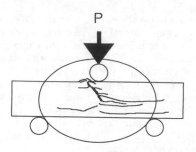

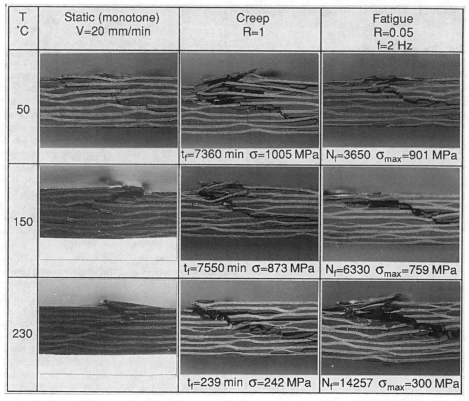

T °C	Static (monotone) V=20 mm/min	Creep R=1	Fatigue R=0.05 f=2 Hz
50		t_f=7360 min σ=1005 MPa	N_f=3650 σ_{max}=901 MPa
150		t_f=7550 min σ=873 MPa	N_f=6330 σ_{max}=759 MPa
230		t_f=239 min σ=242 MPa	N_f=14257 σ_{max}=300 MPa

1mm

Figure11 The side-view of flexural failure under static, creep and
fatigue loadings at various temperatures for T400/3601

were carried out for the matrix resin 3601 over the frequencies, 0.015 ~ 8 Hz and at temperatures from T=50 to 260°C with the step of 10°C; the maximum strain in each test was kept at 0.06 %. Figure 12 shows the storage modulus E'(t, T) versus time period t=1/f for various temperatures on the left side and the master curve E'(t', T_o) for the reference temperature of T_0=50°C on the right side. Since a smooth curve is produced on the right of this figure by shifting horizontally the curves on the left, the reciprocation law of time and temperature hold for the matrix resin. In Figure 13, the solid circles represent the shift factors a_{T_o}(T) versus the inverse temperature T in Kelvin of 3601 for various temperatures used to obtain the smooth master curve.

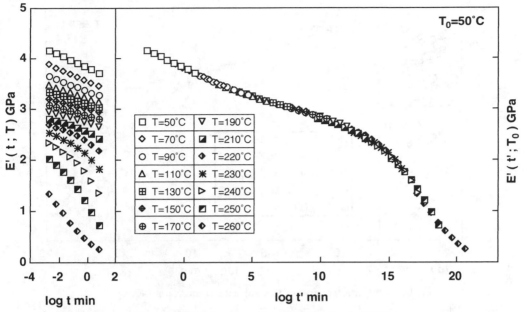

Figure 12　Master curve of storage modulus for 3601

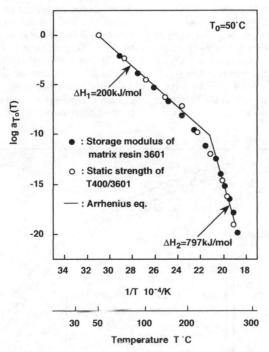

Figure 13　Time-temperature shift factors for 3601
and T400/3601

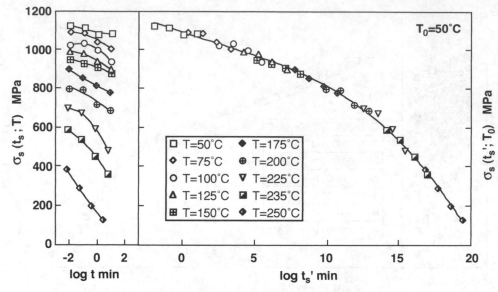

Figure 14   Master curve of flexural static strength for T400/3601

3.4 *Flexural static strength*

In the left side of Figure 14, the dependence of flexural static strength σ_s for the CFRP laminates T400/3601 upon the time to failure t_s is presented at various temperatures where each point for a temperature corresponds to one of the four deflection rates V=0.2, 2, 20, 200 mm/min. Shifting the curves in this figure horizontally, the master curve for the flexural static strength against the reduced time t_s' at the reference temperature $T_o=50°C$ is obtained as shown in the right

side of this figure. The shift factors for T400/3601 used are shown by the open circles in Figure 12 where the solid circles represent the shift factors for the matrix resin 3601 and the solid lines represent Arrhenius equation. The experimental shift factors are very close to Arrhenius equation with $\Delta H=200kJ/mol$ for T<239°C and with $\Delta H=797kJ/mol$ for T>239°C. The two shift factors, one for the matrix and the other for T400/3601 experimentally determined for completely different responses, agree well with each other and the both with Arrhenius equations. This agreement

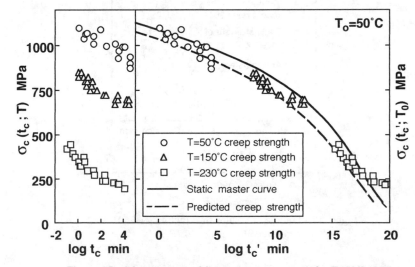

Figure 15   Master curve of flexural creep strength for T400/3601

demonstrates the fact that the time-temperature dependencies of the static failure of T400/3601 is controlled by the viscoelastic behavior of the constituent matrix resin 3601.

3.5 Flexural creep strength

The three-point bending creep tests were carried out for T400/3601 with several stress levels at temperatures T=50, 150, and 230°C. The deflection increases slowly with time until the sudden and instantaneous failure. Figure 15 displays the flexural creep strength σ_c versus time of failure t_c; the left side shows the experimental data at the temperatures T=50, 150, 230°C, while right side exhibits the data shifted to the reference temperature T_o=50°C using the shift factors for the viscoelastic behavior of matrix resin. The right side of this figure also displays the master curve for the static strength in solid curve and that for creep strength

in dashed curve which is calculated based on the master curve for the static strength and the linear cumulative damage law. The predicted creep strength agrees well with the results obtained experimentally. Therefore, it is confirmed that the creep strength can be obtained based on the master curve of static strength and the linear cumulative damage law.

3.6 Flexural fatigue strength for zero stress ratio

To confirm the validity of the reciprocation law of time and temperature for the flexural fatigue strength of T400/3601, we carried out the experiments for a fixed stress ratio R=0.05 closed to zero stress ratio and combinations of frequency and temperature listed in Table 2.

Curves of the flexural fatigue strength σ_f versus the number of cycles to failure N_f, S-N_f curves, are shown in Figure 16 where σ_f at N_f=1/2 is represented by σ_s

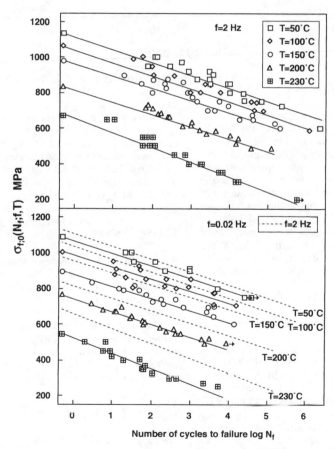

Figure 16   S-N_f curves for stress ratio R=0.05 and frequencies
f=2, 0.02Hz at various temperatures T for T400/3601

185

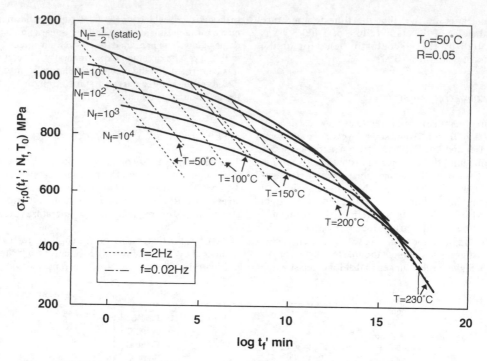

Figure 17 Master curves of flexural fatigue strength for stress ratio R=0.05 for T400/3601

at t_s=0.5/f. The top graph shows the case of f=2 Hz, while the bottom the case of f=0.02 Hz in which the S-N_f curves for f=2 Hz are included in dotted curves. The slope of the S-N_f curves are the same and also constant for all temperatures and frequencies tested over the six decades of N_f. Furthermore, since the reciprocation law of time and temperature holds for σ_s we consider that the same reciprocation law holds for the fatigue strength.

Based on S-N_f curves, Figure 16, we can construct the master curve of fatigue strength σ_f against the reduced time to fatigue failure t_f' for an arbitrary number of cycles to failure N_f at R=0.05 as shown in Figure 17.

3.7 Flexural fatigue strength for arbitrary stress ratio

Figure 18 shows experimental data of σ_f-t_b for several stress ratios at f=2 Hz and T=50, 150 and 230°C and the dashed lines are those predicted by Equation (5) using known $\sigma_{f,0}(t_b;f,T)$ and $\sigma_c(t_b;T)$ plotted in solid lines. As can be seen, the predictions correspond well with the experimental data.

As stress ratio R increases, the flexural fatigue strength at a fixed time to failure increases at T=50°C while it decrease at T=230°C. This reversion of dependence on R indicates that the strength at lower temperature is influenced strongly by the stress

amplitude while, near the glass transition temperature, by the mean stress. We observe that the predicted curves capture the trend of experimental values adequately except for T=230°C which is very close to the matrix glass transition temperature. We can predict σ_f-t_b relation for given R, f and T from Equation (5) if $\sigma_c(t_b,T)$ and $\sigma_{f,0}(t_b;f,T)$ are known.

4 CONCLUSION

We first explained the proposed procedure for the long term prediction method of static creep and fatigue strengths of polymer composites at an arbitrary frequency, temperature and stress ratio using the modified reciprocation law of time and temperature and the linear cumulative damage law. Data from the testing of a satin woven CFRP were used to verify this method. It was confirmed that the proposed prediction method for long term static, creep and fatigue strengths of CFRP composites is reliable.

ACKNOWLEDGMENT

The author wishes to thank Professor Rokuro Muki, Professor Richard Christensen and Dr. Michael K. McMurray for many helpful discussions.

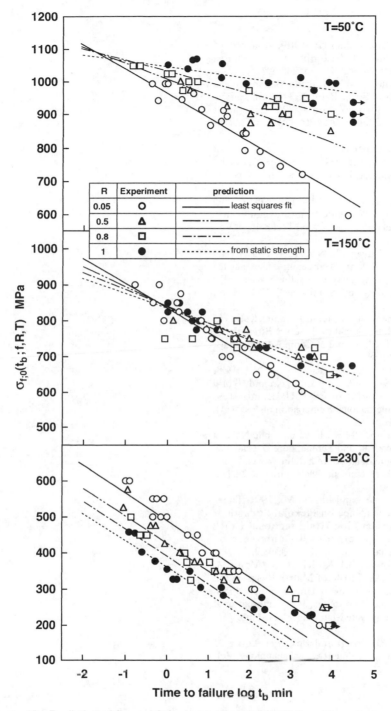

Figure 18   Prediction of flexural fatigue strength $\sigma_f(t_b;f,R,T)$ for various stress ratios R at frequency f=2Hz and various temperatures T for T400/3601

REFERENCE

Aboudi J. and Cederbaum G., 1989, Analysis of Viscoelastic Laminated Composite Plates, Composite Structures, Vol. 12, pp. 243-256.

Enyama J., McMurray M.K., Nakada M. and Miyano Y., 1993, Effects of Stress Ratio on Flexural Fatigue Behavior of a Satin Woven CFRP Laminate, Proceedings of 3rd Japan SAMPE, Vol. 2, pp. 2418-2421.

Gates T., 1992, Experimental Characterization of Nonlinear Rate Dependent Behavior in Advanced Polymer Matrix Composites, Experimental Mechanics, pp. 68-73.

Karakaya M. and Kurath P., 1994, Deformation and Fatigue Behavior of Woven Composites Laminates, Journal of Engineering Materials and Technology. Vol. 116, pp. 222-232.

McMurray M.K., Enyama J., Nakada M. and Miyano Y., 1993, Loading Rate and Temperature Dependence on Flexural Fatigue Behavior of a Satin Woven CFRP Laminate, Proceedings of 38th SAMPE, Vol. 38 No. 2, pp. 1944-1956.

Miyano Y., Kanemitsu M., Kunio T., and Kuhn H., 1986, Role of Matrix Resin on Fracture Strengths of Unidirectional CFRP, Journal of Composite Materials, Vol. 20, pp. 520-538.

Miyano Y., Amagi S., and Kanemitsu M., 1988, Flexural Fracture Behavior of Carbon/Aramid Hybrid Unidirectional Reinforced FRP Laminates, Proceedings of International Symposium on FRP/CM, pp. 7.C.1-7.C.11.

Miyano Y., McMurray M.K., Enyama J. and Nakada M., 1994, Loading Rate and Temperature Dependence on Flexural Fatigue Behavior of a Satin Woven CFRP laminate, Journal of Composite Materials, Vol. 28, No. 11, pp. 1250-1260.

Mohri M., Miyano Y., and Suzuki M., 1991, Time-Temperature Dependence on Flexural Strength of Pitch-Based Carbon Fiber Unidirectional CFRP Laminates, Proc. International Conference on Composite Materials 8, pp. 8. 33.B.1-33.B.9.

Nakada M., McMurray M. K., Kitade N., Mohri M. and Miyano Y., 1993, Role of Matrix Resin on the Flexural Fatigue Behavior of Unidirectional Pitch Based Carbon Fiber Laminates, Proc. International Conference on Composite Materials 9, Vol. 4, pp. 731-738.

Sullivan J., 1990, Creep and Physical Aging of Composites, Composite Science and Technology, Vol. 39, pp. 207-232.

Progress in Durability Analysis of Composite Systems, Cardon, Fukuda & Reifsnider (eds)
© *1996 Balkema, Rotterdam. ISBN 90 5410 809 6*

Time and temperature dependencies of the tensile strength in the longitudinal direction of unidirectional CFRP

Masayuki Nakada & Yasushi Miyano
Materials System Research Laboratory, Kanazawa Institute of Technology, Japan

Takao Kimura
Graduate School, Kanazawa Institute of Technology, Japan

ABSTRACT: The tensile static tests for a carbon fiber/epoxy composite strand (CF/Ep strand) were performed at various temperatures and loading rates. The characteristics of acoustic emission (AE) for the fracture process were evaluated. The time and temperature dependencies of the tensile static behavior of CF/Ep strand was discussed from the viewpoint of the viscoelastic behavior of the matrix resin. As results, the tensile strength of CF/Ep strand depends on time to failure and temperature. The reciprocation law of time and temperature for the creep compliance of matrix resin holds for the tensile strength of CF/Ep strand. Since the characteristics of AE at low and high temperatures are clearly different, it is considered that the cumulative mechanism of damage in the tensile fracture process of CF/Ep strand changes with temperature.

1 INTRODUCTION

In recent years materials that possess high specific strength and specific modulus were developed to fulfill the need for advance structures. Carbon fiber reinforced plastics (CFRP) are materials that have these properties, and are being used to extended the life of structures, in not only secondary structures but also primary structures.

It is well known that the mechanical behavior of polymer resins exhibits time and temperature dependence, called viscoelastic behavior, not only above the glass transition temperature T_g but also below T_g. Thus, it can be presumed that the mechanical behavior of CFRP using polymer resins as matrices also significantly depends on time and temperature. It has been confirmed that the viscoelastic behavior of polymer resins as matrices is a major influence on the time and temperature dependence of the mechanical behavior of CFRP.

In this paper, a carbon fiber/epoxy composite strand (CF/Ep strand) was chosen as the specimens. The tensile static tests were performed at various temperatures and loading rates. The characteristics of acoustic emission (AE) for the fracture process were evaluated. The time and temperature dependencies of the tensile static behavior of CF/Ep strand was discussed from the viewpoint of the viscoelastic behavior of the matrix resin.

2 EXPERIMENTAL PROCEDURE

2.1 *Preparation of specimens*

A carbon fiber/epoxy composite strand (CF/Ep strand) consists of high strength carbon fibers, T400-3K (TORAY, brand name TORAYCA) and a general purpose epoxy resin (YUKA SHELL EPOXY, brand name EPIKOTE 828) is produced by filament winding molding process. The glass transition temperature T_g of the epoxy resin is 112 °C. The CF/Ep strand is cured at 70 °C for 12 hours and postcured at 150 °C for 4 hours and then at 190 °C for 2 hours. The fiber volume fractions of CF/Ep strand was approximately 60 %.

2.2 *Test procedures*

The tensile static tests for the CF/Ep strand were conducted using an Instron type testing machine with a constant temperature chamber. The nominal dimensions of the test specimens are shown in Figure 1. The tests were conducted at 6 points between the temperatures of 50 and 150 °C. The tensile load was applied at the end tabs. The cross-head speeds were 0.01, 1 and 100 mm/min. Two acoustic emission (AE) sensors (200 kHz, resonant type) were attached to the end tabs. The characteristics of AE for the fracture process were evaluated.

The tensile strength σ_s of CF/Ep strand is defined as follows.

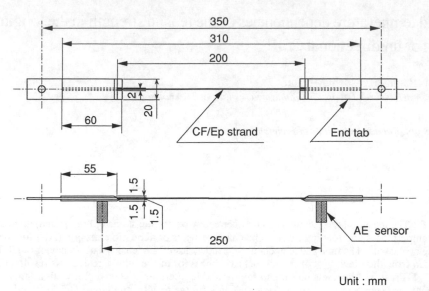

Fig.1   Configuration of CF/Ep strand specimen

$$\sigma_s = P_s \frac{\rho}{te} \qquad (1)$$

where: P_s: maximum load [N]

ρ : density of fiber [g/cm³]

te : tex of fiber [(g/1000m)×10⁻³].

3 RESULTS AND DISCUSSION

3.1 *Viscoelastic behavior of the matrix epoxy resin*

Figure 2 shows the master curve of creep compliance Dc versus reduced time t' at a reference temperature of T_0=50 °C for the matrix resin. The master curve was constructed by shifting Dc horizontally and vertically at various constant temperatures until they overlapped. The horizontal time and temperature shift factor $a_{T_0}(T)$ and the vertical temperature shift factor $b_{T_0}(T)$ are defined as follows.

$$a_{T_0}(T) = \frac{t}{t'} \qquad (2)$$

$$b_{T_0}(T) = \frac{Dc(t, T)}{Dc(t', T_0)} \qquad (3)$$

where: t : time [min]

t' : reduced time [min]

T : testing temperature [K]

T_0: reference temperature [K].

Since Dc at various temperatures can be superimposed so that a smooth curve is created, the modified reciprocation law of time and temperature is applicable for the stress-strain relation of the matrix resin.

Figures 3 and 4 show the time-temperature shift factor $a_{T_0}(T)$ and the temperature shift factor $b_{T_0}(T)$ obtained experimentally for the master curve of Dc. The shift factor $a_{T_0}(T)$ is quantitatively in good agreement with Arrhenius' equation by using two different activation energies ΔH:

$$\log a_{T_0}(T) = \frac{\Delta H}{2.303 R} \left(\frac{1}{T} - \frac{1}{T_0} \right) \qquad (4)$$

where: ΔH: activation energy [kJ/mol]

R : gas constant 8.314×10⁻³ [kJ/(Kmol)].

The temperature at the knee point of two Arrhenius' equations corresponds closely to the glass transition temperature of the matrix resin.

The horizontal temperature shift factor $b_{T_0}(T)$ can be described by a straight line from the following equation:

$$\log b_{T_0}(T) = K (T - T_0) \qquad (5)$$

where K is a material constant [K⁻¹].

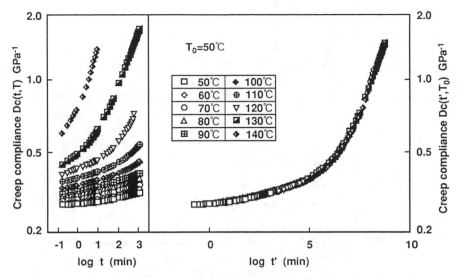

Fig.2 Master curve of the creep compliance for the matrix resin

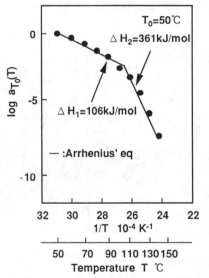

Fig.3 Time-temperature shift factor $a_{T_0}(T)$ for the matrix resin

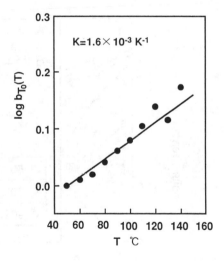

Fig.4 Temperature shift factor $b_{T_0}(T)$ for the matrix resin

3.2 Tensile strength of the CF/Ep strand

The tensile strengths σ_s of CF/Ep strand versus temperature at three steps of loading rate are shown in Figure 5. From this graph it is shown that the tensile strength σ_s of CF/Ep strand depends on loading rate and temperature.

3.3 Tensile strength dependencies on time and temperature of the CF/Ep strand

The tensile strengths σ_s of CF/Ep strand versus time to failure t_s at various temperatures are shown in the left side of Figure 6, obtained from Figure 5. Where t_s is defined as the time period from initial loading to

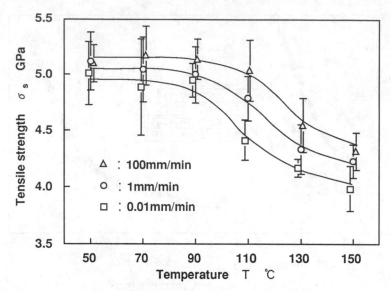

Fig.5   Tensile strength of CF/Ep strand versus
temperature at three steps of loading rate

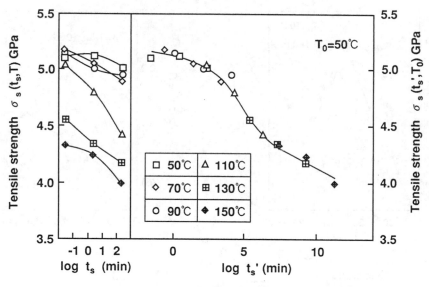

Fig.6   Master curve of the tensile strength for the
CF/Ep strand

maximum load P_s for constant loading rate test. The master curve of σ_s versus reduced time to failure t_s' at a reference temperature $T_0=50$ °C as shown in the right side of Figure 6 was constructed by shifting σ_s at various constant temperatures along the log scale of time to failure t_s until they overlapped each other. The vertical shift applied in the case of the stress-strain relation of the matrix resin is not needed for the strength of CF/Ep strand because the applied load is mostly transferred to the fiber of CF/Ep strand. Since tensile strength at various temperatures can be superimposed smoothly, the reciprocation law of time and temperature also holds for the tensile strength of CF/Ep strand.

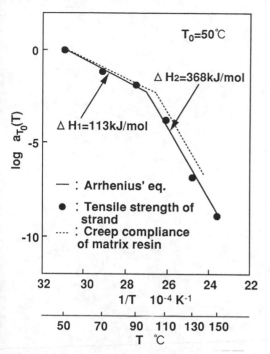

Fig.7 Time-temperature shift factors for the tensile strength of the CF/Ep strand

The time-temperature shift factors $a_{T_0}(T)$ obtained experimentally for the master curve of the tensile strength of CF/Ep strand are compared with that of the creep compliance of the matrix resin in Figure 7. Since the shift factors for CF/Ep strand and matrix resin almost agree with each other, it can be considered that the time and temperature dependencies of the tensile strength of CF/Ep strand are controlled by the viscoelastic behavior of the matrix resin.

3.4 *The characteristics of acoustic emission (AE)*

The load, AE cumulative event count and total AE energy versus elongation diagrams of CF/Ep strand at 50 °C and 150 °C are shown in Figure 8. AE event generation begins at 40 % of the maximum load of 50 °C. However, AE events are not seen until the vicinity of the maximum load of 150 °C. Therefore, it is considered that the cumulative mechanism of damage in the tensile fracture process of CF/Ep strand changes with temperature.

4 CONCLUSION

The tensile static tests for a carbon fiber/epoxy composite strand (CF/Ep strand) were performed at

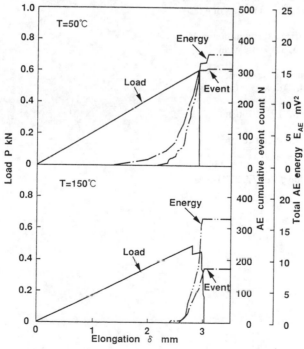

Fig.8 Load, AE cumulative event count and total AE energy versus elongation diagrams

193

various temperatures and loading rates. The characteristics of acoustic emission (AE) for the fracture process were evaluated. As results, the tensile strength of CF/Ep strand depends on time to failure and temperature. The reciprocation law of time and temperature for the creep compliance of matrix resin holds for the tensile strength of CF/Ep strand. Since the characteristics of AE at low and high temperatures are clearly different, it is considered that the cumulative mechanism of damage in the tensile fracture process of CF/Ep strand changes with temperature.

REFERENCES

Aboudi, J. and G. Cederbaum 1989. Composite Structures, Vol 12, 243-256.
Gates, T. 1992. Experimental Mechanics, 68-73.
Miyano, Y., M. Kanemitsu, T. Kunio, and H. Kuhn 1986. Journal of Composite Materials, Vol 20, 520-538.
Miyano, Y., S. Amagi, and M. Kanemitsu 1988. Proceedings of International Symposium on FRP/CM, 7.C.1-11.
Mohri, M., Y. Miyano, and M. Suzuki 1991. Proceedings of International Conference on Composite Materials 8, 33.B.1-9.
Nakada, M., M. McMurray, N. Kitade, M. Mohri, and Y. Miyano 1993. Proceedings of International Conference on Composite Materials 9, Vol 4, 731-738.
Miyano, Y., M. K. McMurray, J. Enyama and M. Nakada 1994. Journal of Composite Materials, Vol 28, 1250-1260.
Miyano, Y., M. K. McMurray, N. Kitade, M. Nakada, M. Mohri 1994. Advanced Composite Materials, Vol 4, 87-99.
Schwarxl, F., A. Staverman 1952. Journal of Applied Physics, Vol 23, 838-43.
Sullivan, J. 1990. Composite Science and Technology, Vol 39, 207-232.
Williams, M., R. Landel, J. Ferry 1955. Journal of American Chemical Society, Vol 77, 3701-3706.

Progress in Durability Analysis of Composite Systems, Cardon, Fukuda & Reifsnider (eds)
© *1996 Balkema, Rotterdam. ISBN 90 5410 809 6*

Durability evaluation on soft membrane composites under combined loading (mechanical and environmental)

P. Mailler, D. Bigaud & P. Hamelin
Laboratoire Mécanique et Matériaux, Université Lyon I, IUT A Génie Civil, Villeurbanne, France

ABSTRACT: Amongst numerous families of composites' materials, soft composites associating thermoplastic matrix with textile reinforcement are particularly used in civil engineering and building constructions in the shape of tensile membranes. The development of the used of those materials is conditioned by a better knowledge of their long term behavior under combined prompting (mechanical and environmental).
To models that behavior we use rheological models like "Zener biparabolic". Those models' parameters are identified by the mean of viscoelastic measurements made on a DMTA analyzer (Dynamic, Mechanic, Thermal, Analysis). With those models we can then express creep and relaxation functions of our material (PES coated with PVC). Those functions are obtained with approach inversion methods. We also use relaxation tests at several temperatures on large dimension specimens. Those tests are made on a biaxial testing apparatus developed by the Mechanical and Material Laboratory in collaboration with the French Textile Institute.
With those tests by applicating time - temperature principle we build a master curve that describes the long term behavior of our material.
So those tests permit us to compare rheological models with experimental results in the aim to determine the limits of the models against long term behavior.

1 INTRODUCTION

Amongst numerous families of composites' materials, soft composites associating thermoplastic matrix with textile reinforcement are particularly used in civil engineering and building constructions in the shape of tensile membranes.
To answer to the permanent used of those structures, it is necessary to know stiffness and failure behaviors of those materials, and also to apprehend their long term behaviors under combined prompting. So we must develop experimental characterization technics and rheological models taking into account the specifies of those materials and their long term behaviors. Those technics are based on microscopic tests on the constituent of the materials and on macroscopic tests on the materials.
To know that behavior we carry out viscoelastic tests by two means. Either by rheological models or by time temperature superposition in order to have a master curve for our material. This work will be the second part of my paper and is essentially numerical and previsional. Or by real time tests like creep and relaxation tests. This part is the third one and is mainly experimental.

2 VISCOELASTICMEASUREMENT ELABORATION OF RHEOLOGICAL MODELS

2.1 *Studied material*

We study a polyester coated with PVC (Polyvinyle chloride) fabric. The main characteristics are:

- fibers:	1100 dTex,
- texture:	9x9 tafetta,
- thickness:	0,6 mm,
- weight:	800 g/m².

2.2 *Experimental process*

To define previsional behavior, from simple viscoelastic tests

The results are shown on fig 1

we define rheological models. The first one uses assemblies in serie or/and parallel of springs and dash pots and is called Zener model, the second one uses assemblies in serie or/and parallel of fractional element. From those models by operational calculus we obtain time dependent expressions for relaxation and creep modulus. The viscoelastic properties of

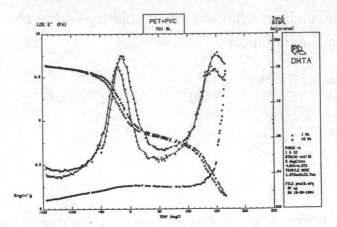

Fig1 : Evolution of E' and tangent δ with temperature

the materials are measured with a DMTA (Dynamic Mechanical Thermal Analysis). The test conditions are:

- tensile test,
- frequencies: 1 Hz and 10 Hz,
- temperature: -150 °C to +200 °C,
- Speed temperature rate: 3 °C/min.

2.3 Rheological models

To choose a well-suited model, we draw with the experimental results Cole - Cole diagrams [1] and we compare those diagrams with model ones.
The complete modulus E* can be written:

$$E^* = E' + iE'' \qquad (1)$$

where E' is associated to elasticity: E' is called conservation modulus and E" is associated to viscoelastic behavior: E" is called waste modulus.
Cole - Cole diagrams are made putting E" modulus as a function of E' modulus. The complex plane is made with E" in Y axis and E' in X axis.

2.3.1 Zener models'

Our experimental results can be approached by two types of models. This is Zener parabolic and Zener biparabolic. [2], [3]
When the Cole - Cole diagrams show a dissymetric behavior, the viscoelastic behavior description is made with a biparabolic model. [4], [5]

The model equation is:

$$E^* = E_0 + \frac{E_\infty - E_0}{1 + (i\omega\tau_1)^{-k} + (i\omega\tau_2)^{-h}} \qquad (2)$$

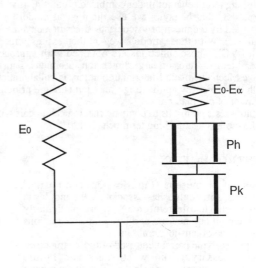

Fig 2: limited biparabolic Zener

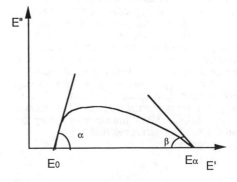

Fig 3: Cole - Cole diagrams

where:

$$k = \frac{2\beta}{\Pi} \quad h = \frac{2\alpha}{\Pi} \qquad (3)$$

E_0 is the static modulus,

$E_\alpha - E_0$ is the mechanical dispertion,

Ph is for high temperature (low frequencies)

Pk is for low temperature (high frequencies).

2.3.2 Fractional models

Fractional calculus is used to build models to describe the viscoelastic materials' properties in a mathematical manner.

A fractional model is an assembly in serie or/and in parallel of springs and fractional calculus elements. (springs pot)

The basic model is a model of one mechanism (spring pot) which constitutive law satisfies: [6], [7]

$$K.D_t^\alpha.x(t) = F(t) \qquad (4)$$

where α is the order of fractional derivation ($0 \le \alpha \le 1$). A spring pot has the characteristics of both a spring ($\alpha=0$) and a dashpot ($\alpha=1$).

As we see before, the model corresponding to experimental results is Zener biparabolic which constitutive law satisfies:

$$x(t) = \frac{F_1(t)}{E_\infty - E_0} + \frac{1}{K_k} D_t^{-k} F_1(t) + \frac{1}{K_h} D_t^{-h} F_1(t) \qquad (5)$$

with

$$\begin{cases} -0 < k < h < 1 \\ -D_t^{-k} F_1(t) = (F * \frac{t^{k-1}}{\Gamma(k)})(t) \\ -D_t^{-h} F_1(t) = (F * \frac{t^{h-1}}{\Gamma(h)})(t) \end{cases} \qquad (6)$$

D_t^{-k}, D_t^{-h} are the operators of fractional derivation of k and h orders.

* is the convolution product.

$$F_2(t) = E_0.X(t) \qquad (7)$$

$$F(t) = F_1(t) + F_2(t) \qquad (8)$$

The Laplace transform of equations (5),(7), (8) are:

$$\overline{X}(p) = (\frac{1}{E_\infty - E_0} + \frac{1}{K_k} p^{-k} + \frac{1}{K_h} p^{-h}).\overline{F_1}(p) \qquad (9)$$

$$\overline{F_2}(p) = E_0.\overline{X}(p) \qquad (10)$$

$$\overline{F}(p) = \overline{F_1}(p) + \overline{F_2}(p) \qquad (11)$$

Combination of (9),(10), and (11) gives:

$$\overline{F}(p) = (E_0 + \frac{E_\infty - E_0}{1 + (p\tau_1)^{-k} + (p\tau_2)^{-h}}).\overline{X}(p) \qquad (12)$$

the complex modulus is given by:

$$E^*(i\omega) = E_0 + \frac{E_\infty - E_0}{1 + (i\omega\tau_1)^{-k} + (i\omega\tau_2)^{-h}} \qquad (13)$$

The values of the parameters h, k, E_0 are determined by enlarging the externals parts of Cole - Cole diagrams.

In order to determinate the values of E_∞, we devide every part of Cole - Cole diagram in two zones. The first corresponding to high frequencies is assimilated to an arc of circle which center is under the real axis, the second corresponding to low frequencies is assimilated to an arc of circle that the center is under the real axis and cuts it in E_∞.

The equation of every circle is:

$$(E' - E'_c)^2 + (E^* - E'_c)^2 = R^2 \qquad (14)$$

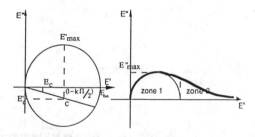

Fig 4:Determination of E_0 and E_∞

Where E'_c and E''_c are given by:

$$E'_c = R\cos(1-k)\frac{\pi}{2}$$
$$E^*_c = -R\sin(1-k)\frac{\pi}{2} \qquad (15)$$

The average radius of the circle is obtained by combination of (14) and (15),

$$R_m = \frac{1}{n}\sum_{i=1}^{n}\frac{1}{2}\frac{E_i'^2 + E_i^{*2}}{E_i' \sin k\frac{\pi}{2} - E_i^* \cos k\frac{\pi}{2}} \qquad (16)$$

where n is the number of points used to calculate this average. The expression of E_∞ is given by:

$$E_\infty = 2R_m \cos(1-k\frac{\Pi}{2}) \qquad (17)$$

To determinate the value of δ, we give arbitrary values to the term $\omega\tau$ (for example $\omega\tau = 10^{-4}$ to 10^{10}), we adjust the value of δ until the calculated and experimental values suited.

We suppose that the ratio of relaxation times τ_1 and τ_2 is independent of temperature, relations (2) or (13) become:

$$E^* = E_0 + \frac{E_\infty - E_0}{1 + \delta(i\omega\tau)^{-k} + (i\omega\tau)^{-h}} \quad (18)$$

with:

$$\delta = \mu^{-k} \text{ and } \mu = \frac{\tau_1}{\tau_2} \quad (19)$$

where:

$$A = \delta(i\omega\tau)^{-k}\cos(k\frac{\pi}{2}) + (i\omega\tau)^{-h}\cos(h\frac{\pi}{2}) \quad (20)$$

$$B = \delta(i\omega\tau)^{-k}\sin(k\frac{\pi}{2}) + (i\omega\tau)^{-h}\sin(h\frac{\pi}{2}) \quad (21)$$

then:

$$E^* = E_0 + \frac{E_\infty - E_0}{1 + A - iB} \quad (22)$$

$$E^* = E_0 + \frac{E_\infty - E_0(A+1+iB)}{(1+A)^2 + B^2} \quad (23)$$

$$E' = \frac{E_0(A^2+B^2) + A(E_\infty + E_0) + E_\infty}{(1+A)^2 + B^2} \quad (24)$$

$$E^* = \frac{B(E_\infty - E_0)}{(1+A)^2 + B^2} \quad (25)$$

Table 1 gives the values of the parameters of the model.

Table 1: parameters of the model

	Zener		fractional	
	1	2	1	2
E_0 (Pa)	8 10^6	8,8 10^8	8 10^6	9,4 10^8
E_α (Pa)	1,2 10^9	4,9 10^9	1,15 10^9	5,1 10^9
k	0,13	0,17	0,13	0,17
h	0,148	0,22	0,148	0,22
δ	1,8	1,5	1,3	1,66

Fig 6 shows a comparison between the curve of E' given by equation (24) and results obtained by viscoelastic tests.

2.4 Creep and relaxation functions

From equation (18) of Zener model, it is possible to find temporal expressions (relaxation modulus E(t), or creep modulus J(t)).
Operational relaxation and creep modulus can be expressed with the help of Carlson - Laplace transforms'. [8], [9]

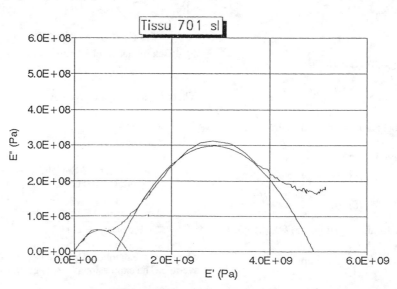

Fig 5: experimental/theoretical Cole - Cole

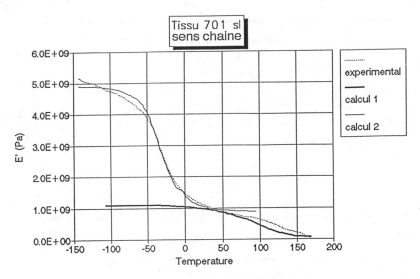

Fig 6: E', comparison model / experimental

$$E^*(p) = p\int_0^\infty e^{-pt}E(t)dt \quad (26)$$

and

$$J^*(p) = p\int_0^\infty e^{-pt}J(t)dt \quad (27)$$

Laplace integral can be write in the modified form:

$$E^*(p) = p\int_0^\infty (e^{-pt}t)\frac{E(p)}{t}dt \quad (28)$$

To calculate the integral we use Ter Haar approximation. [10]
$e^{-pt}.t$ can be approached by a Dirac δ function.

$$te^{-pt} = \frac{\delta(t-t_0)}{p^2}dt \quad (29)$$

that gives

$$E^*(p) = p\int_0^\infty \frac{E(p)}{t}\frac{\delta(t-t_0)}{p^2}dt \quad (30)$$

$$E^*(p) = \int_0^\infty \frac{E(p)}{t}\frac{\delta(t-t_0)}{p}dt \quad (31)$$

The integral is calculated at the pole $t_0 = \frac{1}{p}$ and with the used of Dirac function properties we obtain:

$$E^*\left(\frac{1}{p}\right) = E(p) \quad (32)$$

finally

$$E(t) = E_0 + \frac{E_\infty - E_0}{1+\delta\left(\frac{t}{\tau}\right)^k + \left(\frac{t}{\tau}\right)^h} \quad (33)$$

Using the same way, we calculate the creep function J(t).

$$J(t) = \frac{1+\delta(\frac{t}{\tau})^k + (\frac{t}{\tau})^h}{E_0\left[1+\delta(\frac{t}{\tau})^k + (\frac{t}{\tau})^h\right] + E_\infty - E_0} \quad (34)$$

In a second time, we determine the time variation of E(t) of our material by applicating time temperature superposition. [11]
The test are made on a DMTA between -150 °C and 150 °C at five frequencies (0,3 Hz to 30 Hz).
The superposition of data curves permits us to obtain a master curve for a reference temperature of 25 °C. (figure 9)

3 Test on large dimensions specimen

The Mechanical and Material Laboratory of University Claude Bernard Lyon I in collaboration with the French Textile Institute have developed a biaxial testing apparatus (with the support of Région Rhône Alpes) to carry out test on soft or stiff composites ([12]). A thermal enclosure permits us to realize test between –40 °C to 200 °C. The maximal sizes of the tested specimen are 1 m x 1 m in the aim to minimize the scale effects du to material heterogeneities considered at the scale of fibers or textile textures.
We make a serial of relaxation tests in the following conditions:

199

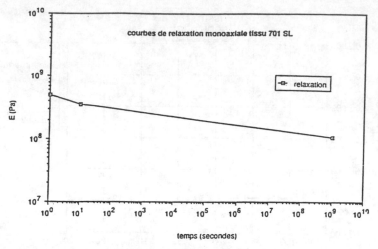

Fig 7: relaxation at 25 °C equation (33)

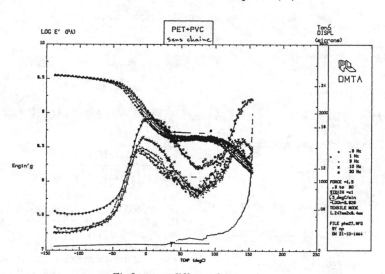

Fig 8: test at different frequencies

specimen 1 m large and 0,3 m wide,

displacement 26 mm at the jaws,

relaxation time 15 minutes,

temperature 25 °C to 130 °C step 15 °C.

From relaxation curves obtained (fig 10), applicating time temperature superposition principle we build a master curve corresponding to our material for a temperature of 25 °C.

We note two kinds of relaxation. The first one between 0 to 10 second and is due to geometrical reorganization and after a second one due to the fibers relaxation.

4 Synthesis and conclusion

We have determined for a given material (tissue polyester coated with PVC) the relaxation modulus variations in function of time by three different means. From:

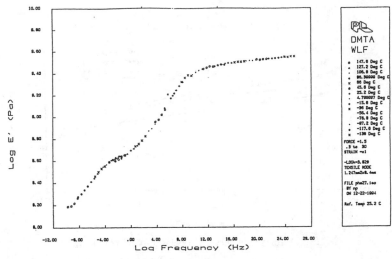

Fig 9 master curve at 25 °C

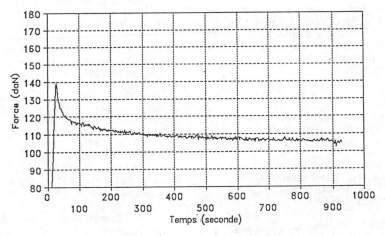

Fig 10: monoaxial relaxation curve

- a rheological model based on viscoelastic tests (either by Carlson Laplace transforms or by fractional derivatives),

- a master curve based on viscoelastic tests at different frequencies,

- a master curve based on relaxation tests in arge dimensions specimen.

We observe two kinds of relaxation. The first one between 0 to 10 second and is due to geometrical reorganization and after a second one due to the fibers relaxation.

We note finally a decrease of about 50 % of E' after 10 years.

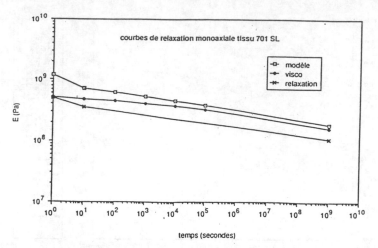

Fig 11: relaxation curve: viscoelasticity, model, tests

ACKNOWLEDGEMENTS

This work was realized with the support of the French Textile Institute (LYON) and especially Mrs Jarrigeon and Mr Nemoz, and with the support of the Région Rhône Alpes for the realization of the biaxial testing apparatus.

REFERENCE

Cole K.S., Cole R.H. "Dispersion and absorption in dielectrics I. Alternating current characteristics" T. Chem. Phys., vol 9, 1941, p. 341.

Huet C. "Etude par une méthode d'inpédance du comportement viscoélastique des matériaux hydrocarbones"Thèse de docteur ingénieur.Faculté des Sciences Université de Paris

Huet C., Bourgoin D., Richemond S. "Rhéologie des matériaux anisotropes" Compte rendu du 19ème colloque national annuel, Paris Cepadues ed. 591p.

Chateauminois. "Comportement viscoélastique et tenue en fatigue statique de composites verre-epoxy. Influence du vieillissement hygrothermique." Thèse UCB Lyon I 1991

Jeanne P. "Contribution à l'étude des zones interfaciales dans les composites polyepoxyde-fibres de verre. Thèse EC lyon 1986

Koller R. C. "Application of fractional calculus to the theory of viscoelasticity" Journal of Applied Mechanics, june 1984 vol.51 pp. 299-307

Jarbouh A. "Appliction de la dérivation fractionnaire à la modélisation du comportement des matériaux viscoélastiques" Thèse de doctorat U.C.B Lyon I 1988

Doan. "Contribution à l'étude du comportement dynamique et au choc des matériaux et des structures composites Thèse INSA Lyon 1989

Boufera, Doan, Hamelin "Comportement viscoélastique des matrices polymères" Comportement des composites à renforts tissus. Comportement dynamique des composites. Ed Pluralis 1990 pp 47-68

Cost T. "Approximate Laplace transform inversion in viscoelastic stress analysis" AIAA journal 1964 vol.2 pp. 2157-2166

Tsai S. W. Composites design 4th Edition Think composites 1988

Williams. American Chemistery Society Vol. 77,3701 1955

Mailler P., Hamelin P., Jarrigeon M., Nemoz G. "Comportement en grandes déformations des matériaux composites souples soumis à un chargement biaxial. JST AMAC Besançon janvier 1995

Progress in Durability Analysis of Composite Systems, Cardon, Fukuda & Reifsnider (eds)
© *1996 Balkema, Rotterdam. ISBN 90 5410 809 6*

Creep damage characterisation of CFRP laminates

O.Ceysson, T.Risson, M.Salvia & L.Vincent
Ecole Centrale, Lyon, France

ABSTRACT: In order to promote the industrial development of CFRP, it is essential to have sufficient knowledge of their durability. It is therefore necessary to study the mechanical behaviour of these materials and more particularly the evolution of various types of damage which appear during loading.
The measurement of the variation of the electrical resistance, in connection with Acoustic Emission analysis, was an accurate technique for the detection and the identification of damage mechanisms in CFRP laminates.
CFRP samples : UD (0°) and (±45°) laminates were subjected to monotonic and creep tests under three-point and post-buckling bending conditions.
The in-situ monitoring of electrical resistance variation and AE activity allowed the detection and the identification, at different stages, of various types of damage mechanisms (fibre fractures but also matrix cracks, debondings and delaminations).

1 INTRODUCTION

Carbon fibre reinforced plastics (CFRP) offer high specific mechanical properties (performance vs weight ratio). Thus during the last decade, they have been increasingly used as components of structures having essentially mechanical functions, particularly in aeronautical applications.

It is therefore imperative to detect, evaluate, and analyse the various types of damage propagation caused by both static or cyclic loads and also by environmental effects.

Although traditional non-destructive techniques (Ultrasonics (A-SCAN or C-SCAN), X-ray radiography, Infrared thermography, Holographic interferometry, Eddy currents,...) enable the a posteriori detection of damage at successive stages of the life of these materials, it seems more difficult to monitor in-situ the evolution of internal damage nucleation and growth, especially in CFRP (opaque materials).

Since carbon fibres are electrical conductors ($\rho = 2.10^{-5}$ Ω.m), the measurement of the variations of electrical resistance appears to be a valuable technique for this purpose.

In the case of CFRP samples, conductivity is not isotropic but depends on the orientation of the carbon fibres.

The electrical conduction of (0°) unidirectional (UD) CFRP parallel to the fibres is due to the current flow along the fibres. This can be modelled using the parallel resistance approach.

The resistance of the composite R, may be written as follows :

$$R = \frac{\rho_f L}{b \, d \, V_f} + R_c$$

Where ρ_f : the fibre resistivity; V_f : volume fraction of unbroken fibres; L : length between the electrodes; b and d : specimen width and thickness respectively;, and R_c : the contact resistance between the sample and the electrodes.

Fibre fractures will cause V_f to decrease hence increasing suddenly the sample electrical resistance R (Figure 1).

Very few previous works (1-5) have used electrical resistance measurements to detect fibre breakage in longitudinal unidirectional CFRP laminates caused by monotonic and cyclic tensile loading.

In the case of transverse unidirectional CFRP, the conduction behaviour depends only on fibre-to-fibre contacts between neighbouring fibres (Epoxy matrix is electrically insulating : $\rho = 10^{13}$ to 10^{15} Ω.m).

The same phenomenon appears in (±45°) laminates in such a way that conduction occurs through continuous paths between carbon fibres along and between the plies of the laminates.

Below some critical values of fibre volume fraction (V_f), there are insufficient fibre-to-fibre contact paths and conductivity is insufficiently large. But with current V_f values in engineering composites (50-60%), contact between fibres is always achieved.

Hence conductivity variations also allow the detection of transverse and longitudinal intraply matrix cracking and delamination by modifications to the fibre-fibre conduction paths and the gradual cutting of the resistive tracks.

Figure 2 illustrates how matrix cracks and delaminations can cause changes in transverse conduction.

Models have been used to develop relationships between crack length and resistance for use in delamination crack length measurements (6,7).

In conclusion, the conduction paths are formed in 3 directions. The current flow in all the volume of the sample.
* In longitudinal direction, by carbon fibres.
* In transverse direction, by fibre-to-fibre contact paths between neighbouring fibres.
* In the thickness of the sample, by contacts between plies of the laminates.

Consequently, the variation of the electrical conductivity can be taken as an indicator of the evolution of various types of damage in CFRP laminates (not only fibre fractures but also matrix cracks and delaminations).

On the other hand, the Acoustic Emission (AE) technique is now used to detect and possibly to identify damage mechanisms in CFRP.
This is achieved by analysis of AE parameters like the amplitude event, and, to a lesser degree, the energy and the duration of the event. But comparison between amplitude values found during different tests and with different samples is sometimes difficult. Beside, it is critical to achieve a relationship between amplitudes and rupture mechanisms.
In spite of this, several authors (8-12) agree that low amplitudes are correlated with matrix cracking, medium amplitudes with delamination and high amplitudes with fibre breakage.

Following these preliminary remarks, the monitoring of both the variations of electrical resistance and the Acoustic Emissions will allow the in-situ detection and identification of various types of damage mechanisms.

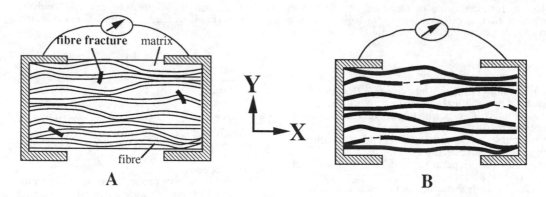

Figure 1. (A) Damage mechanisms (fibre fractures) and (B) its electrical analogue

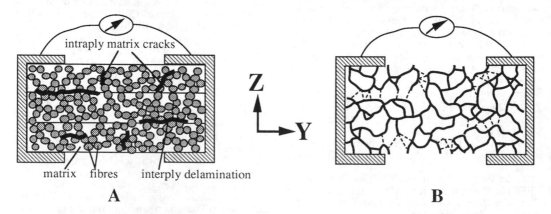

Figure 2. (A) Various types of damage mechanisms and (B) its electrical analogue

2 EXPERIMENTALS

2.1 Materials and machines testing

The material used was a carbon fibre/epoxy laminate of HTA200 - M18/1 (Fibre volume fraction V_f was about 60%).
CFRP samples were subjected to monotonic and creep tests. Two types of monotonic tests were realised with a constant cross-head speed of 2 mm/min: three-point and post buckling bending tests. These monotonic tests allow the determination of the ultimate stress. Then CFRP samples were subjected to creep tests at different load levels (x % of ultimate stress).

1. (0°) UD subjected to monotonic tests under post-buckling bending condition.
Three-point bending tests induce local stress concentration beneath the loading nose. And, in the case of (0°) UD laminates, this may promote premature failure on the compressive side of the sample (13,14). This always results from a microbuckling of the fibre (kink band) due to the local stress concentration. Such tests do not constitute a characterisation - even in compression - of the material, because of the complex stress state in the failure zone on the compressive side of the sample (15,16,17). To overcome the difficulties bound with the compressive effects of the loading nose, a new bending test method was used (14,18,19). This test consists in applying an axial compression to a rectangular cross section sample. The specimen is gripped at each end and can accommodate the local deformation at each end by in-plane rotation. Axial compression induces Euler-buckling of the bar (at the critical load Pc). In a post-buckling configuration, the sample is submitted to a bending strain as shown in Fig 3, where the parameters P, L, d, α, f are defined.
In order to secure a quasi-pure bending strain it is necessary that both the contributions of the shear strain and pure compression strain become negligible. For the strain range considered, this condition is fulfilled if span to depth ratio (L/h) is greater than 50.
So, the sample are about 100 mm long, 10 mm width and 2 mm thick with a [(0°)6]s lay-up.

2. (±45°) laminates subjected to monotonic and creep tests under three-point bending condition.
In this configuration (Fig.4), the material was a 24-ply carbon fibre/epoxy laminate with a [(±45°)6]s lay-up. The CFRP sample had a width of 25 mm, a length of 100 mm and a thickness of about 4 mm. The span length to thickness ratio (L/h) was kept at 20.

2.2 Electrical resistance measurement system

Figure 5 shows schematically the assembly used for the measurement of the electrical resistance.
A constant electric current of about 50mA was introduced into the specimen by electrical cables soldered to U-shape copper plates.

The value of current intensity was deliberately kept very low to avoid heating the specimen.
Electrical contact between carbon fibres and the copper plates was ensured with a silver adhesive.
To increase this contact and therefore allow better current transfer, locally, the specimen ends and side surfaces were polished (suppression of insulator epoxy matrix which are softer than carbon fibre).
Owing to the Ohm's law, the electrical resistance can be determined.

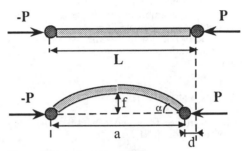

Figure 3. Post-buckling bending test

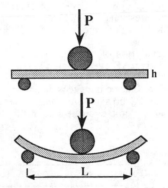

Figure 4. Three-point bending test

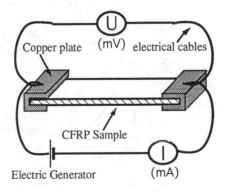

Figure 5. Schematic illustration of assembly used to monitor in-situ the electrical resistance

2.3 Acoustic Emission system

Acoustic emissions signals were acquired and processed with an AE acquisition and analysing system LOCAN Jr of Physical Acoustics Corp.
The AE signals during the tests were detected by a piezo-electric sensor (type R15) attached to the sample by a silicon grease and an adhesive tape. Its operating frequency range is between 50 and 200 KHz, and its resonant frequency is 150 KHz. This transducer was particularly recommended to detect matrix cracking and delamination.
The signals were amplified by a pre and a post-amplifier with a total gain of 70dB.
The threshold of AE system is 45dB, therefore events with an amplitude of lower than 45 dB were not detected.

3 RESULTS AND DISCUSSION

3.1 Monotonic tests in post buckling bending (Configuration (0°) UD)

The observation of fibre fracture is not possible during a test (as the materials are opaque) except when fibres split off the specimen surface. However, the variation in electrical resistance during monotonic loading is a valuable indicator of fibre fracture in UD laminates.
Figure 6 shows the evolution of load P, electrical resistance and cumulative AE event counts versus displacement d.
Failures are progressive. The composite is damaged gradually by failures of fibres (and/or bundles), followed by delaminations on the tensile side of the sample.

Indeed, compressive failures obtained in three-point bending (20) were therefore artifacts that masked intrinsic behaviour. This led to the conclusion that under bending loading conditions, the weak point of carbon epoxy studied composites is not , as it is often said, the behaviour in compression but, on the contrary, the resistance in tension.

The first increase of electrical resistance (after about 3 mm displacement) is caused by the first fibre fracture. This first fibre fracture appears on the free sides of the sample (due to the defects of the cut). Then, the electrical resistance stays at a constant value until a fibre bundle breaks. This may only appear on the tensile side of the specimen.
Owing to this, each fibre bundle breakage occurs regularly (for an increase of displacement of about 1 mm) inducing a step-shaped increase in electrical resistance.
All these fibre bundle fractures are indicated, but with a much lower resolution, on the load-displacement curve by several losses of stiffness and increases in AE activity. When each fibre bundles break, the load P suddenly decreases. The amplitude of this drop is directly proportional to the number of broken fibres on the tensile side. Indeed only fibres which remain undamaged allow the load transfer.
In the same way, as the current flows along the fibres, the electrical resistance increases in proportion to the number of broken fibres.
In both cases, the calculations allow the determination of the number of broken fibres for each fibre bundle breakage. Similar results are obtained if considering the loss of stiffness or the increase of the electrical resistance (for example, the first fibre bundle fracture corresponds to a decrease of useful thickness of about 10%).

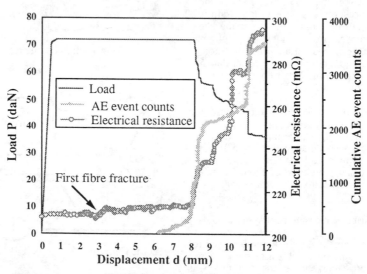

Figure 6. Post-buckling test of a (0°) UD laminate, showing the load/displacement curve and dependent electrical resistance and cumulative AE event counts

3.2 Monotonic tests in three point bending (Configuration (±45°) laminates)

Figure 7 shows the evolution of bending stress, electrical resistance and cumulative Acoustic Emission events counts versus strain.
Three zones can be identified.
They correspond to various types of damage mechanisms.

1. Zone I : (0% < Strain < 1.5%) : Transverse intraply matrix crack nucleation (Fig.8a)
This damage appears as soon as the load is applied, in the outer tensile layers and mainly at the free surfaces of the samples. In this case, the electrical resistance increases slowly due to gradual nucleation of matrix cracking and the subsequent reduction in the number of fibre-to-fibre conduction paths.
The activity of acoustic emission appears as soon as the load is applied and increases consistently. Energy, event duration and amplitude values are low. For the amplitude, one distribution is centred around 50 dB (Fig.9).
Initially during this first stage, the stress vs strain curve is non-linear. This non linearity can be explained by the nucleation of damages (matrix cracks).

2. Zone II : (1.5% < Strain < 2.5%) : Percolation of these matrix cracks led to delaminations (Fig.8b)
At the beginning of the second zone, transverse matrix cracks propagated and multiplied rapidly.These cracks collapsed and caused delaminations (inter-ply cracks) particularly in the middle of the loaded beam (where shear stresses are maximum). There is an effect of mutual induction between various types of damage: intraply matrix cracks induce delamination which induces other intraply cracks.
The mutual coupling of matrix cracking with delamination led to damage growth throughout the whole depth of the sample. Coupling of damage mechanisms arises rapidly and only during the second zone.The presence of intraply matrix cracks modify the conduction paths and the current flow in all the plies of the sample. So, when delaminations occur, contacts between plies are not achieved and the area available for current flow decreases. Therefore, the electrical resistance increases suddenly.
In the same time, the AE activity increases greatly. Amplitude values are higher than those of zone I. There are two distributions centred around 50 dB and 62 dB corresponding respectively to intraply matrix cracking and delamination (Fig.9).

3. Zone III : (2.5% < Strain < 7.5%) : Opening of delamination and slip between plies
This result is linked to friction between plies (fibre/fibre and fibre/matrix). Friction between existing fracture surfaces does serve as a continuous source of AE.
The electrical resistance fluctuates but the mean remains unchanged.
The bending stress reaches a maximum value and becomes constant. Fracture of (±45°) laminates subjected to three point bending tests is never achieved.
All of these assumptions were confirmed by microscopic examination of tested samples. Different microscopic tests were carried out on the samples at the end of each zone. The microscopic views of each specimen are presented in figure 8.

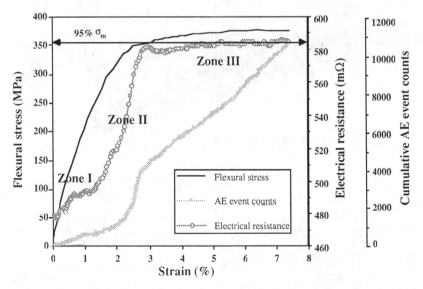

Figure 7. Flexural test of a (±45°) laminate, showing the stress/strain curve and dependent electrical resistance and cumulative AE event counts

A	B

Figure 8. Side view at differents stages of damage of (±45°) CFRP laminates

A : Zone I : Transverse intraply matrix cracks.
B : Zone II : Delaminations and mutual coupling of damage mechanisms

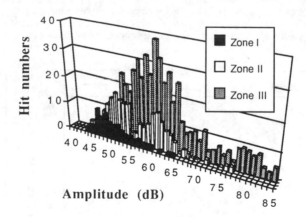

Figure 9. Acoustic emission amplitude distributions for the three zones

3.3 Creep tests in three point bending configuration ((±45°) laminates)

Creep tests were conducted at high load levels (95% of the ultimate stress), at a temperature of 23°C and under a controled relative humidity of 50%.

Figure 10 shows the evolution of creep strain, electrical resistance variations, and cumulative acoustic emission event counts versus time. Three zones can be identified. They correspond to various types of damage mechanisms.

1. Zone A (20 s.< t <45 s.) : Emergence of transverse intraply matrix cracks and delaminations.

This first zone corresponds to the loading of the

sample. Considering the high load level at which creep tests were conducted (See Figure 7), it is easily understandable that both intraply matrix cracks and delaminations appear during this loading zone.

In this case, the electrical resistance quickly increases due to sudden nucleation of matrix cracks and subsequent reduction in the number of fibre-to-fibre conduction paths.

The acoustic emission activity increases too. Three distributions are observed for the amplitude. The first one is centred around 50 dB : this corresponds to intraply matrix cracks. The second one is centred around 62 dB : during loading, delaminations are created. Friction between existing fracture surfaces may explain the third distribution for amplitudes higher than 75 dB (See Figure 11).

2. Zone B (45 s.< t <80 s.) : Multiplication of transverse intraply matrix cracks.

In this zone, The rate of matrix crack apparition decreases. The nucleation and propagation rates of these damages, as well as the creep rate, finally reach a minimum limit value and remain constant : the evolution of strain with time is then linear. At the end of this zone, the number of intraply matrix cracks reaches a maximum limit value.

A slight increase of the electrical resistance is

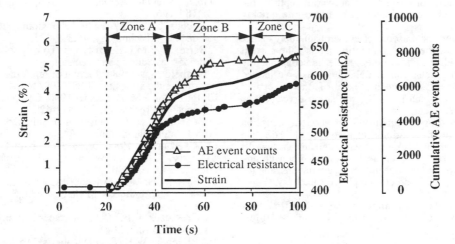

Figure 10. Flexural creep test of a (±45°) laminate, showing the stress/strain curve and relative electrical resistance and cumulative AE event counts

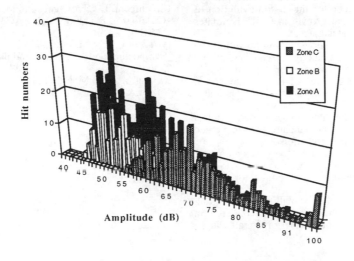

Figure 11 : Acoustic emission amplitude distributions for the three zones

observed due to the modification of the conduction paths.

Concerning the acoustic emission event counts, a similar evolution is obtained. One distribution is centred around 55 dB (characteristic value for intraply cracking-See Figure 11).

3. Zone C (80 s. < t < 100 s.) : Mutual coupling of the two types of damage mechanisms.

Due to the great amount of transverse intraply matrix cracks in the sample tested at the end of the second zone, the probability for such cracks to meet delaminations is significant. This implies coupling of these two types of damage mechanisms which led to damage growth throughout the whole depth of the sample. As a consequence, the creep rate quickly increases.

Moreover, the presence of numerous intraply matrix cracks modifies the conduction paths : the current flows in all the plies of the sample, and consequently allows the detection of delaminations. We indeed observed a sudden increase of the electrical resistance, which is similar to that observed for the creep rate.

Concerning the amplitude of acoustic emission events, one distribution is centred around 70 dB, and an other, around amplitude values greater than 80 dB (See Figure 11) : this can be explained considering the growth and the opening of delaminations, and the subsequent friction between plies respectively.

All of these assumptions were confirmed by microscopic examination of the samples.

4 CONCLUSION

Measurement of the electrical resistance and the analysis of Acoustic Emission parameters provide accurate means to monitor the in-situ evolution of damage nucleation and growth in CFRP laminates particularly for internal damage.

These methods enable the detection as well as the identification of various damage mechanisms : fibre fractures but also intraply matrix cracks and interply delaminations. These types of damage appear at different stages for (0°) and (±45°) CFRP laminates subjected to monotonic and creep tests.

Currently, these techniques are being used to detect the in-situ evolution of internal damage during creep tests under hygrothermal conditions.

Moreover, creep tests at low load levels are under way : the aim is to study the creep behaviour of samples which remain undamaged or quasi-undamaged after loading.

The industrial perspective is to allow the detection, in real time and with low cost, of damage in structures such as helicopter blades or aircraft wings.

REFERENCES

1. Prabhakaran, R. 1990. Experimental Technique. 14:16-20.
2. Schulte, K. & Ch.Baron 1989. Composites Science and Technology. 36:63-76.
3. Schulte, K.1993. Journal de physique IV. 3:1629-1636.
4. Thiagarajan, C. & I.Sturland, D.Tunnicliffe, P.E.Irving 1994. 2nd Eur. Conf. on Smart Structures and Materials. Session 3:128-131.
5. Curtis, P.T. & N.J.Williamson, R.J.Kemp 1994. 6th Int. Conf. on Fibre Reinforced Composites. Paper 17.
6. Moriya, K. & T.Endo 1990. Transactions of Japan Society for Aeronautical and Space Science. 32:184-196.
7. Fischer, C. & F.J.Arendts 1993. Composites Science and Technology. 46:319-323.
8. Barré, S. & M.L.Benzeggagh. To be published. Composites Science and Technology.
9. Berthelot, J.M. 1988. Journal of reinforced plastics and composites. 7:284-299.
10. Berthelot, J.M. & J.Rhazi 1990. Composites Science and Technology. 37:411-428.
11. Komai, K. & K.Minoshima, T.Shibutani 1991. JSME International Journal. 34:381-388.
12. Awerbuch, J. 1988. Drexel University.47-58.
13. Parry, T.V. & A.S.Wronski 1981. Journal of Science. 16:439-450.
14. Fukuda, H. 1990. Advanced Composite Materials, C.Bathias & M.Uemura edit.,SIRPE Pub 171-176.
15. Greszczuk, L.B. 1969. Interfaces in composites, ASTM STP. 452:42-58.
16. Binienda, W.K. & G.D.Roberts, D.S.Papadopoulos 1992. SAMPE q. 23:20-
17. Uemura, M. & H.Iwai 1990. Advanced Composite Materials, C.Bathias & M.Uemura edit. 134-139.
18. Vincent, L. & A.Chateauminois, P.Fournier, O.Pelissou, B.Toumi 1992. EACM, ECCM-CTS. 235-244.
19. Toumi, B. & M.Salvia, L.Vincent 1994. ASTM Symposium on fiber, matrix and interface properties Phoenix.
20. Toumi, B. & M. Salvia, L. Vincent 1994. Mater.& Tech. 6-7:45-49.

Progress in Durability Analysis of Composite Systems, Cardon, Fukuda & Reifsnider (eds)
© *1996 Balkema, Rotterdam. ISBN 90 5410 809 6*

Determination of master bending creep curves of CFRP

Rui Miranda Guedes & António Torres Marques
University of Porto (U.P.), Faculty of Engineering (F.E.U.P.), Portugal

Albert Cardon
University of Brussels (V.U.B.), Faculty of Engineering (T.W.), Belgium

ABSTRACT: A research program was established to determine the long term behaviour of composite materials. This paper is a brief description of a first attempt to validate some experimental procedures. In design, when creep is considered, it is usual that the item should remain in service for an extended time, usually longer than it is practical to run creep experiments on the material to be employed. Thus it is necessary to extrapolate the information obtained from relatively short time laboratory creep tests to predict the behaviour in service.
Bend tests do not produce uniform stresses and strains in the material and introduce, simultaneously, creep in bending and in-plane stress relaxation. Unfortunately it is difficult to perform customary tensile creep tests with current technology. The four point bending test offers an alternative to tensile tests.

1 THEORETICAL MODELS

The Boltzmann Law gives us the relation between the strain and the stress for linear viscoelastic materials at constant temperature.

$$\varepsilon(t) = \int_{-\infty}^{t} S(t - \tau) \frac{d\sigma(\tau)}{d\tau} \, d\tau \qquad (1.1)$$

For vibratory stress which varies sinusoidally with frequency f,

$$\sigma(t) = \sigma_{max} \cdot e^{i2\pi ft} \qquad (1.2)$$

If we define a new variable $\xi = t - \tau$, and if creep compliance vanishes for t<0, after some mathematical manipulation applying the Fourier Transform we obtain the following frequency-time transformation,

$$S(t) = \int_{0}^{t} \left\{ F^{-1}\left[j^*(f) \right] \right\} dt + J(0) \qquad (1.3)$$

where F^{-1} is the inverse Fourier Transform and $j^*(f)$ is the complex compliance, so that

$$\varepsilon(t) = j^*(f) \cdot \sigma(t) \qquad (1.4)$$

The Fast Fourier Transform (FFT) algorithm and numerical integration are employed to carry out the transformations.

From dynamic tests the complex compliance's were obtained for each temperature level. Before using the FFT algorithm it is necessary to calculate the regression curves for the real and imaginary parts of the complex compliance

$$j^*(f) = j_1(f) + i \cdot j_2(f) \qquad (1.5)$$

In this case it was used a variation of the Generalised Kelvin model to fit the storage and loss compliance.

$$j_1(w) = S_0 + \sum_{k=1}^{5} \overline{S}_k \frac{\lambda_k^2}{\lambda_k^2 + w^2}$$

$$j_2(w) = \sum_{k=1}^{5} \overline{\overline{S}}_k \frac{w \cdot \lambda_k}{\lambda_k^2 + w^2}$$

with $w = 2\pi f$ and $\qquad (1.6)$

$\lambda_1 = 0.01$, $\lambda_2 = 0.1$,

$\lambda_3 = 1.0$, $\lambda_4 = 10.0$, $\lambda_5 = 100.0$

where $S_0, \overline{S}_k$ and $\overline{\overline{S}}_k$ are given in units of GPa^{-1} and λ_k are in units of s^{-1}.

Finally applying the TTSP one can determine the shifting factors to obtain the master curve for the creep compliance in the time domain.

2 DYNAMIC TESTS

The test specimens were cut out from panels made of FIBREDUX-920-C-TS-5-42 unidirectional graphite-epoxy prepreg tape with [90°] orientation. The dynamic test specimens had a thickness of 1.8mm, width of 12.0mm and the distance between the supports of 22.0mm.

The dynamic tests were carried out in 3 point bend. For each temperature level, the frequencies varied from 0.01Hz to 30Hz. The maximum imposed deflection was 64μm. The temperatures levels were within the range of 40°C to 110°C.

The softening of FIBREDUX 920 composite happened at a relative low temperature.

After the frequency-time transformation, the short-term compliance curves in time domain were obtained.

From the short-term compliance curves the temperature-time scale factor, a_T, was measured. The parameter was plotted to the reciprocal absolute temperature and the data followed two straight lines and a discontinuity on slope happened around 60°C.

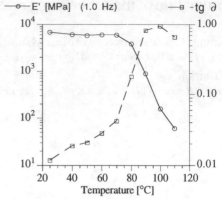

Fig. 2.3 - Temperature dependence of modulus and tg ∂.

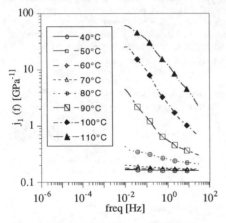

Figure 2.1 - Storage Compliance of [90°] specimens.

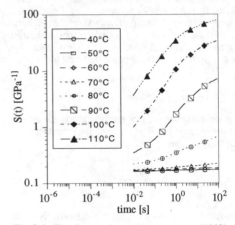

Fig. 2.4 - The short-term compliance curves of [90°] specimens obtained from FFT transformation.

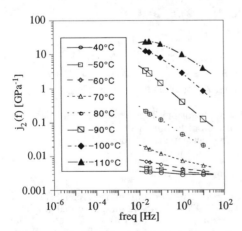

Fig. 2.2 - Loss Compliance of [90°] specimens.

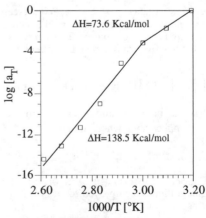

Fig 2.5 - Temperature-time scale factor a_T versus reciprocal absolute temperature of [90°] specimens.

212

This suggests that the a_T function of FIBREDUX 920 composite follow the Arrhenius equation as

$$\log a_T = \frac{\Delta H}{2.303\,R}\left(\frac{1}{T} - \frac{1}{T_0}\right) \qquad (2.1)$$

where ΔH is the activation energy for the relaxation process, R is the universal gas constant: $R=1.98$cal/mol and T_0 is a parameter with dimension of temperature.
The master compliance curve was obtained from the previous data.

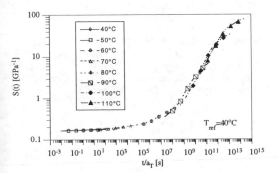

Fig. 2.6 - Master bend transverse compliance curve of FIBREDUX 920.

3 CREEP TESTS

The creep tests were carried out at room temperature. The test specimens were cutted from panels made of FIBREDUX 920 unidirectional graphite-epoxy prepreg tape. The test specimens had a thickness of 1.8mm and a width of 25.0mm.
The loads applied were chosen so that the maximum stress is 30% of the rupture stress in the tensile test (65MPa). For that stress level one should expect a viscoelastic linear behaviour of the composite.

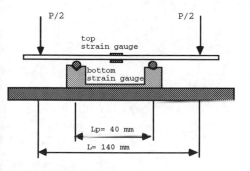

Fig. 3.1 - Four point bend test.

One strain gauge was glued to the top surface (tension) and the other to the bottom surface (compression). The signal from the specimen with no load was used to compensate temperature effects. The maximum stress was calculated as

$$\sigma_{max} = \frac{P\left(L - L_P\right)}{8I}\,h \qquad (3.1)$$

where h is specimen thickness.

The tests were carried during three weeks, which represents around 1.8×10^6 seconds.

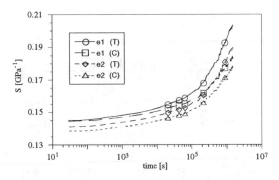

Fig. 3.2 - The compliance for the two [90°] specimens, e1 and e2.

The plotted compliance for the two specimens, e1 and e2, in tension (T) and compression (C), indicates slight differences between them.

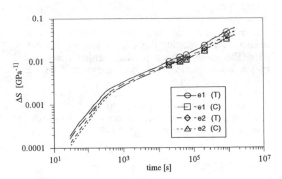

Fig. 3.3 - The transient compliance for the two [90°] specimens, e1 and e2.

On a log-log plot, the transient compliance $\Delta S = S - S_0$, shows two different straight lines where the discontinuity on slope happened around 400s. This indicates that two distinct velocities of deformation take place. At the same time we can see that the compression and tension have a similar behaviour. The two specimens have similar behaviour in tension an compression.

4 LINEAR MODEL

We adopted the general power law to model the experimental data as

$$S(t) = S_0 + \frac{S_\infty - S_0}{1 + \left(\dfrac{\tau_0}{t}\right)^n} \qquad (4.1)$$

To fit the data we used the Levenberg-Marquardt method. It should be noted that this method depends strongly on the initial parameters.

Table 4.1 - Fitted parameters.

	S_0 (GPa^{-1})	$S_\infty - S_0$ (GPa^{-1})	τ_0 10^{13} (s)	n
master	0.147	100.6	1.83	0.485
e1 (T)	0.147	108.1	0.76	0.489
e2 (T)	0.144	112.9	0.74	0.505
e1 (C)	0.147	87.2	1.25	0.486
e2 (C)	0.141	90.0	1.20	0.490

The model fitted the experimental data within a relative small error, i.e. lower than 2.0%.

5 DISCUSSION

The information obtained from the master curve was necessary to compute the parameters for the creep data. The results show a good agreement between the master curve computed from the dynamic tests and the 4 point bend ones.

The creep in bending seems to be driven by the section in tension that forces the section in compression to follow the same path.

ACKNOWLEDGEMENTS

We are grateful to JNICT for providing financial support of this work; to the COSARGUB (VUB) where the test specimens were produced.

REFERENCES

Bracewell, Rolnad Newbold, 1987. The Fourier Transform and its Aplications, International Edition, McGraw-Hill.

Ferry, John D. , 1980. Viscoelastic Properties of Polymers, Third Edition John Wiley & Sons.

Gibson, R. F. Gibson, et al., 1990. Characterization of Creep in Polymer Composites by the Use of Frequency-Time Transformations, Journal of Composite Materials, Vol.24.

Ha, Sung K. and Springer, George S., 1989. Time Dependent Behaviour of Laminated Composites at Elevated Temperatures,Journal of Composite Materials, vol. 23-November.

Press, William H.,et al., 1988. Numerical Recips in C, The Art of Scientific Computing, Cambridge University Press.

Randal, R. B., 1987. Frequency Analysis, 3rd Edition, Brüel & Kjær.

Tuttle,E. Mark, Pasricha, Arun and Emery, Ashley, 1993. Time-Dependente Behaviour of IM7/5260 Composites Subjected to Cyclic Loads and Temperatures, AMD-vol.159, Mechanics of Composite Materials:Nonlinear Effects,ASME.

Xiao, X. R. and Cardon, Albert H., 1989. Temperature Dependence of the Viscoelastic Behaviour of Peek Resin and Peek Composite, Proceedings of ATMAM'89, Canada.

5 Influence of water on the properties

Progress in Durability Analysis of Composite Systems, Cardon, Fukuda & Reifsnider (eds)
© *1996 Balkema, Rotterdam. ISBN 90 5410 809 6*

Sea water effects on the fatigue response of polymeric composites

L.V.Smith & Y.J.Weitsman
The University of Tennessee, Knoxville & Oak Ridge National Laboratory, Tenn., USA

ABSTRACT: This article concerns the effects of sea water on the fatigue life of graphite/epoxy polymeric composites. Experimental data indicate a substantially shorter life of saturated coupons fatigued while immersed in sea water, when compared to dry coupons fatigued in air. This result counters the observation of comparatively longer life of saturated coupons fatigued in air.

The foregoing distinct values of fatigue lives are explained by analytical and computational models, which also account for observed differences in failure modes.

1. INTRODUCTION

The effects of fluids on polymeric composites were investigated extensively over the past several decades. For the sake of brevity these researchers will not be detailed here. Suffice it to say that in several circumstances fluids have detrimental effects on the strength, stiffness, time-dependent response and durability of polymeric composites and that these effects are particularly pronounced when fluid sorption is associated with synergistic mechanisms. An extensive listing of references, which also includes a list of review articles on the subject, is given in a recent review (Weitsman 1995).

This article concerns the effects of sea water on the fatigue life of polymeric composites. It is demonstrated that the coupling of mechanical fatigue and sorption of sea water give rise to a synergistic mechanism that accelerates the progression of damage within the composite, resulting in a significantly shorter fatigue life.

It is worth noting that, in contrast with the extensive literature on the fatigue of composite materials, very few investigations were conducted on "wet fatigue" (Dewimille et al 1980, Friedrich and Karger-Kocsis 1990, Phillips et al 1987, Sandifier 1982, Sumsion 1976, Yang et al 1992). The afore-mentioned articles documented the response of pre-soaked polymeric composites that were fatigued in air. In contrast, the current work concerns the fatigue behavior of pre-saturated coupons fatigued while immersed in sea water and compares it with the response of pre-saturated, and of dry, coupons fatigued in air. It is shown that the case of immersed fatigue is the most severe because it gives rise to a detrimental synergism associated with capillary flow.

2. EXPERIMENTAL

The experiments subjected cross-ply $\left[0°/90°_3\right]_s$ gr/ep AS4/3501-6 coupons to tension/tension mechanical fatigue at a frequency of 5 Hz., with $R = \sigma_{min}/\sigma_{max} = 0.1$. Fatigue life data were collected at four levels of σ_{max}, namely $\sigma_{max}/\sigma_{ult} = 0.74$, 0.79, 0.84, and 0.89. For the above lay-up $\sigma_{ult} = 540$ MPa.

For purpose of comparison, fatigue experiments were performed under three distinct exposure regimes as follows:

(*i*) dry coupons fatigued in air.
(*ii*) pre-saturated coupons fatigued in air.
(*iii*) pre-saturated coupons fatigued under an immersed condition.

Each experimental circumstance was replicated between five and seven times.

The results, together with the scatter bands are exhibited in the S-N plots shown in Fig. 1. Note that the shortest fatigue life occurs in circumstance (*iii*), while the longest fatigue life occurs in case (*ii*). The fatigue life of dry coupons falls in the intervening range.

Pre-saturation was achieved by immersing dry coupons in a bath of simulated sea water at a temperature of approximately 35°C. Periodic weighing indicated that saturation, at a weight gain level of approximately 1.7%, was reached in about three months.

Since the most pronounced effect that exists during immersed fatigue, and is absent otherwise, is the ingress of water into the mechanically induced fatigue cracks by capillary action, additional experiments were conducted to estimate the rate of capillary flow. Therefore, both dry and presaturated

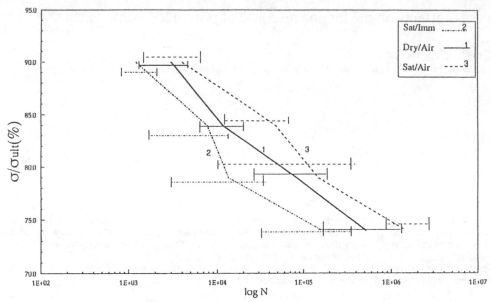

Fig. 1:   S-N data for $\left[0°/90°_3\right]_s$ gr/ep AS4/3501-6 coupons fatigued at 5 Hz., R = 0.1
(·——·——· saturated-immersed, ——— dry, - - - - pre-saturated in air).

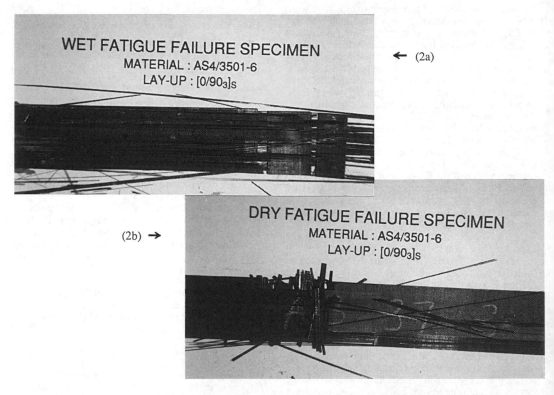

Fig. 2:   Fatigue failures under (2a) immersed and (2b) dry conditions. Note the extensive delaminations in the immersed case and the higher density of transverse cracks in the 90° ply-group in the dry case.

cross-ply coupons were subjected to mechanical loads that caused transverse cracks within the 90° ply group. These cracks spanned the entire width of the specimens. The specimens were subsequently positioned with the foregoing transverse cracks oriented vertically, inside a shallow bath of slightly acidic water and with pH sensitive paper attached to the upper side of the specimens. The rate of capillary climb was determined by monitoring the time required to observe coloration of the pH sensitive paper. The variability or constancy of the above rate was assessed by monitoring cross-ply specimens of various widths. It was found that the rate of capillary motion within the damaged coupon was between 0.3 and 1.1 cm/min.

Typical observations of failed coupons are exhibited in Fig. 2. Note the extensive inter-ply delaminations that occur during immersed fatigue, shown in Fig. 2a, which is contrasted with the profusion of transverse cracks in the 90° ply group that develop under dry fatigue as seen in Fig. 2b. It appears that dry fatigue and immersed fatigue give rise to distinct failure mechanisms.

Furthermore, the phenomena observed herein occur at three highly disparate time scales, these are:

(i) diffusion-to-saturation time $t_s \sim 3$ months, i.e. $\sim 15 \times 10^4$ minutes.

(ii) capillary flow time $t_c \sim 1$ to 5 minutes.

(iii) fatigue cycle time $t_f \sim 3 \times 10^{-3}$ minutes.

It is thus obvious that capillary flow can be separated from the water diffusion process and that, at the present frequency of fatigue cycling, the flow of water within the capillaries is much too slow to respond to the load fluctuations during mechanical fatigue.

3. ANALYSIS

The purpose of the analytical and computational results presented herein is to provide a quantitative explanation for the disparate fatigue lives and distinct failure modes observed for all the immersed and "in-air" fatigue cases. The analysis assume quasi-static behavior, discarding inertia effects associated with fatigue.

Immersed fatigue is typified by the presence of incompressible sea water, drawn by capillary action, within the fatigue-induced micro-cracks of the the cross-ply specimens. In view of the above mentioned disparity between t_c and t_f, this water cannot be expelled during the down-loading stages of the fatigue cycle and therefore imposes compressive normal traction on the surfaces of the micro-cracks at all levels of applied stress σ_a below σ_{max}.

The following analysis consists of two parts. The purpose of the first part (sub-section 3.1) is to demonstrate that in the immersed case an externally applied tension-tension fatigue results in internal tensile-compressive stress fluctuations.

The second part (sub-section 3.2) of the analysis indicates that immersed fatigue presents a more favorable opportunity for inter-ply delaminations when compared with "fatigue in air".

3.1 Shear-lag analysis of cross-ply laminates with transverse cracks. "Immersed" and "in-air" cases.

Consider a cross-ply $\left[0°_m/90°_n\right]_s$ laminate subjected to uni-axial tensile load N_x, with transverse cracks within the inner 90° ply group at distances 2L apart as shown in Fig. 3. Denote the average value of a function $f(x,z)$ with respect to z by $\bar{f}$, namely

$$\bar{f} = \frac{1}{h} \int_0^h f(x,z)\, dz.$$

Employing standard notation assume $\sigma z = 0$, $\partial w/\partial x = 0$ and consider $\bar{\varepsilon}_y$ consistent with $N_y = 0$. Also let the index i (i = 1, 2) refer to the 90° and 0° plies, respectively.

In the sequel we adopt the shear lag model of Nuismer and Tan (1988), which is traceable to the kinematic assumption of parabolic distribution of displacements u_i, namely

$$u_1 = u_1(x, z_1) = a_1(x) + c_1(x) z_1^2 \tag{1}$$

$$u_2 = u_2(x, z_2) = a_1(x) +$$
$$c_1(x) h_1^2 \left[1 + \frac{G_{TT}}{G_{LT}} \left(2\frac{z_2}{h_1} - \frac{z_2^2}{h_1 h_2} \right) \right] \tag{2}$$

In view of the assumption $\partial w/\partial x = 0$ expressions (1) and (2) satisfy the boundary conditions $\tau_1(x, 0) = 0$ and $\tau_2(x, h_2) = 0$ as well as the continuity conditions $u_1(x,h_1) = u_2(x,0)$ and $\tau_1(x,h_1) = \tau_2(x,0) = \tau^*(x)$. Obviously, τ denotes τ_{xz}.

Upon averaging of u_i one obtains

$$\tau^* = G\, \frac{\bar{u}_2 - \bar{u}_1}{h_1 + h_2} \tag{3}$$

where $G = \dfrac{3(h_1 + h_2) G_{LT} G_{TT}}{h_1 G_{LT} + h_2 G_{TT}}$

The main accomplishment of the shear-lag model is its employment of average values $\bar{u}_1$ and $\bar{u}_2$ in place of detailed displacement distributions.

Turning to equilibrium, it suffices to employ by "global" considerations, whereby

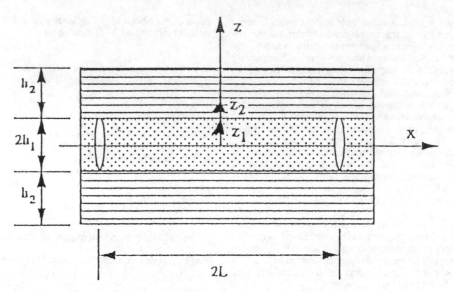

Fig. 3: The characteristic geometry of a cross-ply laminate with a cracked inner 90° ply group.

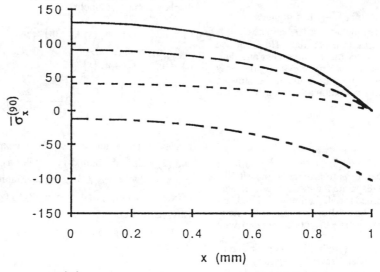

Fig. 4: The stresses $\overline{\sigma}_x^{(90)}$ vs. x corresponding to $\overline{\sigma}_x^{max}$ and $\overline{\sigma}_x^{min}$ in the case of dry fatigue and in the circumstance of saturated-immersed fatigue (——— dry 450 MPa, - - - - dry 45 MPa, — — — saturated immersed 450 MPa, — – — – saturated immersed 45 MPa).

$$\overline{\sigma}_x^{(1)} h_1 + \overline{\sigma}_x^{(2)} h_2 = \overline{\sigma}_x (h_1 + h_2) \qquad (4)$$

$$\overline{\sigma}_x^{(1)'} = -\tau^*/h_1 \qquad (5)$$

$$\overline{\sigma}_x^{(2)'} = \tau^*/h_2 \qquad (6)$$

Employment of laminate analysis yields the following stress-strain relations

$$\overline{\sigma}_x^{(i)} = Q_{11}^{(i)} \overline{\varepsilon}_x^{(i)} + Q_{12}^{(i)} \overline{\varepsilon}_y - \sigma_{xR}^{(i)}$$

$$\overline{\sigma}_y^{(i)} = Q_{12}^{(i)} \overline{\varepsilon}_x^{(i)} + Q_{22}^{(i)} \overline{\varepsilon}_y - \sigma_{yR}^{(i)}$$

$$(i = 1,2) \qquad (7)$$

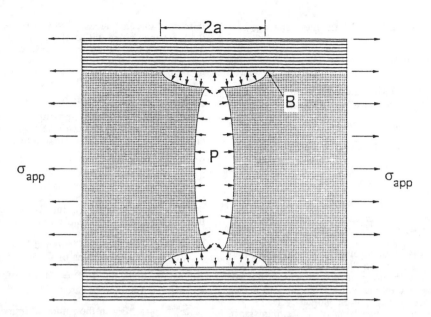

Fig. 5: A sketch of delaminations along the 0°/90° interface emanating from a transverse crack in the 90° ply group. All crack surfaces are subjected to an internal pressure P.

where $\sigma_{xR}^{(i)}$ and $\sigma_{yR}^{(i)}$ denote residual hydro-thermal stresses and $\bar{\sigma}_x = N_x/2h$ is the average value of the applied stress.

Omitting intermediate steps, which are detailed in the article by Nuismer and Tan (1988), the foregoing expressions can be shown to yield

$$\tau^*(x) = A_0 \sinh(\alpha_0 x)$$

$$\bar{\sigma}_x^{(1)}(x) = \frac{A_0}{h_1 \alpha_0}[\cosh(\alpha_0 L) - \cosh(\alpha_0 x)] \qquad (8)$$

$$\bar{\sigma}_x^{(2)}(x) = \frac{\bar{\sigma}_x(h_1 + h_2)}{h_2} - \frac{\bar{\sigma}_x^{(1)}(x) h_1}{h_2}$$

where $\alpha_0^2 = \dfrac{G}{h_1 + h_2}\left(\dfrac{1}{h_1 Q_{11}^{(1)}} + \dfrac{1}{h_2 Q_{11}^{(2)}}\right)$

The determination of A_0 requires the evaluation of $\bar{\varepsilon}_y$, which is established from the condition

$$N_y = \int_0^L \left(h_1 \bar{\sigma}_y^{(1)} + h_2 \bar{\sigma}_y^{(2)}\right) dx = 0.$$

Therefore, A_0 is determined iteratively.

Expressions (8) can be used directly to compute $\tau^*(x)$, $\bar{\sigma}_x^{(1)}$ and $\bar{\sigma}_x^{(2)}$, with the obvious boundary conditions $\sigma_x^{(1)}(L) = 0$, $\sigma_x^{(2)}(L) = \dfrac{h_1 + h_2}{h_2}\bar{\sigma}_x$, for all values of externally applied stresses $\bar{\sigma}_x$ and internal residual stresses, *except in the case of immersed fatigue*.

As already noted, in the circumstance of immersed fatigue the crack faces at $x = L$ cannot contract during the down-loading stages of the fatigue cycle due to the presence of incompressible sea water, resulting in $\bar{\sigma}_x^{(1)}(L) < 0$.

The corresponding stress field is obtained by a superposition of the solution listed in eqns. (8) for $\bar{\sigma}_x^{max}$ with the stress field for an *uncracked* laminate subjected to a compressive stress $\bar{\sigma}_c = \bar{\sigma}_x - \bar{\sigma}_x^{max}$ and without residual stresses. The latter stresses are obtained from a straight forward laminate analysis.

Computations were performed for a $\left[0°/90°_3\right]_s$ AS4/3501-6 gr/ep with the following material properties:

$E_1 = 139$ GPa, $E_2 = E_3 = 11$ GPa, $\nu_{12} = \nu_{13} = 0.27$, $\nu_{23} = 0.34$, $G_{12} = G_{13} = 4.8$ GPa, $G_{23} = 4.1$ GPa. Coefficients of thermal expansion (per 1°C): $\alpha_1 = -0.9 \times 10^{-6}$, $\alpha_2 = \alpha_3 = 31 \times 10^{-6}$. Coefficients of moisture expansion (per 1% weight gain): $\beta_1 = 0.01 \times 10^{-2}$, $\beta_2 = \beta_3 = 0.32 \times 10^{-2}$. Ply thickness $t = 0.125$ mm. Saturation moisture weight-gain 1.7%, temperature drop due to cool-down $\Delta T = -125°C$ and half crack spacing $L = 1mm$.

Results are shown in Fig. 4, where $\sigma_x^{(90)}$ is plotted vs. x for the dry and immersed circumstances under $\overline{\sigma}_{max} = 450$ MPa and $\overline{\sigma}_{min} = 45$ MPa. Note that in the dry case, as well as in the immersed case under $\overline{\sigma}_{max}$, the entire region is in a state of tension and $\overline{\sigma}_x^{(90)} (L) = 0$. However, in the immersed case with $\overline{\sigma}_{min}$ the entire region of the 90° plies is subjected to compression.

3.2 *Finite element analysis of delamination fracture energy. Immersed and "in-air" cases.*

The sharp distinction between the failure modes shown in Figs. 2a and 2b suggest that the water entrapped within the transverse cracks in the 90° plies is squeezed into growing delaminations during the down loading stages of the fatigue cycle.

A quantitative evaluation of this circumstance was obtained by a finite element analysis employing the ABAQUS code.

The computations employed the same material properties as noted previously, except that the half-spacing between transverse cracks in the 90° ply group was taken to be $L = 5mm$.

In addition, a symmetric configuration of delamination cracks of length a were considered to emanate from the place where the foregoing transverse cracks impinge on the 0°/90° interface as shown in Fig. 5.

Solutions were computed that corresponded to applied stresses ranging between $\overline{\sigma}_x^{max} = 450$ MPa and $\overline{\sigma}_x^{min} = 45$ MPa for both dry and saturated-immersed circumstances. The incompressibility of the water contained within the combined region of transverse and delamination cracks was accounted for by ascertaining the constancy of the cross-sectional area bounded by the crack surfaces at all load levels. Consequently, the stresses during the down-loading stage, namely for all $\overline{\sigma}_x < 450$ MPa, were determined by imposing the additional requirements that (*i*) the cross-sectional area of the cracked region remains identical to its magnitude at $\overline{\sigma}_x^{max}$ and (*ii*) that all crack boundaries be subjected to a constant normal compressive traction p.

The finite element results were subsequently utilized to calculate J-integral values around the tip of the delamination crack (point "B" in Fig. 5) and thereby the levels of available delamination fracture energy.

Results are shown in Fig. 6 and 7.

Fig. 6 contrasts values of G vs. a growing delamination length a at $\overline{\sigma}_x^{max} = 450$ MPa for both dry and saturated-immersed cases with the circumstance of $\overline{\sigma}_x^{min} = 45$ MPa in the saturated-immersed case. Note the preponderance of the effect of internal pressure, that enhances the tendency for crack growth in the saturated-immersed case (it attains its largest magnitude at $\overline{\sigma}_x^{min}$), over the role of external loads.

The consistently higher values of G that occur in the saturated-immersed case at $\overline{\sigma}_x^{min} = 45$ MPa explain the observed delaminations in Fig. 2b.

Additional insight is provided by the plots drawn in Fig. 7. In that figure the available energy for delamination G is plotted vs. the applied stress $\overline{\sigma}_x$ ranging between $\overline{\sigma}_x^{min} = 45$ MPa and $\overline{\sigma}_x^{max} = 450$

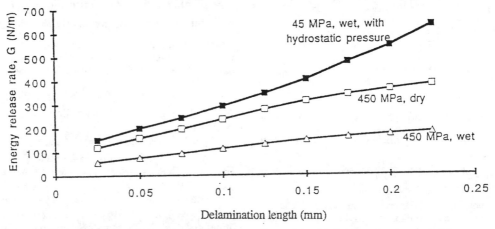

Fig. 6: Fracture energy G at the tip of delamination cracks vs. delamination length a (computed for transverse cracks spaced 10 mm apart).

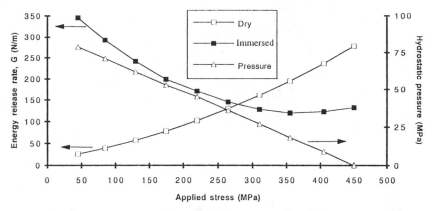

Fig. 7: Delamination energy release rate (left scale) and internal pressure P (right scale) vs. applied stress $\overline{\sigma}_x$ for a delamination of length 0.125 mm and transverse crack spacing of 10 mm.

MPa, with a fixed delamination length a = 0.125 mm and transverse cracks spaced 10 mm apart. Note the increase of G during down-loading in the saturated-immersed case, which contrasts the trend computed for the dry case. The aforementioned increase is associated with an accompanying rise in the internal pressure, as indicated on the right-side scale of Fig. 7.

4. CONCLUSIONS

It has been shown that immersed fatigue may present the most severe limitation on the fatigue life of polymeric composites. The coupling between fatigue induced microcracks and the accompanying ingress of water into the thus formed capillaries enhances the formation of damage in a synergistic manner, which accelerates the failure of the composite. At sufficiently high frequencies the characteristic fatigue-cycle time is much to short to allow for a responsive capillary motion of the nearly incompressible fluid located within the microcracks, thereby squeezing the fluid into weak locations within the composites - such as interlaminar boundaries.

The resulting extensive delaminations, which are especially pronounced in the case of immersed fatigue, bring about early material failures.

This article provides experimental evidence to contrast immersed and "in-air" fatigue responses, which is explained by appropriate analyses.

5. ACKNOWLEDGEMENT

This work was performed under Contract N00014-90-J-1556 from the Office of Naval Research.

REFERENCES

Dewimille, B., J. Thoris, R. Mailfert & A. R. Bunsell 1980. Hydrothermal aging of an unidirectional glass-fibre epoxy composite during water immersion. A. R. Bunsell, C. Bathias, A. Martrenchar, D. Menkes & G. Verchery (eds.), *Proceedings of the Third International Conference on Composite Materials*: 597-612.

Friedrich, K. & J. Karger-Kocsis 1990. Fracture and fatigue of unfilled and reinforced polyamides and polyesters. J. M. Schults & S. Fakirov (eds.), *Solid State Behavior of Linear Polyesters and Polyamides*: 249-322. Prentice-Hall.

Nuismer, R. J. & S. C. Tan 1988. Constitutive relations of a cracked composite lamina. *Journal of Composite Materials*. 22: 306-321.

Phillips, D. C., J. M. Scott & N. Buckley 1978. The effects of moisture on the shear fatigue of fibre composites. B. Noton, R. Signorelli, K. Street & L. Phillips (eds.), *ICCM/2 Proceedings of the 1978 International Conference on Composite Materials*: 1544-1559.

Sandifier, J. P. 1982. Effects of corrosive environments on graphite/epoxy composites. T. Hayashi, K. Kawata & S. Umekawa (eds.), *Proceedings of the Fourth International Conference on composite Materials*: 979-986.

Sumsion, H. T. 1976. Environmental effects on graphite-epoxy fatigue properties. *Journal of Spacecrafts and Rockets*, 13:150-155.

Weitsman, Y. 1995. Fluid effects on polymeric composites - a review. University of Tennessee, Department of Engineering Science and Mechanics, Technical Report No. ESM 95-3.0 CM.

Yang, B.-X., M. Kasamori & T. Yamamto 1992. The effect of water on the interlaminar delamination growth of composite laminates. C. T. Sun & T. T. Loo (eds.), *Proceedings of the Second International Symposium on Composite Materials and Structures*: 334-339.

Progress in Durability Analysis of Composite Systems, Cardon, Fukuda & Reifsnider (eds)
© *1996 Balkema, Rotterdam. ISBN 90 5410 809 6*

Evaluation of the integrity of composite tubes

R. Baizeau, P. Davies & D. Choqueuse
IFREMER, Brest, France

J. Le Bras
EDF, Les Renardières, France

ABSTRACT: This paper presents results from an experimental study of filament wound tubes for cooling water applications. Tests aimed at characterizing the influence of matrix resin on composite toughness are first described. Quasi-static internal pressure loading of tubes to failure is then presented and critical loads for damage onset and weeping are determined. Water-composite interactions and aging are then discussed and results from immersion of tubes are presented. Finally, results from long term (up to one year) internal pressure loading of tubes are then given and extrapolation to 40 year lifetimes is discussed. Overall the preliminary results suggest that these composite tubes have good long term stability and are well-suited to cooling water pipework system applications. Further tests are underway to confirm these results.

1. INTRODUCTION

Composite tubes are finding increasing applications in areas where corrosion resistance and long term stability are essential. Many authors have studied the behaviour of composite tubes under internal pressure (see for example Bax 1970, Spencer & Hull 1978, Soden et al. 1989, Rawles et al, 1990, Maire 1992, Thiebaud 1994). These authors have presented experimental data and proposed damage and failure criteria for short term loading of tubes, but the long term behaviour of such structures has received much less attention. Long term behaviour and life prediction of composites have been the subject of many studies for the aerospace industry, and a recent book reviews the subject (Reifsnider 1990). Nevertheless, the estimation of the durability of composite pipework systems is still largely empirical, as there are important gaps in the data which are required for design on a more rational basis. These include a lack of appropriate short term material data on damage and fracture properties, to enable the damage tolerance of fibre-resin combinations to be improved. A second problem is the unavailability of long term data which would allow predictive models to be checked, particularly with respect to behaviour under load in contact with water. Existing viscoelastic models have been reviewed recently (Dillard 1990) but few examples of their application can be found. For cooling water circuits in power stations, or offshore firewater circuits, where high reliability is required over periods up to 40 years, this lack of data is reflected in high design safety factors.

This paper presents a summary of results from a four year study to examine the long term mechanical behaviour of such tubes. Four aspects are treated. First, the influence of matrix resin on the fracture resistance of composite materials for tube applications is discussed. Little information is available to allow a rational material choice to be made, and data are presented for epoxy and vinyl ester resins. In the second part of the paper the short term failure of glass/epoxy tubes subjected to internal pressure loading, with and without end effects, is presented. These tests have allowed the safe operating conditions to be established, with respect to damage thresholds and weeping criteria. Aging of tubes in the presence of water is then treated. In the final part of the paper data from long term internal pressure tests on tubes are described and the extrapolation of results from accelerated tests lasting several months to in-service conditions and a lifetime of 40 years is discussed.

2. MATERIALS

The short and long term internal pressure tests described below were performed on E glass reinforced epoxy tubes wound at ±55°, of 150 mm internal diameter. The epoxy is the Ciba LY 556 bisphenol A based resin with HY917 anhydride hardener. None of the tubes tested had liners. For fracture resistance tests 4 mm thick plates were wound at 0° and tubes of internal diameter 160 mm and wall thickness 5 mm were wound at ±30°, ±45° and ±60°. Both the epoxy (LY556) and a vinyl ester (Dow 411-45) resin were used and 13 μm thick aluminium foil defects were placed at mid-thickness. Some unreinforced resin plates were also cast and post-cured.

3. RESIN & COMPOSITE FRACTURE

One of the areas of uncertainty with respect to filament wound structures is their sensitivity to defects. The application of a fracture mechanics approach offers a means of establishing critical defect sizes in these composite structures under different loadings, as was shown over 30 years ago (Kies & Bernstein 1962). This could allow improvements in both material selection and NDT reliability. In order to produce data for such calculations and also to assess the influence of resin on fracture behaviour, a number of tests have been performed involving two resins, an epoxy and a vinyl ester. First, the K_C and G_C values of unreinforced resins were measured. Then filament wound unidirectional specimens were tested under mode I, mode II and mixed mode loading to determine failure envelopes. Finally, specimens taken from tubes containing defects were tested to establish whether flat plates were suitable to obtain values appropriate to defects in tubes. Results from the resin and mode I tests have been presented in detail elsewhere (Davies & Rannou 1995) and initiation values (defined as deviation from linearity of the load-displacement plot according to ASTM D5528, are summarized in Table 1. All delaminations were propagated directly from the insert, without precracking. A comparison of mixed mode results from flat unidirectional specimens of the two materials is shown in Figure 1. End notch flexure (ENF) and asymmetrically loaded double cantilever beam (ADCB) specimens used for mode II and mixed mode. The use of these tests on the curved specimens taken from tubes requires detailed analysis of mode separation, and will be presented elsewhere.

These results indicate that there is little difference in initiation values for the two materials, although some differences have been noted for propagation

Table 1. Mean mode I initiation results for plates and tubes, J/m² unless stated, (std. deviations)

	Epoxy	Vinylester
Resin		
K_C MPa√m G_C J/m²	0.70 (0.08) 175	0.78 (0.11) 195
Composite		
Flat UD plate 0° G_{Ic}	247 (13)	214 (28)
Tube ±30° G_{Ic} ±45° G_{Ic} ±60° G_{Ic}	216 (31) 256 (99) No propagation	229 (41) 335 (88) No propagation

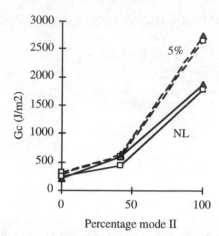

Figure 1. Mixed mode initiation envelopes for unidirectional composites (from insert), using non-linearity (full lines) and 5% compliance change (dashed lines) to define initiation. The open squares indicate epoxy, closed triangles vinylester.

values. It is also interesting to note the correlation between initiation values for flat unidirectional and curved specimens taken from ±30° wound tubes.

Data from such tests are useful in the selection of matrix resins for composite pipes. For comparison, published orthophthalic polyester resin composites values for mode I and mode II may be less than half these values, (Davies & Brunellière 1993). In the remainder of the paper only the glass/epoxy composite will be discussed as this is the combination selected for the power station cooling water application targeted here.

4. QUASI-STATIC INTERNAL PRESSURE LOADING

Tubes wound at ±55° were tested under internal water pressure loading. They were 1.18 metres long and of 150 mm internal diameter, and instrumented with 0°/90° strain gauges and two acoustic emission transducers. Two types of end condition were employed, fixed ends and free ends, such that the stresses resulting are either biaxial or uniaxial :

Fixed ends : $\sigma_\theta = Pr/t$, $\sigma_z = Pr/2t$

Free ends : $\sigma_\theta = Pr/t$, $\sigma_z = 0$

where σ_θ is the hoop stress, σ_z the axial stress, P the internal pressure, r the radius and t the wall thickness of the tube.

Tests were performed with loading-unloading cycles, with 4 minute hold times at each stress level (of the type recommended in the CARP procedure 1986, see Fowler et al 1989), and 2 minutes hold time between cycles. One test thus took up to 18 hours to perform. Residual strains were measured two minutes after each unloading. The following results were obtained from these tests : longitudinal and circumferential stress-strain curves, stress vs residual strain on unloading, and acoustic emission recordings including Felicity ratio. These measurements have enabled three levels of damage to be identified. These are the first damage threshold, which is indicated by a change in slope of the hoop stress-strain plot, residual strain on unloading, and the first acoustic emissions of amplitude greater than 70 dB. As the pressure is increased an accumulated damage threshold is reached at which a second change in slope of the stress-strain is noted and there is a change in slope of the stress-residual strain plot. This corresponds to a significant Felicity effect, as the ratio of stress at which acoustic emission is detected during reloading to the stress before unloading becomes very small (typically 0.2). The correlation of mechanical and acoustic data allows these damage levels to be clearly identified. The third level of damage corresponds to failure, either by "weeping", for fixed end conditions, when the tube starts to leak and drops of water are observed to appear on the outside surface of the tube, or by bursting (free ends). Examples of the results of these tests are given in Figures 2 to 4 and Table 2.

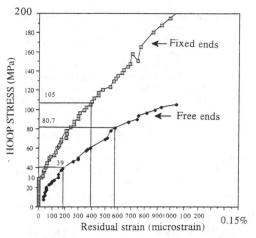

Figure 3. Hoop stress v residual strain 2 minutes after unloading, for free and fixed end conditions

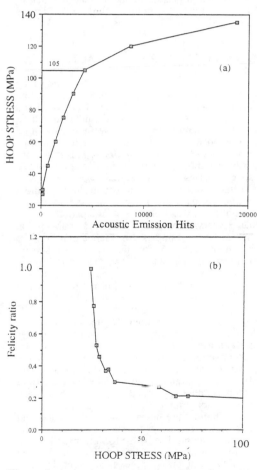

Figure 4. Acoustic emission recordings for tube with fixed ends under internal pressure loading. a) Hits recorded, b) Felicity ratio on reloading

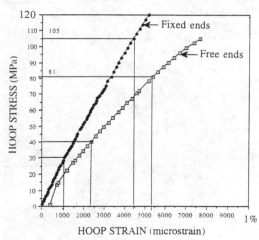

Figure 2. Hoop stress v strain , σ_θ v ε_θ, for free and fixed end conditions (first part of curves).

Table 2. Levels of hoop stress σ_θ and strain ε_θ corresponding to different damage thresholds.

	Free ends	Fixed ends
First damage	39 MPa 0.23%	30 MPa 0.11%
Accumulated damage	81 MPa 0.54%	105 MPa 0.44%
Weeping or Burst	420 MPa	197 MPa

227

The effect of the fixed end condition is to lower the damage hoop strain thresholds due to the additional axial stress. However, the types of damage involved also differ for the two cases. From calculation of ply loading it can be shown that the failure of the tube in this case is dominated in the ply by transverse tension, which combines with shear to cause matrix cracks. Interlaminar shear, then promotes delaminations as shown by fractography (LeBras 1995). Two acoustic emission transducers have been used to locate the sources of noise and it was observed that damage was much more localized in tubes with free ends.

The importance of these results is that they can be used in damage models to determine design limit levels for short term loading (Perreux et al. 1995). At present these are based on the first damage threshold, but this is conservative and it may be that the accumulated damage threshold can be used. In order to examine this it is necessary to establish the influence of stress levels on long term loading, and this is discussed in Section 6 below. The effect of the environment will first be addressed.

5. WATER-COMPOSITE INTERACTION

One of the areas of uncertainty concerning long term application of composite tubes in permanent contact with water is the effect of diffusion and subsequent degradation of matrix and fibre-matrix interface. Many studies on the effect of moisture on composite behaviour exist (see for example, Springer, 1981, Weitsman 1990), but little work has been published on tubes. In order to study this aspect a series of tube sections were filled with distilled water at different temperatures, while a second series were fully immersed in water. Weight gains were followed over a period of more than 3 years. Examples of mean water absorption curves, (three samples for each condition) are shown in Figure 5, together with results from immersed filament wound panels (angle ±55°).

It is apparent from these plots that the diffusion of water through the inside wall of the tube is much slower than that through the external wall, and is negligible even after more than 3 years at 40°C (and 60°C). This is encouraging as far as the durability of the tube is concerned but requires explanation. Two possible factors are the inhomogeneous structure of the tube, and the existence of internal stresses due to winding. Microscopic examination revealed that the fibre content is rather higher at the inner wall than on the outer surface, Figure 6.

If the fibre-matrix interface is good and does not offer easy access to moisture then diffusion is controlled by the local resin content, so the low resin content at the inner wall may be beneficial. Compressive pre-stress might also slow moisture uptake, and cutting strain gauged tubes indicated compressive residual strains but at a low level (10's of microstrain).

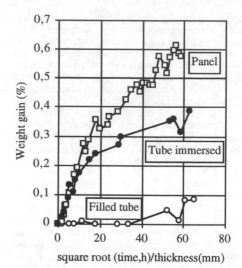

Figure 5. Water absorption curves for filled and immersed tube sections and immersed panels, 40°C for 3 years.

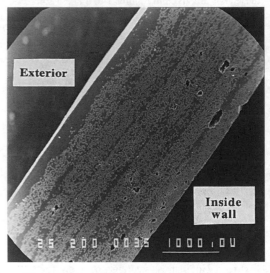

Figure 6. Photo of section through tube wall.

Overall these studies and parallel studies on filament wound plates suggest that the aging effect of water will be a secondary consideration in the long term durability of these tubes, although the effect of resin-rich liners also needs to be checked.

6. LONG TERM PRESSURE TESTING

Two series of tubes have been tested to date, and both series were loaded with free end condtions, as it was believed that such tests would be easier to interpret than the biaxial stress state induced in tubes with fixed ends. However, the latter is closer to real loading conditions, and in a third series of creep tests currently underway fixed end loading conditions are being applied. Only the results from tests with free ends will be presented here.

Tubes were placed in temperature cabinets controlled at 25, 40 and 55°C. They were instrumented with at least six strain gauges, three or four circumferential and two or three axial, and loaded to different pressure levels, as summarized in Table 3.

Table 3. Conditions for twelve tubes tested under long term pressure loading, free end conditions.

	Pressure (σ_θ) MPa	Temp. °C	Duration
Series 1			
	2.7 (38)	25 40 55	12 months creep+1 m recovery
	4.0 (56)	25 40 55	9 months creep+1 m recovery
Series 2			
	2.7 (38)	40 (2) 55 25	6 months creep + 2.5 mths
	5.4 (76)	55 (2)	recovery

The extensometry used was based on strain gauges, after the first 24 hours readings were taken every 12 hours and stored using a data acquisition system. It is clearly of importance to establish at the outset :
- the reliability of the strain gauge measurements over long periods at temperature, (see Tuttle & Brinson 1984),
- the scatter in strain gauge measurements along a tube, as these measurements give local values.
For the former reference gauges were bonded to unloaded tube sections in the temperature cabinets, and their response was recorded during the tests. There was very little drift in values. This is believed to result from the discontinuous supply to the gauges which are only excited by 1 mA when they are scanned (once every 12 hours) so there is minimal heating of the composite material. In order to study the level of scatter in measurements one tube was instrumented with 11 circumferential gauges. Values measured during the first 12 hours are shown in Figure 7 for a hoop stress of 38 MPa at 40°C and scatter along the tube is quite low, (coefficient of variation 2.5%).

An example of data obtained from two tubes under the same test conditions over 3 months is shown in

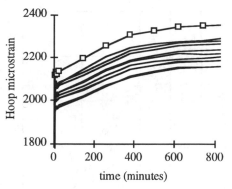

Figure 7. Readings from 11 strain gauges on one tube, 27 bars 40°C.

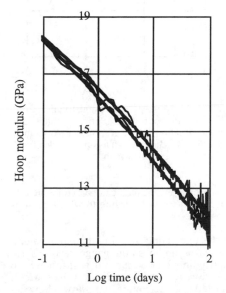

Figure 8. Effective hoop modulus versus time plots for two tubes under 76 MPa hoop stress, at 55°C. Linear regressions fitted to data.

Figure 8. The measured data are plotted together with a smoothing function applied to facilitate the construction of master curves.

Results for the two tubes are very similar, mean hoop strains are within 4% of each other after 100 days. Data from all the tubes were then combined by shifting horizontally to form a master curve at 25°C and 38 MPa hoop stress. Figure 9 shows data from series 1 and 2, nine tubes in total, and at first sight a reasonably unique master curve is found. Performing tests at a hoop stress level of 76 MPa rather than 38 MPa leads to a shift of 1.1 decade, while raising temperature and stress (76 MPa at 55°C) gives a horizontal shift of 2.4

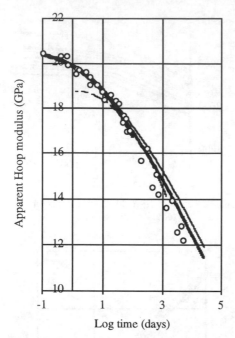

Figure 9. Creep master curve construction for 25°C, 38 MPa hoop stress, Series 2, obtained using 76 MPa hoop stress response at 25°C and 55°C. Series 1 data shown as open circles for comparison.

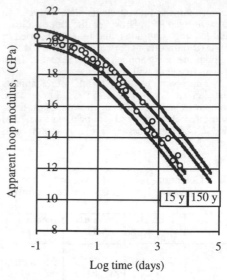

Figure 10. Influence of uncertainty of ±2.5% in initial modulus on long term predictions.

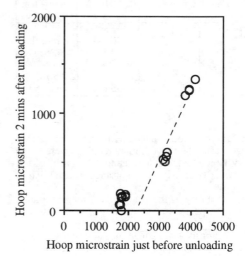

Figure 11. Instantaneous creep strain recovery (tubes series 1).

decades. However, the data can be extrapolated to 14 years or 66 years according to which curve is taken.

Given the uncertainty in initial modulus values, (2.5 % variation in strain gauge readings for different points on one tube) it is instructive to examine how uncertainty in initial values affects the shifted value. Figure 10 shows how an uncertainty of ±2.5% on initial moulus will affect the extrapolated values, and it is apparent that this can modify the extent of the time extrapolation considerably from 15 to 150 years. Thus, while the time-temperature superposition shifts can still allow a useful extrapolation to longer times than those employed for testing, the uncertainty in initial measurements will seriously affect the accuracy of long term predictions.

The tubes were allowed to recover at their creep temperature after unloading. Figure 11 shows examples of instantaneous strain recovery, and it is apparent that a creep strain of 0.2% is completely recovered immediately.

Higher creep strains are recovered more gradually, as shown in Figure 12 below, and a creep strain of 0.32% is completely recovered after 2 weeks. A creep strain of 0.41% also continues to recover slowly, while a creep strain of 0.48% (not shown) remains above 0.17% even after a 3 month recovery period.

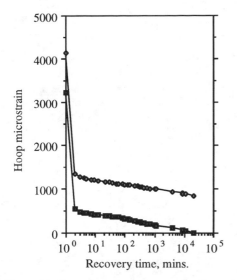

Figure 12. Examples of recovery curves for two tubes.

7. CONCLUSIONS & FURTHER WORK

This paper presents first results from a study of the behaviour of composite tubes for cooling water applications. The work is continuing, but the following preliminary conclusions may be drawn :

■ Fracture tests indicate few differences between epoxy and vinyl ester resins for the composites studied here.

■ Tubes containing water absorb much less than would be expected from results for tubes or panels fully immersed. This is believed to be related to the material gradient across the wall thickness, and should result in reduced aging of the composite in cooling water applications. This result illustrates the difficulty in relating artificial environmental aging of panels to tube behaviour in service.

■ Static internal pressure tests provide allowable damage levels for design. The quasi-static first damage threshold strain for the free end condition is around 0.23%. This corresponds closely to the maximum recoverable hoop strain 2 minutes after long term loading. After longer recovery times a creep hoop strain of 0.33% is completely reversible.

■ Extrapolation of data from temperature accelerated tests lasting several months to long term behaviour of tubes under internal pressure loading involves considerable uncertainties. Scatter in initial moduli values, within variations for measurements along one tube (±2.5%), can cause order of magnitude differences in the range of extrapolations.

These results are now being confirmed by further tests. Other tests are also underway on tubes with fixed ends, and on different types of connections. The assembly of tubes in a pipework system is often the critical operation for long term reliability and both adhesively bonded and mechanical joining methods are being evaluated.

8. REFERENCES

Bax J, Plastics & Polymers Feb. 1970, pp27-30.

CARP test procedure, ASTM E1118-86, 1986.

Davies P & Brunellière 0, J. Mat. Sci. Letters 12, 1993, pp427-9.

Davies P & Rannou F, Applied Comp. Materials, 1, 1995 p 333-349.

Dillard D, Chapter 8 in "Fatigue of Composite Materials", Ed. KL Reifsnider, Composite Materials Series Vol 4, Elsevier 1990.

Fowler TJ, Blessing JA, Conlisk PJ, Proc AECM3, Paris June 1989, pp16-27.

Kies JA, Bernstein H, Proc. 17th SPI conf. on Reinf. Plastics, 1962, 6B, p1-8.

LeBras J, Proc. Journées AMAC, Besançon Feb. 1995.

Maire JF, "Etudes théorique et expérimentale du comportement de matériaux composites en contraintes planes" PhD thesis Université de Franche Comté, Besançon, 1992.

Perreux D, Varchon D, LeBras J, ENERCOMP, Proc. Montreal, May 1995, pp 818-26.

Rawles JD, Roscow JA, Phillips MG, Proc.1990 Pressure Vessels & Piping conf., Nashville June1990, ed. D. Hui ASME Publications p 17-21.

Reifsnider KL, Editor, "Fatigue of Composite Materials", Composite Materials Series Vol 4, Elsevier 1990.

Soden P, Kitching R, Tse PC, Composites 20, 2, March 1989, pp 125-135.

Spencer B & Hull D, Composites, 9, 1978 pp263-271.

Springer GS, Editor, "Environmental effects on composite materials", Technomic 1981.

Thiebaud F, "Modélisation du comportement global en sollicitations quasi-statiques d'un composite straifié verre-époxy", PhD thesis Univ. Franche Comté, Besançon,1994.

Tuttle ME, Brinson HF, Experimental Mechanics, March 1984, pp54-65.

Weitsman Y, Chapter 9 in "Fatigue of Composite Materials", Ed. KL Reifsnider, Composite Materials Series Vol 4, Elsevier 1990.

Progress in Durability Analysis of Composite Systems, Cardon, Fukuda & Reifsnider (eds)
© *1996 Balkema, Rotterdam. ISBN 90 5410 809 6*

Effect of plasticization with water on the behaviour of long-term exposed filament wounded composites

Philippe Castaing
Polymers & Composites Department, CETIM, Nantes, France

Hubert Mallard
Department of Mechanical Calculation, CETIM, Nantes, France

ABSTRACT : Glass fibre reinforced pipes are widely used in various agressive combined environments like moisture or chemical products with high temperature. These materials offer a quite good long term behaviour during their life time, considering the unloaded aging. However, during the transient absorption of water, they are very sensitive to the plasticization and to differentiel swelling, leading to a significant but partially reversible loss of properties. The present paper focuses on the importance of determining accurately the microstructure of the pipes, and taking into account the decrease of properties in order to predict the failure envelopes during the non steady state of sorption.

1- INTRODUCTION

The use of glass fibre reinforced composites is widely increasing as structural materials in various fields like offshore oil industry, chemical industry or heating system pipes, mainly due to their intrinsic chemical-mechanical propertics compared to traditional materials.

The glass fibre reinforced polyester and epoxy are generally exposed during their life time to an aggressive environment like moisture, high temperature, chemical products...

Composites offer a quite good long term behaviour in various aggressive environments and high strength - stiffness to weight ratios. However, it appears that these reinforced plastics are more or less sensitive to water and moisture through an absorption of water molecules, leading first to physical degradations, like plasticization of the matrix with water and the differential swelling between fibres and resin or between plics.

At longer times, the uptake of water can also induce chemical degradations like hydrolysis of the matrix and the glass sizing, leading to delamination. Consequently, the residual mechanical properties depend on these degradations, and for a mean life time of 20 years, the main effects to these materials are the plasticization of the resin (reversible loss of properties) and the differential swelling (cracks appearing at interfaces fibre-resin). So, it appears to be important to focus on these latter phenomenons.

The aim of this paper is to present mechanical and physico-chemistry results about the behaviour of filament wounded pipes, subjected to accelerated ageing tests in distilled water.

Before ageing tests, it is necessary to determine accurately the physical properties and the microstructure (data determined by image analysis) in order to optimize the mechanical calculations.

It is shown how the different chemical formulations of the laminate resins account for their different behaviour. The reversible phenomenon of plasticization has to be taken into account for the evaluation of the life time of the composites regarding the failure envelope.

If several workers (1,2,3..) have been studying the problem of moisture absorption by composites, few of them take interest in the plasticization itself of the matrix and its effect on the failure envelope.

2- MATERIALS STUDIED AND PROPERTIES MEASURED

The material studied are of three types : a glass fibre - orthophthalic polyester (ORTHO), a glass fibre - isophthalic polyester (ISO) and a glass fibre - DGEBA epoxy pipes (EPOXY), diameter : 120 mm.

These composites are manufactured by filament winding with theoretically a constant angle of winding equal to 55° for the ISO, EPOXY and for the ORTHO pipes. The reinforcement is a E glass fibre (800 tex) roving. In order to obtain a total thickness of 6 mm, 5 helicoïdal windings at +a and 5 at -a were necessary. The curing was completed in an oven for 2 hours at 80°C in the case of the polyester pipes, for 3 hours at 140°C for the epoxy pipes.

Two series of pipes are manufactured : one series with a low resin ratio ("dried pipes") and another one with a correct resin content ratio.

The interesting properties measured (see Table 1) are the thermal ones (glass temperature NFT 57-501), the physical ones like the weight fibre ratio (according standard NFT 57-102), the porosity rate or the fibre angles, by image analysis.

The latter data are necessary to optimize the calculation of the pipes, and also the mechanical characteristics such as the ultimate stress and strain according a tensile test on rings (Nol tensile test, standard ASTM D 2290-76) and compressive test; the mechanical characteristics are given in next paragraphs (Tables 3, 4, and 5).

Samples are analyzed by Differential Scanning Calorimetry in order to evaluate the advancement of the crosslinking by measuring ΔH (j/g) and the glass temperatures Tg (°C) (v=10°C/mn). The results are closed to the maximal glass temperatures given by the resin suppliers :

Orthophthalic resin	Tg=96°C
Isophthalic resin	Tg=117°C
Epoxy resin	Tg=119°C

3- WATER ABSORPTION

The 0,8 m long pipes are then immersed in distilled water at a temperature of 60°C and ambient pressure for 1200 hours. Referring to previous

Table 1 : Physical and thermal characteristics of the three types of pipes before ageing (4)

Material	Vf (%)	σ (Vf)	Poros. rate P (%)	σ (P)	Fibre angle	σ	Tg (°C)	ΔH (j/g)
ORTHO	46,70	6,30	3,17	1,56	61,1	4,9	95,9	-4,9
ISO	45,20	8,50	3,41	1,16	60,6	2,4	118,9	-4,7
EPOXY	44,60	6,10	3,59	1,32	61,8	4,1	119,0	-2,0

studies (4,5) concerning laminates, the ageing is performed at T=60°C as this test leads to a correct degree of acceleration of the degradation, compared to a natural ageing in water (acceleration factor of about 25 to 30, that is to say that 2000 hours at 60° may be equivalent to about 10 years in fresh water at 20°, or 2000 hours at 80°C equals to 10 years at 40°C); in addition, this temperature prevents from inducing other thermal degradations. The period of 1000 hours of testing corresponds to the plasticization of the matrix, without hydrolytic attack, as shown in a previous study.

The absorption of water in pipes can be well predicted by Fick's model, in polar coordinates for pipes with an internal radius a and an external radius b. Recall that at steady state, for a pipe in total immersion, the general solution is given by (6) :

$$-1- \quad C(r) = c_1.[\ln(b/r) + \ln(r/a)]/(\ln(b/a)$$

where c_1 is the water concentration on both side of the thickness pipe. The water concentration profile C(r) through the thickness of the cylinder is no more linear as for a plane sheet, as the quantity of molecules crossing a surface A at r would cross a surface A+dA at r+dr; A is constant for the plane case. A specific decrease of absorption for $0,6 < M(t)/Ms < 0,9$ is then observed compared to plane sheets.

However, when $(b-a)/a$"1 et $b/a \geq 1$, $b/a \approx 1$, the pipe tends to behave like a plane sheet, which is useful to determine the diffusivity D, and to predict the sorption curves of hollow cylinders the same way as for plane sheets.

The experimental procedure its mechanical ones) at the end of the ageing test. Examples of absorption curves on pipes are displayed on the following figure (Figure 1), the mean absorption of three pipes is determined.

Table 2 : Diffusional parameters of the materials studied

Material	D (mm²/h x 10³)	Ms (%)	D (mm²/h x 10³) low vr%	Ms (%) low vr %
ORTHO	8,6	1,25	13,5	1,52
ISO	7,2	1,05	7,8	1,11
EPOXY	7,5	1,10	7,9	1,20

The ISO and EPOXY pipes tends to reach the saturation state, on the other hand it appears that the ORTHO resin begins being hydrolyzed (decrease of the curve $M(t)$ % as hydrolyzed products are rejected from the materials).

Curiously, D and Ms (%) for materials with low resin rate ("dried fibres") are higher than for the correct materials : a preferential absorption and diffusion along the interfaces fibres-matrix certainly occurs. The diffusivity of water and the saturation rate of the ORTHO resin is higher than for the orther materials, this difference is explained by the chemical hydrophilic behaviour of the orthophthalic acid and by the unsaturation rate lower than for the other polyester resin.

4- VARIATION OF THE CHARACTERISTICS

4-1 Mechanical testing (tensile test)

The mechanical characteristics measured are the ultimate stress and strain according a tensile test on rings (Nol tensile test, standard ASTM D 2290-76)

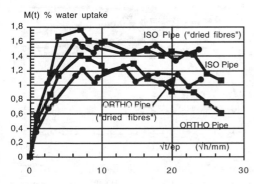

M(t) % water uptake

Figure 1 : Classical absorption curves of ISO pipes (dried fibres or not) and ORTHO pipes (dried fibres or not), in distilled water at 60°C.

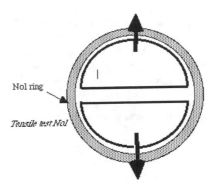

Nol ring

Tensile test Nol

Figure 2 : Tensile test on rings according standard ASTM D 2290-76

Tables 3 : Tensile tests performed on rings at initial state, after ageing (humid state) and after ageing on dried samples

Ortho	Initial state		1000 h Humid state		1000 h Dry state	
	Xr	Ar%	Xr	Ar%	Xr	Ar%
1	232,5	4,01	148,5	3,92	168,7	3,65
2	232,3	4,03	143,0	3,72	171,1	3,58
3	244,1	4,65	129,4	3,33	171,6	3,24
4	244,7	3,71	131,0	3,61	176,7	3,32
mean	238,4	4,10	138,0	3,65	172,0	3,44
σ	6,9	0,42	9,3	0,25	3,37	0,19

ISO	Initial state		1000 h Humid state		1000 h Dry state	
	Xr	Ar%	Xr	Ar%	Xr	Ar%
1	238,7	3,74	206,7	3,77	212,2	3,52
2	248,2	3,54	204,6	3,49	218,1	3,30
3	255,0	4,44	182,5	3,61	205,4	3,22
4	258,4	3,75	188,5	3,85	217,6	3,36
mean	250,1	3,87	195,6	3,68	213,3	3,35
σ	8,7	0,39	11,9	0,16	5,92	0,13

Epoxy	Initial state		1000 h Humid state		1000 h Dry state	
	Xr	Ar%	Xr	Ar%	Xr	Ar%
1	262,7	5,78	211,7	5,69	236,2	4,12
2	239,6	5,91	221,6	5,70	229,6	4,05
3	271,6	5,11	219,1	5,29	241,8	4,16
4	307,7	4,53	217,8	4,88	248,2	4,23
mean	270,4	5,33	217,5	5,39	238,9	4,14
σ	28,26	0,64	4,23	0,38	7,9	0,07

described in Figure 2. The results are presented in the following tables (Table 3). Tests are performed on "humid" samples, just after ageing, and on "dried" samples, after being dried in room conditions for one month.

4-2 Mechanical testing on large rings (compressive test)

The tests are performed according a compressive test described in Figure 3.

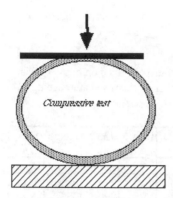

Figure 3 : Compressive test on wide rings according standard ASTM D 2412-77

The results are presented in Table 4.

Table 4 : Compression test on large rings at initial state (I) and after 1200 hours of ageing (A), (dried samples)

Material	σ 1st failure (MPa)	E circonf. (MPa)	σ rupt (MPa)
ORTHO (I)	392,00	11469,05	425,17
ORTHO (A)	246,34	15499,56	286,21
ISO (I)			383,36
ISO (A)	295,24	12471,45	327,21
EPOXY (I)			520,77
EPOXY (A)	321,06	12124,79	336,40

4-3 Interpretation

Generally after an ageing, Ar% increases (2),which is surprisingly, not the case for our pipes : a loss of 10% of Ar% is observed on ORTHO humid samples, 5% for humid ISO

samples, +1% for EPOXY but this latter results is not very significant. Regarding the utltimate stresses Xr, the loss is -42% for the ORTHO, -22% for the ISO and -19% for the EPOXY pipes. Looking at the results obtained for "dried samples", it appears that the ultimate stresses increase compared to the humid state after ageing.

So the % of loss due to the plasticization with water is as follows :
- 15% of loss for Xr (ORTHO pipe)
- 8% of loss for Xr (ISO pipe)
- 7,5% of loss for Xr (EPOXY pipe)

But why the mechanical characteristics don't totally recover ? in fact no chemical degradation is observed (hydrolysis, osmosis...) during the first 1000 h of ageing (except maybe for the ORTHO resin); this has been confirmed by a previous study.

No degradation products was detected until 2000 hours of ageing in distilled water at 60°C. On the following Figure 4, is displayed the evolvement of E_{xx}, E_{yy}, and G_{xy} with time of ageing for a laminate, compared to the uptake of water. The chemical degradation begins initiating a significant loss of properties at t = 2000 hours of ageing. For ultimate stresses, the shape of the evolution is the same (5, 7).

A first decrease is observed from 0 hours to 1000 hours (5% for ISO, 10% for ORTHO), followed by a steady state up to 2000-3000 hours. The properties degradation begin decreasing up to 5000 hours, resulting in a final large decrease with

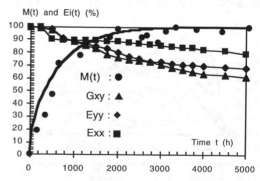

Figure 4 : Evolution of E_{xx}, E_{yy}, G_{xy} and M(t)% for a GFR polyester (quasi UD 90-10) immersed in water at 60°C during 5000 hours. (Modal analysis)

respect to the initial state. The decline of the moduli is more important in the transverse direction compared to the longitudinal direction as expected : the matrix and the glass sizing are preferentially attacked by water in the perpendicular direction, while the longitudinal fibres seems to keep their integrity. The decrease is especially high for ORTHO laminates, whose matrix is more sensitive to water than the ISO one.

The shear modulus G_{xy}, which particularly characterizes the intralaminar bond, also decreases for all the materials, and the fall is more important for ORTHO laminates, which are very sensitive to water.

The first decrease of the moduli is mainly assigned to the plasticization and the differential swelling between fibres and the matrix during the non steady state of water absorption at high temperatures. These two phenomenons are appearing with the uptake of water. Figure 5 represents a picture of cracks appearing at interface fibres-matrix, after 300 hours of ageing in distilled water while no large chemical attack has occurred.

The phenomenon of hygrothermal microcracks by differential swelling (acceleration of the phenomenon with the temperature) is increasing with the local volumic ratio of matrix and happens mainly on laminates and in the external surface of a pipe rather than in the internal one (local fibre ratio higher than on the external side). Under the transient absorption state, stresses of about 30 MPa have commonly been calculated.

However, to explain the irreversible loss of properties, one can suppose in addition that some molecules remain bonded to polar groups inside the polymer : first absorbed molecules of water are

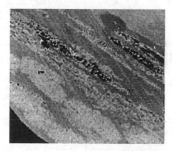

Figure 5 : Cracks appearing during the first 1000 hours of ageing, due to the differential swelling by absorption of water (GFR polyester laminate) - SEM picture

believed to exist in tightly bound state until all hydrogen bounds are replaced by water bridges; at higher concentration, the molecules of water are more weakly bound and serve as diluent, plasticizer (8).

4-4 Thermal properties

The loss of Tg due to water absorption has been widely theoretically studied and can be predicted (2, 9, 10, 11...), but the problem of measuring precisely Tg precisely in plasticized polymers having a Tg greater than the boiling point of water remains. That is to say that the results of thermal analyses exposed in Table 5 do not represent the real values of the Tg which may certainly be lower.

Table 5 : Values of Tg (°C) after ageing, for various pipes

Material	Initial State Tg(°C)	After ageing Tg(°C) "humid state"	After ageing Tg (°C) "dry state"
ORTHO	95,9	90,2	90,9
ISO	118,9	113,1	116,0
EPOXY	119,0	105,2	115,6

The results show that the ORTHO resin begins being hydrolyzed very early compared to the other materials, which recover partly their thermal properties. The dramatic effect of the plasticization appears clearly for the Epoxy resin regarding the thermal properties.

The real Tg may be under 100°C; therefore the max temperature for using those pipes has to be under 80°C, for preventing creep and formation of cracks (2,14), alterations of viscoelastic properties (12, 13).

Indeed, it has been shown that the effect of plasticization with a depression of Tg is equivalent to subjecting the laminate to a higher service temperature with a serious effect on the creep compliance (15). A loss of 10 to 20°C is observed for common epoxys with an uptake of water of 0,6 to 1% (9, 16).

5- PLASTICIZATION AND CALCULATIONS

Because of the difficulty to measure all the mechanical, hygral and thermal properties of the laminates, it's necessary to predict some of them.

For example, the halpin-Tsaï equation which combines parallel and serial springs is a useful and sufficient tool for the prediction of the stiffness and the non mechanical deformations as shrink, thermal and hygral expansion.

Now, these equations are mainly used in linear mechanics but the models can also be used with a non linear behaviour.

An epoxy resin has for example a linear behaviour in tension or even in compression, but its shear behaviour has to be smoothed with a Ramberg-Osgood or a form similar to a Mooney-Rivlin one.

With the models of the basic nonlinear behaviour of the plies, we can also include the effects of the plasticization in time by the evolution of this basic behaviour.

Finally, the classical first and last ply failure will change versus time with the internal stresses and the rate of internal moisture.

These curves are obtained with the Tsaï-Wu criterion for the fist ply failure, then in modifiing the matrix behaviour we calculate the last ply failure with the same criterion or the max strain one. With time, these enveloppes change in shape and dimensions with the plasticization, moisture absorption or desorption.

This non linear behaviour is presently studied in CETIM. In some cases, the last failure envelope is closer to the service conditions after plasticization : for an epoxy pipe, regarding the ultimate stress Xr, this latter one can vary along the axis (●) as shown on Figure 7.

6- CONCLUSION

This study focuses on the effect of water during the non-steady state of diffusion on the properties of a pipe : the main damaging effects are the plasticization (of the laminate resin) - inducing a loss of 10 to 25% of loss of mechanical characteristics- and the differential swelling, increasing internal stresses lead to cracks and to a loss of 5 to 10 % of the characteristics). The presence of water produces a lowering of the glass transition temperature and the loss is greater for the Epoxy material than for the Polyester ones. It remains important to consider that the characteristics of a laminate or a pipe may be affected (loss of properties, recovering of some properties after desorption) during their life-time by the plasticization of the material as soon as the latter one is in contact with water and take into account the non-linear and non steady state for the prediction of the short-term and long term behaviour.

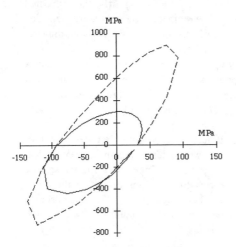

Figure 6 : Example of the first and last ply failure at time t

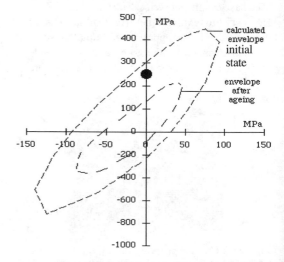

Figure 7 : Variation of the failure envelope due to reversible change of the matrix properties (plasticization with water).

7- BIBLIOGRAPHY

(1) : C.H. Shen, G.S. Springer, *J. Composite Materials,* 1, 11, 2, (1977)

(2) : J. Verdu; in Vieillissement des plastiques, Afnor Editions, (1984)

(3) : P. Bonniau et al, in Composite Structure, Ed. I.H. Marshall, App. Sci. Pub., (1981)

(4) : Ph. Castaing, H. Mallard, *Proc. of ECCM-CTS-2*, p 437, (1994)

(5) : Ph. Castaing, Thesis n° 621, National Polytechnic Institute of Toulouse (INPT) (1992)

(6) : J. Cranck, in The mathematics of diffusion, 2ème Ed, Clarendon Press, Oxford (1976)

(7) : Ph. Castaing et al, *Proc. of Int. Conf. Comp. Mater. N°9*, Sess.A - Env 2, 577, (1993)

(8) : Z.A. Mohd Ishak et al, *Polymer Composites,* 15, 3, 223, (1994)

(9) : A. Chateauminois, Thesis n°24291, University of Lyon I, (1991)

(10) : M.L. William et al, J. *Amer. Chem. Soc,* 77, 3701, (1955)

(11) : H.G. Carter et al, *J. of Comp. Mat,* 11, 1643, (1977)

(12) : CE. Browning, *Polymer Engineering & Sci.*, 18, 1, (1978).

(13) : R.S. Chen et al, *J. of Composite Materials,* 27, 16, (1993)

(14) : P.C. Upadhyay et al, *J. of Reinforced Plastics and Composites*, 13, 1056, (1994)

(15) : E.M. Woo, *Composites,* 25, 6, 425, (1994)

(16) : B. Mouhamath, Thesis, ENSMP, (1992)

Progress in Durability Analysis of Composite Systems, Cardon, Fukuda & Reifsnider (eds)
© 1996 Balkema, Rotterdam. ISBN 90 5410 809 6

Hydrothermal and shear loading of polymer composites

R. D. Adams & M. M. Singh
Department of Mechanical Engineering, University of Bristol, UK

ABSTRACT: A flexibilised epoxy and its carbon-fibre reinforced composite were exposed to humid conditions, and the resulting changes in the shear modulus and loss factor at a temperature well below the glass transition temperature were determined. Shear stress was applied to "as-received" composite samples under ambient conditions, and to samples that were saturated with moisture at 75% relative humidity under the same humid conditions. The moisture uptake in the composite was found to be more per unit mass of matrix material than that in the unreinforced resin. Both absorbed moisture and applied stress each reduced the longitudinal shear modulus of the composite; the application of shear stress to already saturated samples caused further degradation. Prediction of changes in the properties of the composite cannot readily be made from the observed changes in the resin matrix properties under the same conditions.

1. INTRODUCTION

In considering the durability of a composite system, there are two complementary approaches. Firstly, there is the need to be able to predict the long-term properties of the system from results of accelerated or short duration tests, so that its expected life-time can be assessed. The second approach is to find a nondestructive method of determining the state of the material whilst in service, so that its imminent failure can be detected, before any serious damage is caused. The latter is not an easy task, so, at present, the first approach is more usually adopted. With this end in view, the effect of moisture and of applied shear stress on the shear properties of a polymer matrix composite material have been investigated.

2. OBJECTIVES

The aims of this programme were to determine the effects of both moisture and applied shear stress on the shear modulus and loss factor of a carbon-fibre reinforced flexibilised epoxy. To assess the extent to which the properties of the composite could be predicted from the long-term properties of its constituents, samples of the unreinforced matrix resin were also exposed to the same conditions of moisture.

3. MATERIALS

The composite material studied was a flexibilised epoxy system, based on Bisphenol A, reinforced with unidirectional carbon fibre. The fibre volume fraction was approximately 0.4. The glass transition temperature of the unreinforced epoxy was about 28°C.

The composite laminates were between 2.7 mm and 3.2 mm thick, and unreinforced resin samples were of 2 mm nominal thickness. The width of all test-pieces was 12.7 mm, but the lengths varied.

4. EXPERIMENTAL PROCEDURE

4.1. *Conditioning*

Samples were exposed to conditions of 98%, 75%, 50% and 0% relative humidity (R.H.) at 40°C until they reached an equilibrium condition. The humid conditions were produced by placing the samples in a sealed container over the appropriate saturated salt solution. The dry conditions were achieved using self-indicating silica gel. A temperature of 40°C was chosen, rather than ambient temperature, in order to increase the rate of moisture absorption, without causing any damage to the materials. This temperature was maintained by placing the conditioning containers inside an oven.

A shear stress of 1.25 MPa was applied in an atmosphere of 75% R.H. at 20°C to samples that had been conditioned at 75% R.H. and 40°C, and to unconditioned control specimens under ambient conditions (approximately 50% R.H. and 20°C). The applied shear stress was maintained for 30 days.

4.2. *Moisture absorption characteristics*

The moisture uptake was monitored as a function of time by periodically weighing the specimens on a Sartorius lever balance precise to 0.1 mg. The moisture content, M, was given by:

$$M = \frac{m_t - m_0}{m_0} \times 100\% \qquad (1)$$

where m_t, is the mass of the sample after time t, and m_0 the dry mass. The coefficient of diffusion, D, was derived from curves of moisture content against the square root of time, as:

$$D = \pi \left(\frac{h}{4M_\infty}\right)^2 \left(\frac{M_2 - M_1}{\sqrt{t_2} - \sqrt{t_1}}\right)^2 \qquad (2)$$

where h is the thickness of the specimen, M_1 and M_2 are the moisture content after times t_1 and t_2, respectively, during the initial part of the absorption process, and M_∞ is the moisture content at saturation. This value was then corrected using Shen and Springer's (1976) correction factor to allow for the fact that the samples were of finite length and breadth.

4.3. *Shear modulus and loss factor measurement*

The shear modulus was measured statically in a purpose-built static torsion machine, and dynamically with a small torsion pendulum (Adams & Singh 1990). The shear loss factor was determined, using the half-power bandwidth method (Thomson 1965). Since the material was viscoelastic, the dynamic properties were both frequency- amd temperature-dependent. To minimise frequency effects, all the tests were carried out at -30°C, well below the glass transition temperature of the matrix resin. To achieve this temperature, the test rigs were placed in a cryogenic chamber. This was first evacuated, then filled with dry nitrogen gas, after which it was sealed and cooled with liquid nitrogen. The materials were tested firstly after storage under ambient conditions, and secondly after conditioning, when they had reached an equilibrium state. For each conditioning

state, three samples were tested, and mean values of the moduli and loss factor taken.

4.4. *Shear loading*

Shear loads were applied to fibre-reinforced composite samples using a compound specimen illustrated in Figure 1. The composite test-pieces were bonded into grooved stainless steel fixtures, and the assembly was then placed under load in a creep test machine. A stress of 1.25 MPa, 5% of the shear strength of the material, was maintained for 30 days both on as-received samples under ambient conditions, and on samples saturated at 75% R.H. in an environment of the same R.H. To maintain this humidity, test tubes containing saturated sodium chloride solution were fitted into a collar around the loading bar. The collar was pushed into a perspex tube, being sealed with an O-ring, and a similar arrangement sealed the tube at the top.

Measurements of the shear modulus and loss factor were made on the as-received material, then after saturation at 75% R.H., before being stressed,

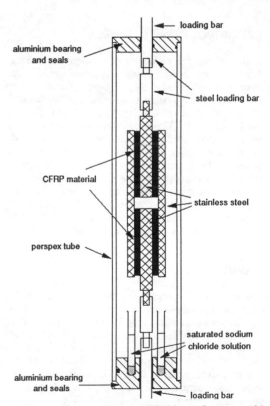

Fig.1. Shear stress application fixture, with environmental chamber

and again after 30 days under load, when the test-pieces were cut from the stressing fixture.

4.5 *Interlaminar shear strength measurements.*

The short beam shear test yielded no meaningful results for the interlaminar shear strengh of these materials, owing to their high ductility at room temperature. The interlaminar shear strength was measured, therefore, after conditioning, using a modified "guillotine" test (ASTM 1970). Three samples in each state were tested, and the mean value taken.

5. RESULTS

5.1. *Moisture absorption characteristics*

The moisture absorption curves of the composites are shown in Figure 2, where the moisture content is plotted against the square root of the exposure time. The lines represent the Fickian diffusion curves that best fit the experimental data. Similar curves were obtained for the unreinforced resin.

As expected, the higher was the R.H., the greater was the equilibrium moisture content.

The measured coefficients of diffusion, D, after applying Shen and Springer's correction, are given in Table 1. Also given are the saturation moisture contents, M_∞, of both the reinforced and unreinforced materials together with the apparent saturation moisture content of the *matrix* in the composite, if it is assumed that all the observed increase in mass is due to moisture absorbed by the matrix.

Table 1. The moisture absorption characteristics of a flexibilised epoxy resin and its carbon-fibre reinforced composite.

Material	Relative Humidity (%)	M_∞ (%)	D (mm²/s)
Unreinforced resin	50	1.18	1.3×10^{-6}
	75	2.35	1.5×10^{-6}
	98	4.7	1.2×10^{-6}
Carbon fibre reinforced composite	50	0.62	7.4×10^{-8}
	75	1.25	1.0×10^{-6}
	98	2.85	1.0×10^{-6}
Composite matrix	50	1.38	
	75	2.78	
	98	5.68	

It can be seen that the moisture content of the matrix in the composite was greater than that of the unreinforced resin under the same conditions.

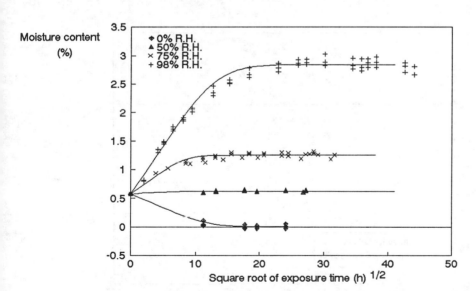

Fig.2. Moisture absorption curves for carbon fibre reinforced flexibilised epoxy at 40°C

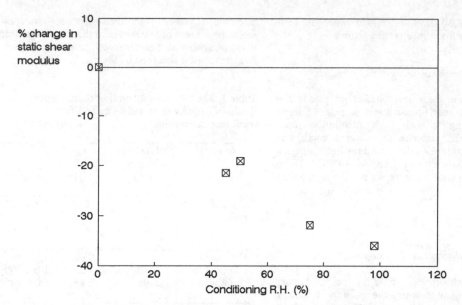

Fig. 3. Changes in static shear modulus at -30°C against conditioning R.H.

5.2. *Changes in shear modulus at -30°C due to moisture*

The changes in the static and dynamic shear moduli, expressed as a percentage of the shear moduli of the dried materials, are shown in Figures 3 and 4, respectively. The static modulus of the dried CFRP was 3.14 GPa at -30°C. At the same temperature, the dynamic modulus of the composite was 3.76 GPa at 38 Hz, while that of the unreinforced resin was 1.30 GPa at approximately 20 Hz. The static modulus of the unreinforced resin was not measured, owing to the difficulty of setting up this extremely flexible material in the static torsion machine.

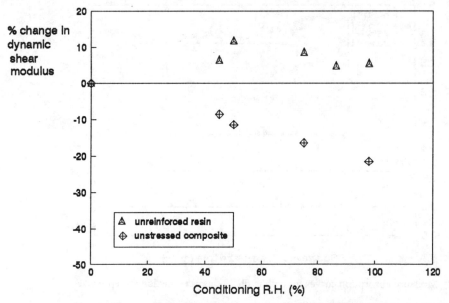

Fig. 4. Changes in dynamic shear modulus at -30°C against conditioning R.H.

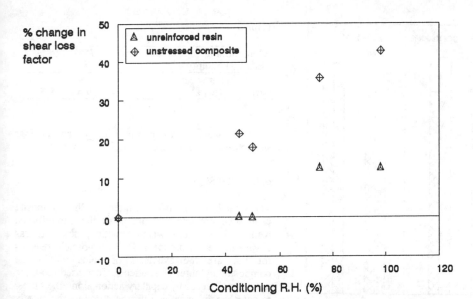

Fig. 5. Changes in the shear loss factor at -30°C against conditioning R.H.

It can be observed that the dynamic modulus of the unreinforced resin was not significantly affected by the absorbed moisture; if anything, there was a small increase. That of the composite, however, decreased by 21.5% when exposed to 98% R.H. The static modulus of the composite was reduced by 36% of its dry value when exposed to these conditions.

5.3. Changes in loss factor at -30°C due to moisture

The changes in loss factor at -30°C are shown in Figure 5, as a percentage of the loss factor of the dried materials, which were 0.048 and 0.028 for the unreinforced and carbon fibre reinforced resin at 20 Hz and 38 Hz, respectively. Again it can be seen that after exposure to 98% R.H. and 40°C, the loss factor of the matrix rose by only 13%, while that of the composite increased by 43%.

5.4. Changes in shear properties at -30°C due to applied stress

The measured values of the shear properties before and after the application of stress (when the applied stress was removed) are given in Table 2, and illustrated graphically in Figures 6 and 7.

It can be seen that the application of stress alone under ambient conditions caused a fall in the shear moduli, of the same relative magnitude as that caused by the ingress of 1.25% moisture (at 75% R.H.). A small fall in the loss factor was observed. When the

Table 2. The effect of applied shear stress on the longitudinal shear properties at -30°C of carbon fibre reinforced flexibilised epoxy.

Condition	Static shear modulus (GPa)	Dynamic shear modulus (GPa)	Loss factor
As received	2.5	3.2	0.035
As received + applied shear stress	2.4	3.1	0.033
After conditioning at 75% R.H.	2.8	3.1	0.038
After conditioning at 75% R.H. + applied shear stress	2.1	2.8	0.040

shear stress was applied in a humid environment, to already saturated samples, the already reduced moduli of the material were decreased further. There was a corresponding slight increase in loss factor. There was also the suggestion, from density measurements, that more water was absorbed when the saturated material was under stress. This has yet to be confirmed, however.

5.5. Changes in interlaminar shear strength due to moisture

The results of the guillotine shear tests performed after conditioning the specimens, are given in Table

245

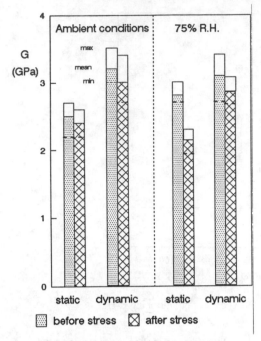

Fig. 6. Measured shear moduli of CFRP before and after the application of shear stress

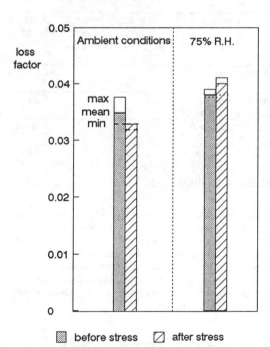

Fig. 7. Measured shear loss factor of CFRP before and after the application of shear stress

Table 3.

Conditioning relative humidity	0%	50%	75%	98%
Interlaminar shear strength (MPa)	25.5	14.4	9.4	8.8

3. It is clear that there was a loss of interlaminar shear strength with increased moisture content.

6. DISCUSSION

The equilibrium moisture content of the composite cannot be predicted from that of the unreinforced resin and the fibre weight fraction, since it was always higher than expected. Two possible reasons for this are the existence of voids within the composite and interfacial effects. Increased moisture uptake may occur by capillary action along the fibres, rather than by simple diffusion through the matrix material. In addition it is possible that there is an "interphase" region surrounding the fibres which is more hygroscopic than the unreinforced matrix material.

The observed degradation of the composite properties is more consistent with the the latter premise. If the extra water merely collected in voids, then it might be expected that, as in the unreinforced material, there would be no significant loss of shear modulus. The loss factor would be likely to be reduced since the water may act as a lubricant between surfaces that, in the dry material, might well introduce frictional losses during small amplitude vibration. The fact, therefore, that there is an increase in loss factor and a reduction in shear modulus when moisture is absorbed by the composite, suggests that this additional moisture is not simply free water. The degradation of the longitudinal shear properties implies a weakening of the fibre-matrix interface.

This is corroborated by the fact that there was a reduction of the interlaminar shear strength when moisture was absorbed.

That the application of shear stress caused a degradation in the shear properties further suggests that it is the interface that is affected. Further experiments are, however, necessary to verify this. Firstly, unreinforced material should also be subjected to shear stresses, and the resultant changes in the shear properties compared with those of the composite. Secondly, the moisture absorption characteristics of the material under stress should be carefully monitored. Thirdly, different levels of stress, sustained for different periods should be applied. In addition, the change in properties of dry material, stressed in dry, rather than ambient, conditions should be investigated, to eliminate completely any effect of moisture on the material.

7. CONCLUSIONS

Comparison of the effects of conditioning in a humid atmosphere on the unreinforced and reinforced materials, suggests that the interface between fibres and matrix plays a signifcant role in the degradation of the composite properties. This means that it may not be possible to predict the long-term behaviour of a composite from a knowledge of the response of its consituent materials to the expected service conditions.

These experiments have indicated that applied shear stress is as damaging to the shear performance of a carbon-fibre reinforced polymer composite as is absorbed moisture. The effects of the design service loading must also be taken into account, therefore, when considering the durability of a composite system.

ACKNOWLEDGEMENTS

The financial support of both the Royal Academy of Engineering and the Defence Research Agency for this work is gratefully acknowledged. The authors would like to thank, in particular, members of staff at the Defence Research Agency at Holton Heath, Dorset, UK for preparing the materials and some of the specimens.

REFERENCES

Adams, R.D. and M.M. Singh 1990. The effect of exposure to hot, wet conditions on the dynamic properties of fibre reinforced plastics. *Proc. Durability '90*. Brussels.

American National Standard ANSI/ASTM D2733-70 1970 (reapproved 1976). Interlaminar shear strength of structural reinforced plastics at elevated temperatures.

Shen, C-H and G.S. Springer 1976. Moisture absorption and desorption of composite materials. *J. Composite Materials* 13: 2-

Thomson, W.T. 1965. *Vibration theory and applications*. London: Allen and Unwin.

Progress in Durability Analysis of Composite Systems, Cardon, Fukuda & Reifsnider (eds)
© *1996 Balkema, Rotterdam. ISBN 90 5410 809 6*

Moisture-matrix interactions in polymer based composite materials

G. Mensitieri, A. Apicella & L. Nicolais
University of Naples 'Federico II', Department of Materials and Production Engineering, Italy

M. A. Del Nobile
National Research Council, Institute of Composite Materials Technology, Naples, Italy

ABSTRACT: Fiber reinforced polymeric composites are currently based both on thermoset polymers like epoxy or polyester resins and on advanced thermoplastic materials like polyetheretherketone (PEEK), polyetherimide (PEI) and polysulfone (PSU). Sensitivity of polymer matrices to the presence of moisture and solvents and to thermally induced stresses is of primary importance in determining the durability of polymer based composite structures. In fact, moisture and solvents can greatly affect not only mechanical properties of the matrix per se, but also its adhesion properties to the reinforcing fibers. The mechanisms through which sorbed moisture and solvents can act are reviewed in this paper. These mechanisms include interaction with specific groups and penetrant induced plasticization and are deeply examined in the case of epoxy, polyester and PEEK resins.

1 INTRODUCTION

Fiber-reinforced polymeric composites are expected to be exposed, over a wide range of temperatures, to mechanical fatigue as well as to extreme environmental conditions. Polymer-based materials such as fiber composites differ from other structural materials in that low molecular weight substances may easily migrate in them even at ambient temperature determining in many cases a degradation of matrix mechanical properties. The environmental degradation of the mechanical properties of the matrix has been reported by several authors [Apicella et al. (1982a), Morgan et al. (1982), Morgan et al. (1977), Apicella et al. (1981a), Chi-hung et al. (1976), Apicella et al. (1984a), Loss et al. (1979), Pogany (1976), McKgue et al. (1975), Apicella et al. (1979), Apicella et al. (1982b)] and, in particular, was associated by Apicella et al. [Apicella et al. (1982a), Apicella et al. (1981a), Apicella et al. (1984a), Apicella et al. (1979), Apicella et al. (1982b)] with the plasticization and micromechanical damage induced by the synergistic effects of temperature, stress and sorbed solvents. Environmental resistance, mainly to hygrothermal fatigue under external loading, is one of the most important matrix properties. Fiber reinforced plastics (GRP) undergo a slow degradation process when subjected to the physical and chemical ageing due to water molecules sorption. The degradation mechanism may act on the fibers, the interfaces and the matrix.

The matrices actually used for polymeric composites are generally restricted to the class of thermosets such as epoxy or unsaturated polyester resins. The epoxy matrices show strong moisture sensitivity due to strong interactions between some groups of the macromolecule and the water molecules leading to strong depression of the glass transition temperature and of the mechanical properties [Moy et al. (1980), Banks et al. (1979), Ellis et al. (1984), Ellis et al. et al. (1983), Ten Brinke et al. (1983)]. Exposure of polyester resins to water at different temperatures induces modifications of the mechanical properties and morphology of the material. This environmental sensitivity of thermoset matrices coupled with their low impact resistance and damage tolerance favored the development of temperature and environment stable thermoplastics. Preliminary investigations [Mensitieri et al. (1989)] on high performance thermoplastics, such as poly(aryl-ether-ether-ketone)(PEEK), proved their high resistance to water plasticization. PEEK is a tough aromatic polymer with attractive high temperature performances. This semicrystalline polymer has a melting point of about 335 °C and a glass transition of about 145 °C [Stober et al. (1984), Blundell et al. (1981), Grayson et al. (1987)]. The morphology and related properties of PEEK are well described in the work of Blundell and Osborne [Blundell et al. (1983)].

In order to elucidate the effect of moisture on material properties, both the thermodynamic aspect of water sorption and kinetic aspects of its diffusion through the polymer have to be analyzed and related to the molecular and morphological characteristic of the polymer. The generally accepted mechanism for penetrant sorption in polymers is an activated sorption-diffusion process [Meares (1957), Vieth et al. (1977), Carfagne et al. (1982), Michaels et al. (1963), Apicella et al. (1985), Apicella et al. (1984b), Apicella et al. (1981b), Apicella et al. (1981c), Carfagna et al. (1983)]. The molecules are first dissolved into the polymer surface and then diffuse through the bulk of the polymer by a series of activated steps.

2 SORPTION EQUILIBRIUM

On the base of thermodynamic grounds the amount sorbed in a polymer can be related to its activity level. The type of relationship depends upon the morphology and state of polymer. The types of equations used to analyze sorption equilibrium are reported in literature [Ellis et al. (1984), Ellis et al. (1983), Ten Brinke et al. (1983), Meares (1957), Vieth et al. (1977), Carfagne et al. (1982), Michaels et al. (1963), Apicella et al. (1985), Apicella et al. (1984b), Apicella et al. (1981b), Apicella et al. (1981c), Carfagna et al. (1983), Mensitieri et al. (1995)].

2.1 *Moisture sorption equilibrium in epoxy systems*

The existence of different possible sorption modes is particularly evident for the epoxy systems, and it can be also argued from the different levels of plasticization as determined by Apicella et al. [Apicella et al. (1981a), Apicella et al. (1984a), Apiccella et al. (1979), Carfagna et al. (1982), Apicella et al. (1983)]. The plasticization effect is illustrated in figure 1a, b. In figure 1a are reported the torsion modulus for an epoxy matrix based on tetraglycidyl 4,4' diamino diphenyl methane-4,4' diamino diphenyl sulfone (TGDDM-DDS) at DDS curing agent concentration of 20 PHR for water saturated (at 20 °C and 60 °C) and dessicated samples. In figure 1b are reported the differential scanning calorimeter (DSC) curves for wet and essicated TGDDM-DDS.

The equilibrium water vapor sorption isotherms [Apicella et al. (1985), Apicella et al. (1984b), Apicella et al. (1981b)], reported in figure 2, have been utilized to investigate the actual mechanisms of sorption. The sorption isotherms of widely utilized epoxy matrix (TGDDM-DDS) equilibrated at low external water activities displayed a region of downward concavity around 1% uptake that can be attributed to the saturation of hydrophilic sites. The phenomenon occurred at higher activities for epoxies

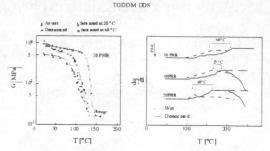

Figure 1: (a) Mechanical characterization of TGDDM-DDS epoxy resin at DDS curing agent concentration of 20 PHR: (o) 'as cast' dry, (Δ) desiccated, (●) saturated at 60 °C, (▲) saturated at 20 °C; (b) Calorimetric characterization of wet and dessiccated TGDDM-DDS epoxy resin at DDS curing agent concentration of 20, 30 and 50 PHR.

containing higher concentration of amino hardeners [Apicella et al. (1985)], leading also to higher plasticization (glass transition temperature depressions $(\Delta Tg) > 80$ °C).

The tendency for sorption by hydrogen-bonding associated with initial negative curvature of the stiffer TGDDM-epoxy systems has been related by Apicella et al. [Apicella et al. (1984a), Apicella et al. (1979)] to the presence of a greater amount of unreacted secondary amines. The equilibrium sorption isotherms experimentally determined for systems in which polymerization and cross-linking exhausted all the primary and secondary amines as in the case of systems based on diglycidyl ether of bisphenolA cured with a stoichiometric ammount of triethylen tetramine (DGEBA-TETA) [Apicella et al. (1981c)], showed in fact, an almost linear initial behavior [Apicella ct al. (1984b)] (see figure 4) and lower plasticization levels (glass transition temperature depressions $(\Delta Tg) \sim 20\text{-}40$ °C).

The increase of the glass transition temperature of

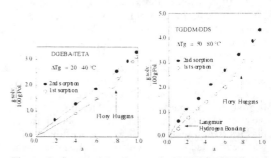

Figure 2: Sorption isotherms in DGEBA-TETA and TGDDM-DDS systems (o)1st sorption, (●) 2nd sorption.

some epoxies is probably related to the formation of strong intermolecular hydrogen bonding which stiffens the polymer network. The higher degrees of plasticization following moisture equilibration of these systems are a consequence of the breakdown of hydrogen bonds (intermolecular hydrogen bond formation) in addition to the ordinary solution formation.

The shape of the sorption isotherms may be, therefore, represented by a dual sorption model, even if the tendency of water to cluster should be considered in order to account for the hygrothermal hysteresis reported in figure 2.

The second sorption cycle, as also reported by Barrie et al. [Barrie et al. (1984), Barrie et al. (1985)], is affected by a significant increase in the regain over the all range of activities. The differences between 1st and 2nd cycle in sorption behavior at the same external conditions have been explained in terms of progressive damage produced in inhomogeneous materials, such as thermosets characterized by locally different cross-linking densities or in composites equilibrated at increasingly higher moisture contents (> 50% R.H.). The most important consequence of such inhomogeneous morphology (from the standpoint of hygrothermal fatigue) is the difference in the rate of moisture diffusion, hence the differential swelling stresses that can arise between regions of the polymer network. A sudden increase in temperature (such as the thermal spikes reported in reference 5) during exposure to humid environments of epoxy based materials will favor clustering of the water molecules initially dissolved in the bulk material [Carfagna et al. (in presss)]. A temperature increase, therefore, produces either an over saturation of the material or an additional micro voiding due to the fast formation of water clusters.

The failure of this excess water to induce further plasticization is discussed in the next section. The amount of water actually dissolved in DGEBA-based systems equilibrated at different temperatures was determined in independent tests [Apicella et al. (1984b)] and has been found to be higher at lower temperatures, whereas micro voiding, which is favored by the higher temperatures progressively increased.

2.2 Thermoplastic PEEK

The amount of water sorbed in thermoplastic PEEK is reported [Mensitieri et al. (1989)] to be much lower than in the case of epoxies and does not involve any significant interaction among polymer backbone and penetrant molecules. Water sorption isotherm reported by Mensitieri et al. [Mensitieri et al. (1989)] at 60 °C for an amorphous PEEK sample equilibrated at different relative humidities is compared in figure 3

with the corresponding behavior of a TGDDM-based high performance epoxy matrix [Apicella et al. (1985), Apicella et al. (1984b)]. The theoretical curves evaluated through Flory-Huggins equation [Mensitieri et al. (1989)] are also reported. The sorption data indicate that for a moisture-saturated environment, the expected equilibrium value is close to 0.70% and is in agreement with the value determined in liquid water sorption. No changes in the matrix morphology such as those observed for the epoxy systems were detected during the sorption and resorption cycles. For semicrystalline PEEK immersed in liquid water, a mean equilibrium value (0.48%) only weakly dependent on temperature, is reported in reference 17.

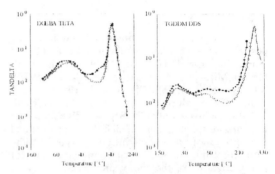

Figure 3: Dynamic-mechanical analysis of dessiccated epoxy systems.

3 SOLVENT SORPTION AND POLYMER PLASTICIZATION

Plasticization is a process of depression of the glass transition temperature and reduction of the mechanical properties associated with the sorption of moisture or, more generally, of a solvent. The plasticization of the network structure depends on the dilution process as well as on intermolecular bond formation. The former is governed by the diluent volume fraction v_d and by the incremental change of the thermal expansion coefficients at the glass transition, α_d and α_p, through the well known expression [Kelly et al. (1961)] based on free volume concepts:

$$T_{gw} = \frac{T_{gp} \cdot \alpha_p \cdot v_p + T_{gd} \cdot \alpha_d \cdot v_d}{\alpha_p \cdot v_p + \alpha_d \cdot v_d} \quad (1)$$

where the subscripts p and d refer to the polymer and the diluent respectively.

Alternative theoretical expressions have been developed by Ellis, Karasz, et al. [Ellis et al. (1984), Ten Brinke et al. (1983)] using classical thermodynamic treatments to describe the compositional dependence of the glass transition in miscible blends, further extended also to the epoxy-water systems. The studies carried out on relatively low glass transition DGEBA-epoxy resins have shown that the plasticization induced by water sorption may be predicted by means of the following equation:

$$T_{gw} = \frac{T_{gp} \cdot C_{pp} \cdot x_p + T_{gd} \cdot C_{pd} \cdot x_d}{C_{pp} \cdot x_p + C_{pd} \cdot x_d} \quad (2)$$

where x is the weight or mole fraction and C_p is the incremental change in the specific heat at the glass transition.

3.1 *Effect of moisture on epoxy matrices. Plasticization mechanisms*

The chemical structure of the prepolymer constituents and the processing conditions have been described by several authors [Apicella et al. (1983), Apiccella (1986), Gillham (1979), Lewis et al. (1979), Mijovic et al. (1981), Morgan et al. (1978), McKague et al. (1978), Mijovic et al. (1985), Antoon et al. (1981), Morgan et al. (1979), Peyser et al. (1981), Delasi et al. (1978)] as influencing the network morphology, and hence the properties and durability, of a cross-linked polymer. The investigation of the actual mechanisms of polymerization of the epoxy prepolymers gives a useful insight into the chemistry of the system, hence into the potential interactions that can arise between the penetrant molecules and the polymer. Chain extension and cross-linking of the epoxy resins are due to the reactions of the epoxidic groups between themselves and with the hydrogens of donor compounds, such as amines and organic acids, which are usually used as hardeners. An additional cause of reduced hygrothermal stability is premature formation of a glassy solid occurring as a consequence of the network becoming denser through further intramolecular cross-linking, which prevents the full curing of the system, since it strongly reduces the mobility of the still unreacted functional groups. The existence of differences in properties between desiccated and soaked samples reinforces the concept that physical modifications are introduced into the polymer network by moisture and thermal aging. Although the exact nature of this change has not been completely identified, we proposed previously an explanation related to the micro voiding and crazing of polymer and assuming a dual state of water in the polymer.

Differential scanning calorimeter tests performed on wet and desiccated TGDDM-DDS samples showed greater plasticizations for systems with a larger amino content, even though the same amount of water was sorbed by the systems (i.e. ≅ 5%) [Morgan et al.

(1982), Apicella et al. (1985)]. Another interesting effect of compositional changes on the plasticization of these systems has also been reported [Apicella et al. (1982b)]. Samples of intermediate composition showed lower degrees of plasticization than systems containing larger or smaller amounts of amino hardeners. An increase of the concentration of the hardener was reported to increase the density or the thermal expansion coefficients of TGDDM-DDS systems [Apicella et al. (1982b), Apicella et al. (1985), Apicella et al. (1983)]; according to these values, one can expect an increase in the hardener concentration to result in systems of higher hygrothermal stability. Conversely, anomalously high plasticization were experimentally observed, especially for the systems richer in hardener. This result indicates that the diffusion and the related plasticization process become less important as the hardener concentration is increased but, as previously discussed, while the network is becoming more tightly cross-linked, a larger number of secondary amines can remain unreacted. For the systems containing lower amounts of amines, it was assumed [Apicella et al. (1984a), Apicella et al. (1982b)] that plasticization occurred primarily by dilution, while the influence of hydrogen bonding on hydrophilic sites was found to have lower relevance. Conversely sorption and plasticization of materials richer in amino-hardener are described to be primarily related to the presence of unreacted amines.

Ellis and Karasz have theoretically calculated from equation [Loss et al. (1979)] glass transition depressions of 10-15 °C for each percent of water uptake for the less cross-linked DGEBA systems and of 25 °C/wt % for the TGDDM-based systems, using a value of C_{pd} for water of 1,94 J/g K [Keenan et al. (1979)] and values ranging from 0,34 to 0,35 J/g K for the epoxies. These expectations were generally confirmed by their experiments on samples equilibrated at high temperatures. High plasticizations are experimentally observed [Apicella et al. (1984a)] for these commonly used epoxy systems. Moisture uptakes of 2-5% [AApicella et al. (1984a), Apicella et al. (1981b)] lead to significant drops of the glass transition of wet samples from 150-200 °C down to 60-80 °C, well inside the application limits of most epoxy-based composites. Some tests on high performance carbon fiber composite laminates obtained from commercial-type TGDDM-based pre impregnated materials have indicated, for samples equilibrated in water at low temperatures (i.e., 5% water uptake), calorimetrically measured T_g depressions higher than 100 °C.

Dynamic mechanical tests clarified the actual mechanism of plasticization and further reinforced the basic assumption of the proposed sorption and plasticization mechanisms. DGEBA-based epoxies and TGDDM-based epoxies exhibit the high and low temperature transitions reported in figure 3, which are associated, respectively, with the sample main glass transition and a quite broad series of low temperature secondary transitions. The broad low temperature transition area indicated a wide spectrum of motion types and activation energies contributing to the transition. The water conditioning of DGEBA and TGDDM resins resulted in an increase in magnitude of the transition and a slight drop in the temperature location of the transition, observed around 20-25 °C, with increasing moisture content. By plotting the areas under the loss compliance transition as a function of moles of water per mole of initial nitrogen content for epoxies equilibrated at different moisture levels, it has been confirmed that the DGEBA system shows the saturation of the available amino-hydrogen bonding sites to occur at lower moisture content than the TGDDM system [Carfagna et al. (in press)]. Dynamic mechanical tests such as those reported in figure 3 hence support the assumption that the sorption by hydrogen bonding is more pronounced for TGDDM systems than for DGEBA systems.

Sorbed moisture may be present in the polymer network in different forms associated with an equilibrium content that is dissolved or bonded in the compact resin and "holes". Assuming that only the former contributes to the plasticization of the network, the higher microcavitation expected to occur at high temperatures will produce higher water uptakes, even though the actual water solubility is a decreasing function of the temperature.

3.2 Thermoplastic PEEK

PEEK has been found [Mensitieri et al. (1989)] to have good moisture and liquid water resistance. In fact, only a slight depression of Tg (2 °C) is reported for samples equilibrated with water at unit activity at 60 °C. While the amorphous PEEK network is essentially formed by the highly rigid repeating units, the epoxy matrices, such as those based on TGDDM resins, present rigid sections cross-linked by more flexible segments. The glass transition of the latter, however is higher ($\sim$ 200 °C) than that of the amorphous PEEK (145 °C), and this difference is probably induced by the presence in epoxy system of strong intramolecular hydrogen bonds, which, however, while stiffening the structure, increase its moisture sensitivity. The PEEK glass transition temperature, on the other hand, is influenced only by the sterically induced structure stiffness, and although PEEK possesses several oxygens that may potentially form hydrogen bonds, it is not moisture sensitive, since these atoms are sterically shielded by the aromatic rings.

4 DIFFUSION MECHANISMS

The possible multimodal sorption modes of a glassy polymer greatly influence the actual kinetic of sorption. Glassy polymers, in fact, generally exhibit a complex mass transfer behavior. Both concentration-gradient-controlled diffusion and relaxation-controlled swelling contribute to the rate and extent of penetrant sorption. The physical phenomena that simultaneously occur are dissolution, diffusion, swelling and relaxation, together with deformation and stress buildup in the matrix. A wide variety of effects can be therefore exhibited, depending on the mutual polymer-penetrant affinity, on the temperature, and the penetrant activity. Varying the temperature and the penetrant activity it is possible to meet the full range of behaviors, from ideal Fickian diffusion to limiting Case II (relaxation controlled) [Hopfenberg et al. (1969), Keenan et al. (1979), Nicolais et al. (1984), Berens (1977), Berens et al. (1978)].where sorption is characterized, respectively, by smooth concentration gradient (Fick's mode) and by the presence of a sharp boundary dividing an unpenetrated glassy core from an outer swollen and solvent-saturated shell (Case II), as also indicated in figure 4. Penetrant diffusion and solvent-induced polymer relaxation, in fact, are temperature activated processes characterized by significantly different activation energies. Figure 4 summarizes the possible

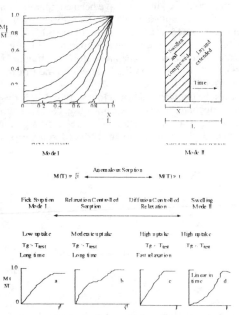

Figure 4:   Possible sorption kinetics in glassy polymers.

253

sorption kinetics reported as a function of square root of time, that can be experimentally observed for glassy systems exposed to environments held at different temperatures, penetrant activities, and types of solvents. As suggested by Crank [Crank (1975)], the complex problems of anomalous transport include time-dependent boundary conditions and diffusion coefficients, polymer relaxation providing the rate-determining transport step, and polymer crazing accompanying penetrant sorption.

4.1 Relaxation-controlled diffusion in amorphous PEEK and in epoxies

To describe the sorption behavior of some glassy matrices at higher activities, the model proposed by Berens and Hopfenberg [Berens et al. (1978), Crank (1975), Enscore et al. (1977)] can be used. This model considers the linear superposition of a Fickian diffusion contribution and a first order relaxation term and assigns to each mechanism a fixed fraction of the total sorbed solvent equilibrium amount. The amount of solvent sorbed is given by:

$$M(t) = M_F(t) + M_r(t) \qquad (16)$$

where $M_F(t)$ and $M_r(t)$ are respectively, the Fickian and relaxation contributions. This model provides a good interpretation of some experimental results obtained at high activities, as shown in figure 5 for epoxy and PEEK samples.

The long-time behavior and sorption equilibrium uptakes for epoxy samples exposed to a relatively low temperature (20 °C) in water have been compared with the equilibrium uptake reached by a sample that has been equilibrated at a high temperature (60 °C) and then brought back to the lower temperature. Figure 6 indicates that the low temperature slow sorption (> 2 years) tends to the same final value obtained in the accelerated test (few weeks).

It must be emphasized that at low activity levels this double mechanism usually is not observed. In such conditions, as a consequence of the reduced amount of sorbed penetrant, solvent-induced relaxations

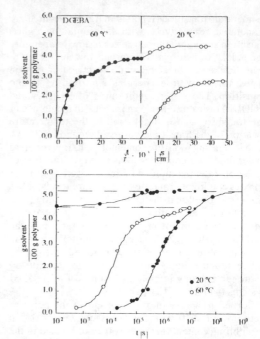

Figure 6: Short-time (diffusion-controlled) and long-time (relaxation-controlled) liquid water sorption behavior for an epoxy system, at 20 and 60 °C.

probably are too slow to be observed in the elapsed experimental time.

The PEEK samples, after exposure to the maximum water activity, were submitted to further sorption experiments at low and medium activities. No hysteresys effects were detected, since the values of equilibrium uptakes were practically the same as those obtained initially at the same activities.

The liquid water sorption experiments performed on semicrystalline PEEK films exhibited a purely fickian behavior. The water uptake values obtained for both amorphous and semicrystalline PEEK confirm the good moisture and liquid water resistance of this kind of high performance thermoplastic polymer.

5 HYGROTHERMAL STABILITY OF POLYMER BASED COMPOSITES

Fiber-reinforced may undergo a slow degradation process when subjected to environmental aging due to the synergism of moisture sorption, temperature and stress. The degradation mechanism acts on the fibers, on the interface and on the matrix. The matrices of composites have been described as taking up more water than the pure resin, especially during equilibration in severe conditions of temperature and humidity. Fiber debonding from the matrix induces

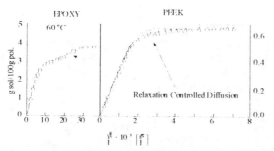

Figure 5: Relaxation-control water sorption kinetic in epoxy and in amorphous PEEK at 60 °C.

localized water entrapment hence apparently higher water uptakes.

Sorbed solvents may promote craze and crack initiation; surface energy reduction and plasticization are involved in void nucleation and growth under the action of the tensile stress. Crazing of a glassy polymer involves significant localized cavitation and fibrillation induced by the dilatational component of the stress [Sternstein et al. (1968)]. Crazing, in fact, cannot occur unless a finite hydrostatic tensile component of the stress exists; conversely, other types of plastic deformations (e.g., yielding) may also occur under conditions of hydrostatic compression. This suggests that as the stress field becomes non-dilatational, crazing should be also inhibited. Solvent-induced osmotic stresses, or differential swelling strains between inhomogeneous regions such as those of different cross-linking densities or around the fibers, should produce the "internal" stress field that induces the cavitation and craze formation responsible of the observed localized fiber debonding. Equilibration in water carried out on uniaxially loaded samples resulted in increases of water uptakes 15% higher than those observed for the unloaded materials even if the applied loads were only 7% of the ultimate stress [Apicella et al. (1984b)]. Conversely samples of the same epoxies equilibrated in water under hydrostatic compression (i.e., 20 atm, which should decrease the tendency of the polymer to fail by crazing) showed reduction of the water uptakes of more than 25%.

Glass and carbon fiber composites based on epoxy matrices, in fact, when exposed to severe environmental conditions (high temperatures and relative humidities) often show a progressive degradation of the mechanical properties associated with anomalous higher sorptions. The weight gain experimentally observed for these composites, if referred to the resin contents, are significantly higher than those relative to the neat resin. Fiber debonding and delamination are observed. The synergistic action of temperature, stress, and moisture is then particularly effective in composites exposed to the actual service conditions. The defects generated during the processing of these materials, such as incomplete and inhomogeneous cure of the matrix or temperature differences generated in the part by uncontrolled processing leading to inter laminar residual stresses, strongly affect the durability of the composite. The characteristics of the curing process and the final properties of thermoset-based polymeric composites are strongly dependent on the thermokinetics and chemorheological properties of the matrix.

On the other hand, according to preliminary investigations, composites based on thermoplastic PEEK tested in the same conditions do not seem to undergo the same type of environmental degradation encountered for the epoxy systems due to the lower affinity to water of this thermoplastic polymer, coupled with high strength and stiffness at elevated temperatures.

Styrenated unsaturated polyesters were found to be incompletely cured in air, due to inhibition of the curing reaction in the presence of oxygen. The system resulting from such a cure had low heat stability, low resistance to hydrolysis, and showed greater swelling in solvents than those cured in absence of air. Postcuring in water or in an inert medium led to further reaction of the residual styrene monomer and anhydrides with radicals fixed in the network. Increase in the glass transition temperature, however, were also observed for polyesters that had been post cured in water when the residual monomer had been completely exhausted. Hygrothermal ageing of commercial grade glass fiber-reinforced bisphenolic polyester resin has been deeply investigated by Apicella et al. [Apicella et al. (1982a)]. The breakdown and leaching observed in such systems is thought to be due to the polyester component, since bisphenolic segments are resistant to chemical attack. Ester linkages are likely the most reactive groups to be involved, as indicated by the good agreement observed between ester content and corrosion. Hydrolysis of ester groups results in the formation of carboxyl groups which have been shown to auto-catalyze further decomposition.

Calorimetric analysis is reported Apicella et al. [apicella et al. (1982a)] on samples of the unfilled resin and its composites, made with a 20% by weight of glass fiber mat aged one month in water at 100 °C. The thermograms of the 'as prepared' system, the unfilled resin and the composite, behave the same way. At temperature above the Tg the 'as prepared' samples show a wide exotherm associated to the residual curing of the unreacted prepolymer components. The exposure in a liquid environment at 100 °C for one month erased the final exotherms and increased the Tg of the flexibilizer present in the system and the resin, both for the neat resin and for the composite. The Tg of the flexibilizer occurred at 70 - 100 °C instead of 45 - 70 °C, while that of the resin occurred at 120 -140 °C instead of 100 - 120 °C. The displacement of the transitions may be related to both the residual curing of the resin achieved during the ageing and to the desorption of partially bonded or free low molecular weight components, which may be initially present in the resin. The residual curing, due to the long exposure to high temperature, however, is not sufficient to increase the Tg significantly; in fact an independent curing carried out at 100 °C in a dry environment gave only a slight change in Tg. Postcuring in water (or in another inert medium) leads to a further reaction of the residual styrene monomer previously inhibited by the presence of the oxygen. Tg increases, however, were described on water post-cured polyester also when the residual monomer was completely exhausted. Thus, a post-cure test carried out in a dry inert environment (nitrogen) should result in the same Tg as for a water post-cured sample, if only oxygen inhibition and residual curing are responsible for the lower value of Tg. The loss of low molecular weight component, conversely, may be sufficient to reduce the degree of plasticization of the resin and the flexibilizer increasing their Tg.

Sorption data on neat resin and GRP are reported in figures 7 and 8, at 20, 40, 60 and 90 °C. A weight gain of 0.5 - 0.7 % was initially reached during the water conditioning then it slowly decreased over the entire range of time. The phenomenon was progressively less evident as the test temperature was decreased. After an equivalent ageing time of 8000 $\frac{\sqrt{s}}{cm}$, the samples were desiccated and weighed in the dry state. Weight losses ranging from 0.38% to 1.12% for the neat resin and from 0.53% to 2.41% for the composites are reported in table 5 for the temperatures investigated. The values of the actual water uptakes calculated by using the effective final dry weight, are compared in table 6 for both types of samples. The actual water uptakes of pure resin at 20, 40 and 60 °C are practically the same (0.78%) which indicates that sorption is athermal; while the value at 90 °C is noticeably higher, i.e. 1.17%. The latter phenomenon could be possibly attributed to microcavitational damage.

Results of tensile tests carried out at temperatures of 20, 40, 60 and 90 °C on samples of the resin and the composite both in the 'as prepared' state and after

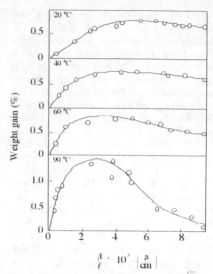

Figure 8: Weight gain as function of $\frac{\sqrt{t}}{l}$ for 'as prepared' GRP kept in water at T equal to 20, 40, 60 and 90 °C.

thermal ageing in water for 15 days show an increase of the elastic moduli and a reduction of the elongation at break after the ageing, which could be attributed weight losses. However, the mechanical properties of the composites are less sensitive to the temperature changes and ageing, due to the strong influence of the fibers on the strength of the samples.

The plasticization due to the sorbed water has been found to have a small effect only at temperature near Tg of the dry resin. The comparison of the elastic moduli at 20 °C and 90 °C of the pure resin, conditioned in water at 90 °C for 15 days in the wet and desiccated state, gave indication of the effect of the dissolved water. At 20 °C the moduli were found to be the same ($3.2 \cdot 10^3$ MPa). To evaluate the actual influence of the weight losses, the mechanical tests were performed at 20 °C far from the glass transition where the dissolved water had no practical influence. Both the pure resin and the composite show an increase of the elastic modulus and a strong reduction of the elongation at break, from 1.7 to 0.5 and from 1.4 to 0.4 respectively. The thermo-calorimetric analysis gave similar information, indicating that the exposure of this flexibilized system to water at high temperatures induces an increase in Tg for both the resin and the flexibilizer.

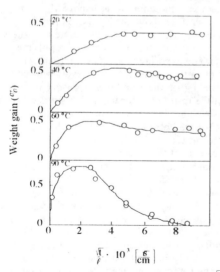

Figure 7: Weight gain as function of $\frac{\sqrt{t}}{l}$ for 'as prepared' pure resin kept in water at T equal to 20, 40, 60 and 90 °C.

6 CONCLUSION

Environmental degradation of the mechanical

properties of both polymer matrices and polymer based composites is associated with plasticization and micromechanical damage induced by synergistic effects of temperature, stress and sorbed moisture.

Both moisture equilibrium sorption and kinetics of diffusion have been reviewed for epoxy, polyester and PEEK matrices.

In the case of epoxy matrices, different mechanism of moisture induced plasticization are involved and the resulting glass transition temperature depression is related to the amount of unreacted secondary amines, to the extent of crosslinking (amount of hardener) and to the level of interaction between water and epoxies.

Much lower water sorption equilibrium uptakes are reported for thermoplastic PEEK. Water induced plasticization is, consequently, negligible.

Some degree of relaxation has been found to slightly affect the moisture sorption both in PEEK and epoxy net matrices.

In the case of polymer based composites, moisture affects not only matrix characteristics, but acts also on fibers and on matrix-fiber interfaces. In fact, debonding from the matrix induces localized water entrapment, hence apparent higher water uptakes.

Defects generated during the processing of these materials, such as incomplete and inhomogeneous cure of the matrix or temperature difference generated in the part by uncontrolled processing leading to interlaminar residual stresses, strongly affect the durability of composites.

Generally, PEEK based composites are more resistant of epoxy-based composites, due to the lower affinity to water, coupled with high strength and stiffness at elevated temperatures.

In the case of polyester neat resin and polyester based composites, plasticization effects are coupled to desorption of low molecular weight compounds, whose is due to incomplete curing. Plasticization due to sorbed water has been found to have a small effect only at temperature near Tg of the dry polyester resin.

REFERENCES

M.K. Antoon, J.L. Koeneing, and T. Serafin, *J. Polym. Sci., Polym. Phys. Ed.*, *19*, 1567, (1981).

A. Apicella, L. Nicolais, G. Astarita, and E. Drioli, *Polymer*, *20*, 9 (1979).

A. Apicella and L; Nicolais, *Ind. Eng. Chem. Prod. Res. Dev.*, *20*, 138 (1981a).

A. Apicella, L. Nicolais, G. Astarita, and E. Drioli, *Polymer*, *22*, 1064 (1981b).

A. Apicella, L. Nicolais, G. Astarita, and E. Drioli, *Polym. Eng. Sci.*, *21*, 18 (1981c).

A.Apicella, C. Migliaresi, L. Nicodemo, L. Nicolais, L. Iaccarino, and S. Roccotelli, *Composites*, *13*, 406 (1982a).

A. Apicella, L. Nicolais, C. Carfagna , C. de Notaristefani, and C. Voto, *Proceedings of the 27th SAMPE Natl. Symposium*, *27*, 753 (1982b).

A. Apicella, L. Nicolais, and J.C. Halpin, *Proceedings of the 28th SAMPE Natl. Symposium*, *28*, 518 (1983).

A. Apicella and L. Nicolais, *Ind. Eng. Chem. Prod. Res. Dev.*, *23*, 288 (1984a).

A. Apicella, R. Tessieri, and C. de Cataldis, *J. Membrane Sci.*, *18*, 211 (1984b).

A. Apicella, L. Nicolais, and C. de Cataldis, *Adv. Polym. Sci.*, *66* (1985).

A.Apicella, "Influence of Chemorheology on the epoxy Resin Properties", in G. Pritchard, Ed., *Development of Reinforced Plastic*, Vol. 5, Applied Science Publishers, London, 1986.

L. Banks and B. Ellis, *Polym. Bull.*, *1*, 377 (1979).

J.A. Barrie, P.S. Sagoo, and P. Johncock, *J. Membrane Sci.*, *8*, 197 (1984).

J.A. Barrie, P.S. Sagoo, and P. Johncock, *Polymer*, *26*, 1167 (1985).

A.R. Berens, *Polymer*, *18*, 697 (1977).

A.R. Berens and H.B. Hopfenberg, *Polymer*, *19*, 489 (1978).

D.J. Blundell, D.R. Beckett, and P.H. Willcocks, *Polymer*, *22*, 704 (1981).

D.J. Blundell and B.N. Osborn, *Polymer*, *24*, 953 (1983).

C. Carfagna, A. Apicella, and L. Nicolais, *J. Appl. Polym. Sci.*, *27*, 105 (1982).

C. Carfagna and A. Apicella, *J. Appl. Polym. Sci.*, *28*, 2881 (1983).

C. Carfagna, E. Amendola, A. D'Amore, and L. Nicolais, *Polym. Eng. Sci.*, in press.

Chi-hung and G.S. Springer, *J. Compos. Mater.*, *10*, 2 (1976).

J. Crank, *The Mathematics of Diffusion*, Oxford University Press, Oxford, 1975.

R. Delasi and J.B. Whitside, in *ASTM STP 658*, 1978, p. 2.

T.S. Ellis, F.E. Karasz and G. ten Brinke, *J. Appl. Polym. Sci.*, *28*, 23 (1983).

T.S. Ellis and F.E. Karasz, *Polymer*, *25*, 664 (1984).

A.R. Berens, *Polymer*, *18*, 1105 (1977).

D.J. Enscore, H.B. Hopfenberg, and V.T. Stannett, *Polym. Eng. Sci.*, *20*, 102 (1980).

J.K. Gillham, *Polym. Eng. Sci.*, *19*, 676 (1979).

M.A. Grayson and C.J. Wolf, *J. Polym. Sci., Poly. Phys. Ed.*, *25*, 31 (1987).

H.B. Hopfenberg and H.L. Frisch, *J. Polym. Sci., Poly. Phys. Ed.*, *7*, 405 (1969).

G.P. Johari, *Phil. Mag.*, *35*, 1077 (1977).

F.N. Kelly and F. Bueche, *J. Polym. Sci.*, *50*, 549 (1961).

J.D. Keenan, J.C. Seferis, and J.T. Quinlivan, *J. Appl. Polym. Sci.*, *24*, 2375 (1979).

A.F. Lewis, M.J. Doyle, and J.K. Gillham, *Polym. Eng. Sci.*, *19*, 687 (1979).

A.C. Loss and G.S. Springer, *J. Compos. Mater.*, *13*, 17 (1979).

E.L. McKague, J.E. Halkias, and J.D. Reynolds,*J. Compos. Mater.*,*9*,2 (1975).

E.L. McKague, J.D. Reynolds, and J. Halkais, *J. Appl. Polym. Sci.*, *22*, 1643 (1978).

P. Meares, *Trans. Faraday Soc.*, *53*, 101 (1957).

G. Mensitieri, A. Apicella, J.M. Kenny, and L. Nicolais, *J. Appl. Polym. Sci.*, *37*, 381 (1989).

G. Mensitieri, M.A. Del Nobile, A. Apicella, and L. Nicolais, *Revue de L'Istitut Francais du Petrole*, *50* (4), 1 (1995).

A.S. Michaels, R.H. Vieth, and J.A. Barrie, *J. Appl. Phys.*, *24*, 1 (1963).

J. Mijovic and L. Tsai, *Polymer*, *22*, 902 (1981).

J. Mijovic and S.A. Weinstein, *Polym. Comm.*, *26*, 1167 (1985).

R.J. Morgan and J. O'Neal, *J. Mater. Sci.*, *12*, 1966 (1977).

R.J. Morgan and J.E. O' Neal, *Polym.-Plast. Technol. Eng.*, *10*, 49 (1978).

R.J. Morgan , J.E. O' Neal, and D.B. Miller, *J. Mater. Sci.*, *14*, 109 (1979).

R.J. Morgan, E.T. Mones, and W.J. Steele, *Polymer*, *20*, 315 (1982).

P. Moy and F.E. Karasz, *Polym. Eng. Sci.*, *20*, 315 (1980).

L. Nicolais, A. Apicella, and C. De Notaristefani, *J. Membrane Sci.*, *18*, 187 (1984).

P. Peyser and W.D. Bascom, *J. Mater. Sci.*, *16*, 75 (1981).

G.A. Pogany, *Polymer*, *17*, 690 (1976).

S.S. Sternstein, L. Ogechin, and A. Silverman, *Appl. Polym. Symp.*, *7*, 175, (1968).

E.J. Stober, J.C. Seferis, and J.D. Keenan, *Polymer*, *25*, 1845 (1984).

G. ten Brinke, F.E. Karasz, and T.S. Ellis, *Macromolecules*, *16*, 244 (1983).

R.H. Vieth, J.H. Howell, and J.H. Hoseih, *J. Membrane Sci.*, *1*, 177 (1977).

Progress in Durability Analysis of Composite Systems, Cardon, Fukuda & Reifsnider (eds)
© *1996 Balkema, Rotterdam. ISBN 90 5410 809 6*

Effect of water absorption on the low energy repeated impact of carbon/epoxy laminates

R. Boukhili, L. Champoux & S. Martin
CRASP, Mechanical Engineering Department, Ecole Polytechnique de Montréal, Qué., Canada

ABSTRACT: Low energy repeated impact (LERI) tests have been performed on a 36 laminas carbon/epoxy composite. The impact test consists of a 12.7 mm diameter hemispherical impactor repeatedly impacting a 76.2 mm square plate clamped over a 51 mm diameter hole. During the test, the increase of the damaged area as function of the number of impacts is monitored using an ultrasonic apparatus. Some of the specimens were immersed in room temperature distilled water (RTC) and others in hot distilled water (ETC) until the reached 0.35 % water content. From the analysis of the damage growth area as function of the number of impacts for conditioned and non conditioned specimens, it appears that compared to non conditioned specimens, RTC improves the LERI resistance by delaying the matrix cracking and delamination stages while the inverse is seen for ETC. From tests on moist-dried specimens, it is concluded that for the level of water content used, ETC do not induce any irreversible damage. The decrease in the LERI resistance in this case is due to the stresses induced by the non uniform moisture distribution through the thickness.

1 INTRODUCTION

Many research efforts have been dedicated to study the environmental effects on the properties and behaviour of composite materials as illustrated by the first ASTM-STP publication fully devoted to this subject (Vinson 1978). Investigations of environmental effects remained an active research area in the 1980's and the most important developments have been gathered in three volumes edited by G.S.Springer (1981, 1984 and 1988). The first volume summarizes work published prior to 1980, while the second volume is comprised of papers published from 1981 through 1983 and the third volume was published in 1988. According to our survey the most recent review paper is that published by E.G.Wolff (1993) and reports more recent developments in this field. Moisture or water absorption can affect all matrix and fiber/matrix dominated properties. All studies agree that moisture effectively plays the role of a resin plasticizer which softens the matrix and lowers the glass transition (Staunton 1982 and Adams 1986). The plasticization process is believed to involve interruption of the Van Der Walls bonds between ethers, secondary amides and hydroxyl groups (Wolff 1993). Generally, it is reported that

moisture effects are: induced internal stresses, plasticization of the matrix, reduced residual stresses, swelling of the matrix, surface cracking and osmotic cracking.

In general, moisture effect may be beneficial or detrimental. This depends on the damage mechanisms governing the fracture and the fracture criteria used. The fracture mechanisms can be divided in three categories: fiber controlled, matrix controlled and fiber/matrix interface controlled. For some loading conditions moisture effect was reported to be negligible. However, such a statement may be misleading since beneficial and detrimental effects may be balanced. As an example, an impact on a defect free laminate involves transverse matrix cracking and delamination. Moisture can delay matrix cracking (ductile matrix) and promote delamination (weak interface). The overall response will depend on the impact energy and the ratio of initiation and propagation energies. Presently, there is no experimental methodology that has the ability to dissociate the matrix cracking and delamination stages in an impact test. However, the Low Energy Repeated Impact method (LERI) developed recently by R.Boukhili et al (1993 and 1994) is believed to have the potential to answer this question. Low energy impact (LEI) or low velocity impact are expressions becoming popular in the field of composite testing. Unfortunately, there are no standard definitions of these expressions and sometimes, what is called LEI by an investigator is considered high energy impact by another. In this investigation, LEI refers to an impact with an energy below that which produces delamination. The smallest energy that produces a delamination in a single impact is called the threshold energy E_o.

In (Boukhili 1994) it was shown that the damage growth under LERI follows a three stage pattern behaviour. A typical curve showing the damage growth area versus the number of impacts is shown in Figure 1 in which three stages are displayed. It was postulated that stage I corresponds to a matrix cracking of the region under the impactor, stage II corresponds to a rapid damage growth rate in the form of delamination and stage III to a slow damage growth rate involving matrix cracking and delamination.

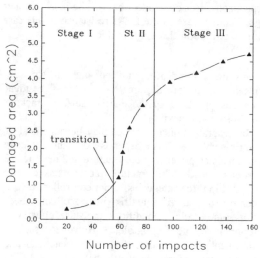

Figure 1 Typical LERI curve showing the damage area as function of the number of impacts

The occurrence of the transition between stage I and stage II (transition I) was found to be described by an impact-fatigue curve in the form of equation (1):

$$E_I = c \log (N_{tf}) + E_0 \qquad (1)$$

where E_I, E_o, c and N_{tf} are respectively the impact energy, the threshold energy that induces a delamination in one impact, a material constant, and

the number of impacts required to induce delamination (transition I). It was suggested that the mentioned impact-fatigue curve may be used as a damage tolerance criteria for composite materials.

2 EXPERIMENTAL PROCEDURES

The material investigated is a Hercules AS4/3501-6 graphite/epoxy laminate supplied by National Defence of Canada in the form of 305 mm wide square plates. The laminate orientation is:

$$[+45_2/-45_2/90_2/0_2/+45/-45/0_2/+45/-45/+45/-45/0_2]_s$$

Ultrasonic C-scan imaging was used to check for defects in the laminates before the tests and during the LERI tests to monitor the damage growth as a function of the number of impacts. The ultrasonic workstation used is an Ultra Pac II apparatus from Physical Acoustics Corporation. The LERI tests were performed at room temperature using a pneumatic instrumented Dynatup (model 8250) drop weight impact machine with a GRC 830-I data acquisition system. To avoid rebounding and extraneous impacts, the cross head is caught automatically after each individual impact. The impact fixture configuration used is that specified by the ASTM D3763 standard, which covers the determination of puncture properties of plastics. Simply stated, a 12.7 mm (1/2 inch) diameter hemispherical impactor strikes a 76.2 mm (3 inches) square plate clamped over a 51 mm (2 inches) diameter hole.

Conditioned specimens at room temperature (RTC) and elevated temperature (ETC) were first dried in an oven at 90°C and their weight measured periodically on an analytical balance until stabilization and then immersed in distilled water. Specimens were periodically removed from the water, wiped, weighted and replaced in the water. The moisture content M expressed in terms of percent weight gain is defined as:

$$M = ((M_i - M_d)/M_d) \times 100$$

where M_d is the weight of dry material and M_i the weight of moist material. It took only one day to reach 0.35% moisture content at 100°C compared to 140 days at RTC and all the $M-\sqrt{t}$ curves are linear in the investigated range. The moisture level achieved after 25 days at 100°C was about 1.6% (Figure 2)

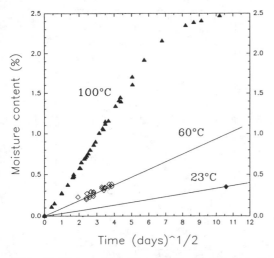

Figure 2 Change of moisture content as function of time for different conditionning temperatures

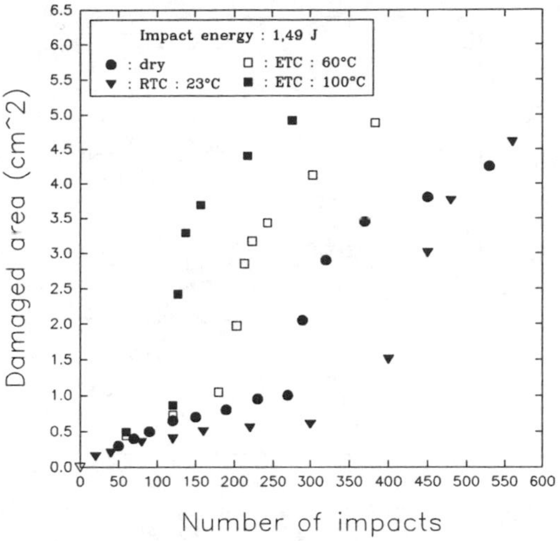

Figure 3 Evolution of the damaged area as a function of the number of impacts for different temperature conditionning and 0.35 % water content

3 RESULTS AND DISCUSSION

Figure 3 is a typical A-N diagram showing the evolution of the damaged area (A) as function of the number of impacts (N) for RTC, ETC and dry specimens repeatedly impacted at 1.49 J. It is clear that the occurrence of transition I from matrix cracking to delamination, i.e. the point in the A-N curve where the damaged area increases rapidly, is delayed to higher number of impacts in the case of RTC and occurs earlier for ETC. It can also be observed from Figure 1 that the slope of stage II increases with the conditioning temperature. Consequently, elevated temperature conditioning promotes matrix cracking as well as delamination.

From LERI tests performed at different impact energies (E_I) it is possible to plot an "S-N" like fatigue curve. The number of impact to failure (N_{If}) corresponds to transition I. The plot of such "S-N" like diagram is shown in Figure 4 in the form of an impact-fatigue curve (E_I-N_{If} curve) from which two observations are made:

1. Compared to non-conditioned dry specimens, the threshold energy is higher in RTC specimens and lower in ETC specimens.

2. Compared to non-conditioned dry specimens, RTC improves the resistance to LERI by delaying the occurrence of transition I, and the inverse is seen for ETC.

It should be mentioned that the analysis of the A-N curves obtained at other impact energies show

that the effect of ETC, which is very pronounced at very low energies, tends to vanish at high energy levels (Figure 5). If this tendency is maintained, the effect of ETC will probably become negligible at high energies for 0.35 % water content. This result shows the ability of LERI to detect small changes in the morphology, which in conventional impact testing will be overlooked. This phenomena is also illustrated by the fatigue curve (Figure 4) from which it can be noticed that the curves at 60°C and 100°C converge at high energies. In Figure 4, the experimental data points corresponding to dry and RTC are replaced by a curve fitting. Obviously, since a semi-log diagram is used, it tends to underestimate the behaviour at high N_{If} values.

At first sight, the increase of LERI resistance at RTC can be explained by the matrix plasticization and the decrease of LERI resistance at ETC by induced matrix microcracking. While such argument may be considered for the 100°C conditioning, it cannot hold for 60°C conditioning. Indeed, almost all studies agree that at this temperature (i.e. 60°C) and particularly for the Hercules 3501-6 matrix, matrix microcracking is not likely to occur. In addition, in this study scanning electron microscope (SEM) observations were made on the specimen sections at very high magnification levels and no microcracks were observed at this moisture level (0.35 %). It is possible to be sceptical and argue that SEM observations do not guarantee that microcracks have not been induced by the elevated

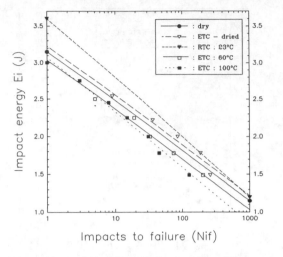

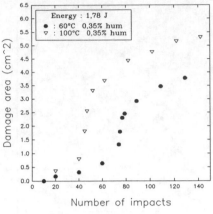

Figure 4 Impact-fatigue behaviour for different temperature conditionning and 0.35% water content

temperature conditioning. However, this uncertainty can be removed indirectly by LERI tests on specimens that have been dried after ETC (moist-dried). If damage has not been induced by ETC, then the material should recover its properties after drying. This is confirmed in the following subsection.

In order to determine the effect of drying on the LERI behaviour of the AS4/3501-6 graphite-epoxy, five plates were conditioned at 100°C until they achieved 0.35% moisture content and dried in a circulating air oven at 90°C until they recovered their initial weight. The plates were LERI tested at energies between 1.48 J and 2.55 J. The results are presented in the form of fatigue E_i-N_{if} curves in Figure 4. As it can be seen, when dried the material recovers its LERI resistance. Actually, the curve fitting of the fatigue curve of moist-dried specimens is very close to that of non-conditioned plates. Consequently, and at least up to 0.35% moisture content, it can be concluded that elevated temperature conditioning does not induce any irreversible damage to the material investigated. It remains that the deleterious effect of ETC is not yet explained since microcracking or damage hypothesis are discarded. One hypothesis that can explain the difference in the LERI behaviour at RTC and ETC can be based on the stresses induced by the non uniform moisture profile distribution as presented by Lee and Peppas [8]. However, this hypothetic explanation which is presented below needs a careful examination.

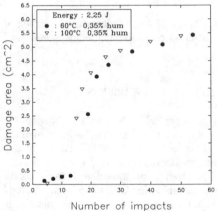

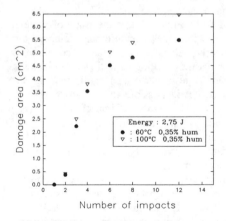

Figure 5 Effect of temperature conditionning ($\bullet$ 60°C and $\triangledown$ 100°C) on the damage area growth for different impact energies (1.78, 2.25 and 2.75 J)

M.C.Lee and N.A.Peppas (1993) investigated a Hercules AS4/3501-6 composite immersed in 100°C water. This is a very interesting case since it is the same material and conditions used in the present project. Many mathematical models are used to describe moisture transport in epoxy composites. These models take into consideration the stresses induced due to moisture and their influence on the transport mechanisms. Lee and Peppas (1993) conclude that: *"All the models showed that absorbed moisture causes transverse directional tensile stresses in the center of the sample and compressive stresses around its surface during the process of moisture absorption. At the same time, longitudinal tensile stresses were induced in the fiber portion whereas compressive stresses were induced in the resin portion..."*. The conclusions of Lee and Peppas (1993) are used in the following manner:

To obtain 0.35% water content it took 140 days at RTC and 1 day at ETC (100°C). The moisture distribution profiles are not uniform in the two cases. However, given the long immersion time at RTC, the induced stresses relaxe and become negligible. On the contrary, given the short time at ETC, the induced stresses are still important and will superimpose with the stresses induced during the LERI test and result in a lower resistance. It is said that the induced stresses will superimpose because for the point impact used, the plate is in a biaxial loading state. This explanation is in agreement with the vanishing effect of ETC at high impact energies (Figure 5). Indeed, when the impact energy is increased the impact induced stresses increase and the relative contribution of the moisture induced stresses decrease.

The analysis developed seems to be rational. However, as a hypothetic explanation, it needs to be deeply investigated and this may constitute a very promising approach to finally understand the real problems encountered by accelerated conditioning. If these problems arc well understood, accelerated conditioning can be optimized on a rational basis and the repercussions can be very important in time and cost of testing concerning moisture problems. For instance, to optimize the conditioning time, ETC and RTC can be applied in an alternate manner. The alternation ETC-RTC allows the induced stresses to relax while the material continues to absorb water. Once the stresses are relaxed a second ETC-RTC cycle can be applied and so forth until the desired moisture level is achieved. Such approach is presently under investigation.

This optimistic note is enhanced by the fact that the composite system used is a real structural lay up which constitutes a very complex case to analyze. Consequently, if the analysis developed in the preceding sections is valid, there is a good chance that it will remain valid for less complex lay ups.

4 CONCLUSIONS

In composite materials water absorption is believed to affect matrix and matrix/fiber interface dominated properties. In conventional impact loading where both matrix cracking and delamination are expected, it is difficult to find out which fracture process has been more affected by water absorption. This investigation has attempted to use the LERI method as a sensitive tool to overcome this difficulty since LERI tests can dissociate matrix cracking and delamination. The effect of RTC on the LERI resistance seems to be beneficial while the inverse is seen for ETC. The deleterious effect of ETC is not due to an irreversible damage (at least up to 0.35% water pick-up) as shown by the impact-fatigue curve of moist-dried specimens. An explanation of this behaviour can be based on the non-uniform water distribution through the specimen thickness and the induced internal stresses.

5 ACKNOWLEDGMENTS

This project has been financed by NSERC OGP0025403 grant and National Defence CDT P1887 project.

6 REFERENCES

ADAMS, D.F., "Environmental Effects on Composite Materials", Seminar Notes, Technomic Publ. Co., Lancaster, PA, 1986.

BOUKHILI, R., BOJJI, C., and GAUVIN, R., "A Feasibility Study on the Impact Fatigue Behaviour of Composites", Final report, Prepared for National Defence, Project CDT P1663, École Polytechnique de Montréal, July 1993.

BOUKHILI, R., C.BOJJI, and R.GAUVIN 1994. Fatigue Mechanisms under Low Energy Repeated Impact of Composite Laminates, Journal of Reinforced Plastics and Composites, Vol. 13, No. 10, p.856-870.

LEE, M.C., and N.A.PEPPAS 1993. Models of

Moisture Transport and Moisture-Induced Stresses in Epoxy Composites, Journal of Composite Materials, Vol. 27, No. 12, p. 1146-1171.

SPRINGER, G.S. 1981, 1984 and 1988. Environmental Effects on Composite Materials, Technomic Publishing Co., Lancaster, PA, Vol. 1, 1981, Vol 2, 1984, Vol. 3, 1988.

STAUNTON, R. 1982. Environmental Effects on Properties of Composites, in Handbook of Composites, Edited by LUBIN, G., Van Nostrand Reinhold, NY.

VINSON, J.R. 1978. Advanced Composite Materials - Environmental Effects, ASTM STP658, American Society for Testing and Materials.

WOLFF, E.G. 1993. Moisture Effects on Polymer Matrix Composites, SAMPE Journal, Vol. 29, No. 3, p. 11-19.

Progress in Durability Analysis of Composite Systems, Cardon, Fukuda & Reifsnider (eds)
© *1996 Balkema, Rotterdam. ISBN 90 5410 809 6*

A family of mathematical models for water sorption/desorption in epoxies

P.J. Shopov
VUB, GOSARGUB, Brussels, Belgium & IM, BAS, Sofia, Bulgaria

P. Frolkovich
FMF, Comenius University, Bratislava, Slovak Republic

W.P. De Wilde
VUB, GOSARGUB, Brussels, Belgium

ABSTRACT: Two classes of mathematical models for penetrant sorption in polymers are proposed, which generalise of the *Carter & Kibler* model. An important feature of the proposed models is that they address the whole set of sorption curves at different RH at ones i.e. they offer unified theory for the sorption behaviour of the material at all RH by means of only few governing parameters. The models are unsteady but they yield equilibrium sorption theories for large times as well. One model has been identified as especially encouraging and it seems to have good perspectives for wide implementation. It gives for large times exactly the classical *Langmuir* theory for the water equilibrium level. Some theoretical criteria to outline the applicability area of each class of models are also suggested. A number of interesting non-linear phenomena have been also observed and discussed.

1. INTRODUCTION

In the present moment we still do not possess complete models for predicting the time-dependent moisture uptake and reduction of the mechanical properties of epoxy matrix and epoxy based composites - Mensitieri et al (1995). There are a number of non-Fickian sorption phenomena that are not covered satisfactory yet:
• *two stage sorption* - long-term effect due to swelling and degradation - microcracking, microvoids
• *preferential sorption* of water by hydrogen bounding
• *sigmoidal sorption* effect for some highly crosslinked epoxies - for small time.
• eventual *chemical reactions* and leakage out

On one hand the classical models are linear and suffer from difficulties to describe a number of non-linear sorption phenomena as preferential sorption, sigmoidal effect, accelerated ageing, double change of the curvature of the swelling graphs and so on.

On the other hand the existing theories are considering every environmental situation separately i.e. the sorption curve is fitted independently at fixed RH and temperature. Consequently the obtained values for the governing parameters (say the diffusion coefficient and at least one parameter for the magnitude of the two stage sorption effect) should be interpolated in some way to yield a prediction for all RH. But as the data often has a rather complex behaviour - e.g. *de Wilde and Shopov*(1992), it is not transparent how to chose this interpolation, nor what will be its mechanical consistency. This problem becomes much more complex if we want to consider also temperature or history effects.

Therefore it seems interesting to try to develop sorption theories which are able to put together the sorption measurements at different RH in one integrated theory. It should be able to give predictions for all RH without interpolation and should include also equilibrium sorption curves.

The classical Fickian model is often practically applied, although clearly neglecting all specifics of the epoxy systems. It assumes that the water is not interacting with the polymer, which is clearly unrealistic. Therefore, the idea is to generalise Fickian model, taking into account the transformation of free water (FW) into another form - it will be conventionally named "bond" water (BW); it is assumed to be immobilised in the epoxy network.

The goal of the present work is thus to offer C- and L-sets of non-linear "two water" models with an analysis of their mechanical background, their properties and performances. The simplest model from this group is the C&KM - Carter and Kibler (1978) and its properties and application area is considered in the next chapter. The direct non-linear generalisations (class C) of the C&K model are considered in the next chapter. Models with finite "bonding" capacity of the epoxy are introduced in the forth chapter - class L. In this class one model is identified as especially encouraging and it is recommended for wider applications. As usually the discussion and conclusions are at the end.

2. CARTER AND KIBLER MODEL (C&KM)

C&KM is very important "two water" model - free water (FW) and bond water (BW); the concentrations are denoted by C_f and C_b.

(2.1) $\dfrac{\partial C_f}{\partial t} = \dfrac{\partial^2 C_f}{(\partial z)^2} - \dfrac{\partial C_b}{\partial t}$

(2.2) $\dfrac{\partial C_b}{\partial t} = \gamma_1\, C_f - \beta_1\, C_b$

(2.3) $C_f(z,0)=0$; $C_b(z,0)=0$; $C_f(0,t)=C_f(2\,l,t)=C_0$
where $C_0 = C_0(RH)$ is the free water concentration at on the surface and l is the half width of the specimen. It depends on 3 governing parameters - (γ_1,β_1,C_0); the first two are "bonding" and "freeing" rates of water and are considered as fixed for the penetrant-epoxy system.

In dimensionless form: $t_0 = l^2 / D$; $C_0 = C_{f,s}$

$t' = t / t_0$; $C_f' = C_f / C_0$; $C_b' = a\, C_b / C_0$; $z' = z / l$

this system becomes

(2.4) $\dfrac{\partial C'_f}{\partial t} = \dfrac{\partial^2 C'_f}{(\partial z')^2} - \dfrac{1}{a}\dfrac{\partial C'_b}{\partial t'}$

(2.5) $\dfrac{\partial C'_b}{\partial t'} = \gamma\,(C'_f - C'_b)$

(2.6) $C'_f(z,0)=0$; $C'_b(z,0)=0$; $C'_f(0,t)=C'_f(2,t)=1$
and depends only on two parameters - the relative rate γ of the "reaction" and the ratio a between the equilibrium levels of FW and BW water.

$$\gamma = \dfrac{\gamma_1 l^2}{D}\,; \quad a = \dfrac{\beta_1}{\gamma_1} = \dfrac{C_0}{C_{b\infty}}$$

Hence a is a parameter, related only with the equilibrium sorption graph and γ is controling the rate of transition to the equilibrium state.

For a fixed penetrant-polymer system the growth of γ corresponds to the growth of the sample thickness l ; please note that the unit time also grows with l .

Limitations of C&KM are:
1. The sorption curve in M_t / M_∞ does not depend on: relative humidity RH and the thickness of the sample l
Therefore, unless all gravimetric data cluster at a single curve, C&K is not applicable. If the relative amplitude of the two-stage sorption is not constant or the qualitative behaviour of the sorption curve change with RH (which is often the case) this is an sufficient condition for inapplicability of C&KM.
The independence of the dimensionless sorption curve on the sample thickness is even more rear situation but such data are rearly available.
2. The *sigmoidal effect* for small times can not be reproduced.
3. The *preferential sorption* of bond water is not covered.
4. The *"apparent" Fickian sorption effect* take place for large bonding rate γ . This leads to *"apparent" diffusion coefficient and "apparent" equilibrium surface concentration;* the amplitude of the deviation can be computed exactly from the analytical solution of C&KM

$$D_{app}= \dfrac{a}{1+a}\,D_{real}; \; C_{0,app}= \dfrac{1+a}{a}\,C_0 \text{ for } \gamma=\infty$$

The potential *bonding water capacity is considered as infinite* and the *bonding rate is independent of the bond water concentration* already achieved.
6. The *finite equilibrium levels could be maintained only due to the freeing* of water.

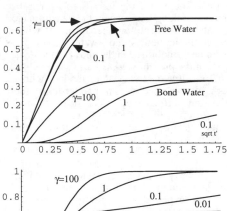

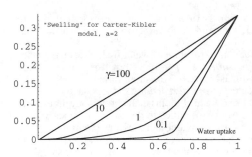

8. Further, we shall introduce the notion "swelling" graphs, which picture the bond water uptake against the total water uptake. The name is suggested because it is really the swelling (up to the scaling of the ordinate) if the swelling is proportional to the bond water concentration .

Normally the swelling curve exhibits non-linear transition to saturation but in C&KM it is linear.

To conclude, although C&KM is able to cover the two stage sorption effect, it exhibits a number of flaws in its qualitative behaviour as well as some conceptual inconsistencies. In some cases it can be

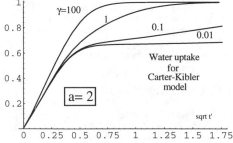

used for trapped water but is compitely unadequate for water bond by polar interactions.

3. NON-LINEAR C&K MODELS (C-MODELS)

It is well known to practising experts that all data for different RH can not be fitted with the same "reaction rates" γ_1, β_1 and that $\gamma_1 = \gamma_1(RH)$, $\beta_1 = \beta_1(RH)$. This means that the "reaction rates" are not a material property of the penetrant-epoxy system but depends also of something else. Examining the original model (2.1-2) we can easily conclude that the relative humidity can not be filled inside the specimen directly but only through the concentrations of "free" and "bond" water. This reasoning leads to the idea that both coefficient should have the form

$$\gamma_1 = \gamma_1(C_f), \; \beta_1 = \beta_1(C_b)$$

i.e. the "bonding" rate should depend of the amount of the available free water to "react" and "freeing" - on the amount of "bound" water to be "freed". This reasoning is common for the theory of chemical kinetics and leads to the generalised C-type model (3.2), where α_1 and α_2 are imperial constants.

$$(3.1) \quad \frac{\partial C_f}{\partial t} = \frac{\partial^2 C_f}{(\partial z)^2} - \frac{\partial C_b}{\partial t}$$

$$(3.2) \quad \frac{\partial C_b}{\partial t} = \gamma_1 \, (C_f)^{\alpha_1} - \beta_1 \, (C_b)^{\alpha_2}$$

Naturally, C-models are more general and we want to know which demerits of C&KM could be surmounted and which not. We shall introduce their dimensionless form in order to study them parametrically

$$(3.3) \quad \frac{\partial C'_f}{\partial t} = \frac{\partial^2 C'_f}{(\partial z')^2} - \frac{1}{a} \frac{\partial C'_b}{\partial t'}$$

$$(3.4) \quad \frac{\partial C'_b}{\partial t'} = \gamma \, (C'_f{}^{\alpha_1} - C'_b{}^{\alpha_2})$$

$$(3.5) \quad \gamma = (\gamma_1 \, l^2 / D) \, (C_0)^{\alpha_1 - 1}$$

$$(3.6) \quad a = (\beta_1/\gamma_1)^{1/\alpha_2} \, C_0{}^{1 - \alpha_1/\alpha_2}$$

Therefore, the number of the governing parameters remains two but their form becomes more complicated in comparison with C&KM. From (3.6) we can obtain the "bond" water equilibrium level in function of "free" water one:

$$C_{b,\infty} = (\beta_1/\gamma_1)^{-1/\alpha_2} \, C_0{}^{\alpha_1/\alpha_2}$$

and clearly it will tend or to infinity or to zero for large surface concentrations. Hence it can never yields the Langmuir type behaviour for the saturation level of "bond" water.

The relations (3.6) could be used to classify C - models:

C.0. $\alpha_1 = \alpha_2 = 1$ - C&KM. The dimensionless sorption graph do not depend on the surface concentration and hence on RH.

C.1. $\alpha_1 = \alpha_2 \neq 1$. The amplitude of the two stage effect do not depend on RH and could be considered as a property of the penetrant-epoxy system as for K&CM. Only the "transition rate" parameter γ changes with RH. In this case both bonding and freeing intensities are decreased in the dimensionless formulation and the dimensionless sorption and "swelling" graphs looks practically identical with C&KM for the same value of dimensionless governing parameters - see the figures for $\alpha_1 = \alpha_2 = 2$.

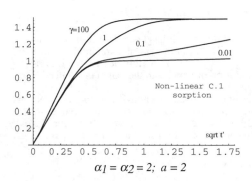

$$\alpha_1 = \alpha_2 = 2; \; a = 2$$

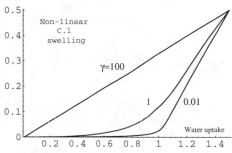

C.2. $\alpha_1 \neq \alpha_2$. Both the relative amplitude of the two-stage sorption effect a and the transition rate parameter γ depend on RH and its behaviour is much more non-linear.

In the case C.2 the relative amplitude of the two stage sorption effect is *increasing* with the growth of RH, if $\alpha_1 > \alpha_2$ and *diminishing* if $\alpha_1 < \alpha_2$. If the relative amplitude a is diminishing with RH, then $\alpha_1 > \alpha_2$ has to be chosen, and $\alpha_1 < \alpha_2$ if a is growing.

In order to study the qualitative behaviour of this model we shall fix $\alpha_1 = 1$ and shall vary α_2.

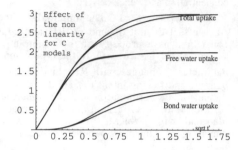

C&KM and A.2M ($\alpha_1 = 1$; $\alpha_2 = 2$; $a=2$; $\gamma = 1$)

So C.2 model exhibits faster transition to saturation than C&KM. This can be explained by stronger bonding of water which leads to increasing penetration of water into the sample for intermediate times.

Higher degree of non-linearity stresses this tendencies and brings qualitative differences in free and bond water sorption curves. But nevertheless, total uptake curve remains Fickian.

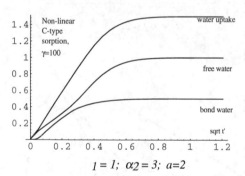

1 = 1; $\alpha_2 = 3$; $a=2$

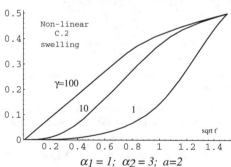

$\alpha_1 = 1$; $\alpha_2 = 3$; $a=2$

In this case the bonding of water is so strong that the sigmoidal effect is shifted towards the origin and after that the "bond" water behaves nearly Fickian.

On the other hand, a lot of "free" water is changing its form and that causes a double change in the curvature of the "free" water sorption curve. But even in this case the total water uptake remains Fickian. This again confirms that the sorption curve is not sensitive to the degree of non-linearity in C-models and their determination with curve-fitting will be difficult.

The swelling curve exhibits s-shape for intermediate γ as observed in the experiments of Adamson (1980).

This investigations permits to conclude that he non-linear Carter&Kibler models are more flexible then the original C&KM and they can cover the case when the relative amplitude of the two stage effect depends on RH. But the integral sorption curve remains qualitatively similar to Carter&Kibler model, although considerable differences in the behaviour of "free" and "bond" water can take place. Therefore, it will be not very easy to determine the degrees of non-linearity in C-model on the base of the usual sorption data.

But probably the most serious theoretical minus is that the Langmuir type behaviour for the bond water equilibrium level can not be reproduced with C-models.

4. NON-LINEAR LANGMUIR TYPE MODELS (L-MODELS)

The fundamental drawback of C&K type models is that they assume the water bonding rate to be proportional only to the free water concentration. In reality there is a some bonding capacity A_1 in the epoxy and the bonding rate is proportional to the saturation of this capacity.

This situation is covered by L-models:

$$(4.1) \quad \frac{\partial C_f}{\partial t} = \frac{\partial^2 C_f}{(\partial z)^2} - \frac{\partial C_b}{\partial t}$$

$$(4.2) \frac{\partial C_b}{\partial t} = \gamma_1 (C_f)^{\alpha_1} (A_1 - C_b)^{\alpha_3} - \beta_1 (C_b)^{\alpha_2}$$

Using the same dimensionless variables as for C-models this system becomes

$$(4.3) \quad \frac{\partial C'_f}{\partial t} = \frac{\partial^2 C'_f}{(\partial z')^2} - \frac{1}{a} \frac{\partial C'_b}{\partial t'}$$

$$(4.4) \frac{\partial C'_b}{\partial t'} = \gamma' [C'_f{}^{\alpha_1}(r - C'_b)^{\alpha_3} - (r - 1)^{\alpha_3} C'_b{}^{\alpha_2}]$$

$$(4.5) \quad \gamma' = (\gamma_1 l^2 / D) (C_0)^{\alpha_1 - 1} C_b{}^{\alpha_3}]$$

$$(4.6) \quad r = A_1/C_{b,\infty} \geq 1$$

This additional parameter indicates that the whole bonding water capacity is only partially saturated because of the presence of the backward "reaction". In the general case the relation between the bond water saturation level and the free water saturation level is explicit

(4.7) $\gamma_1 (C_0)^{\alpha_1} (A_1 - C_{b\infty})^{\alpha_3} = \beta_1 (C_{b\infty})^{\alpha_2}$

In the case $\alpha_1 = \alpha_2 = \alpha_3 = 1$ it gives exactly the Langmuir model for the bond water equilibrium and we shall call it L-model

(4.8) $C_{b,\infty} = A_1 (d C_0)/(1 + d C_0); \quad d = \gamma_1 / \beta_1$

and A_1 is the standard total BW capacity C'_H, d is the affinity constant usually denoted by b in standard textbooks. In this case the governing parameters became functions of C_0.

(4.9) $a = \dfrac{C_0}{C_{b\infty}} = \dfrac{1}{A_1} (\dfrac{1}{d} + C_0)$

(4.10) $r = \dfrac{A_1}{C_{b\infty}} = 1 + \dfrac{1}{d\,C_0}$

Therefore, the relative amplitude of the "two-stage" effect will diminish and the degree of saturation of the total bond capacity will increase for high RH. We shall consider only the L model because it is able to generate the equilibrium sorption theory as a limit case of the sorption kinetics.

If we assume that the *Flory - Huggens* theory holds for the FW saturation level, we can check what it gives for equilibrium sorption

(4.11) $p = B C_0 \exp [(1 + X) - (1 + 2 X) C_0 + X C_0^2]$

where X is a free water - polymer interaction parameter and B is a scaling factor with dimension of pressure.

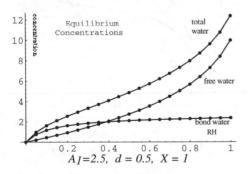

$A_1 = 2.5, \quad d = 0.5, \quad X = 1$

Therefore the L-model is promising for the forms of water which exhibits Langmuir behaviour as water bond by polar interactions and water in "holes" in the epoxy network.

4.1. L.0 - model - No "Freeing" of water

First we shall consider the limit case when the whole bonding capacity is always saturated i.e. there is no "freeing" of water. It corresponds also to high RH, when the whole "bonding" capacity is saturated at equilibrium stage.

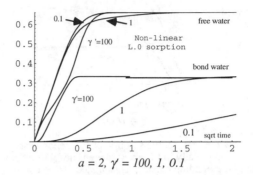

$a = 2, \gamma = 100, 1, 0.1$

From this graph is clear that the bond water capacity is saturated definitely before the free water one for large values of γ i.e. the preferential sorption phenomenon is reproduced.

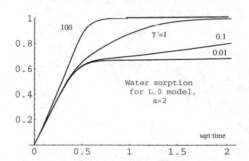

So in general the sorption graph looks like C&KM even when free and bond waters behave different.

Let us now examine the case of large "bonding" rate γ.

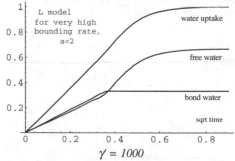

$\gamma = 1000$

The "bond" water increases now linearly and no sigmoidal shape is exhibited. The preferential

269

sorption phenomenon is also visible and the "bond" water concentration is even higher than of the "free" water. But again the sorption curve remains Fickian.

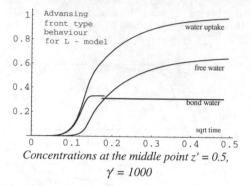

Concentrations at the middle point z' = 0.5,
$$\gamma = 1000$$

Now the preferential sorption phenomenon is very clear, practically in the epoxy there is no "free" water until the bonding capacity is saturated. If we connect the swelling with the presence of "bond" water than this picture corresponds to the moving front sorption mode, when we have a dry core and a swollen epoxy zone.

The bond water graph is always sigmoidal and hence, the sorption curve is sigmoidal, when the bond water capacity is large.

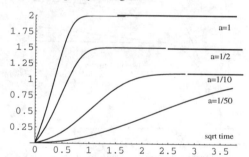

Total water uptake for fixed $\gamma' = 1$

If the "free" water level is higher than the "bond" water one then the Fickian type behaviour prevails and the sigmoidal effect fades out.

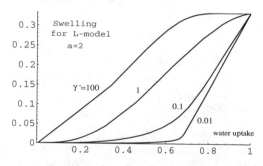

Convex-linear-concave swelling for intermediate γ'

The s-shape behaviour of the swelling curve is well reproduced; it has been experimentally established by Adamson(1980).

4.2. L-model

In the general case the whole "bonding" capacity is only partially saturated due to freeing of water and the degree of saturation depends on RH - see (4.10). The typical phenomena exhibited by the model remains the same and the differences are quantitative.
For illustration we shall consider the case when only 2/3 of the total capacity is saturated

$$C_{b\infty} = 2\, A_1\, /\, 3 \longrightarrow r = 1.5\; ;\; d\, C_0 = 2$$

This corresponds to relatively smaller RH, when not all "bonding" capacity is saturated at equilibrium.

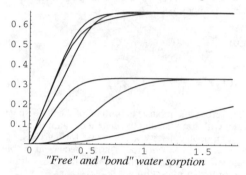

"Free" and "bond" water sorption

The influence of the "bonding" capacity is that BW is increasing faster - compare with L.0 pictures. This effect is so strong, that for large γ and intermediate times BW sorption curve goes slightly above the straight line and the sigmoidal effect is shifted to very low values of time. Also, the water sorption proceeds faster at the transition-to-saturation region.

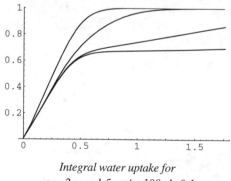

Integral water uptake for
a = 2, r = 1.5, γ'= 100, 1, 0.1

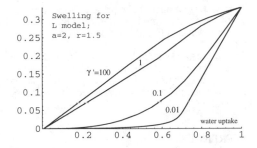

Swelling for L model; a=2, r=1.5

$\gamma'=100$

0.3
0.25
0.2
0.15
0.1
0.05
0

0.2 0.4 0.6 0.8 1

water uptake

Comparing this dimensionless graphs with L.0 case, it is clear that the swelling is higher for small time and the concave regions goes into a linear curve; it well visible for $\gamma'=1$. This is because the total capacity is less saturated now and hence, the bonding rate is higher - see the amplification factor in bonding term of (4.2).

More general values of the degrees of non-linearity in (4.4) will not be considered here because we possess still no experimental information to support their invocation. Also, we have seen with C-model that they do not influence the sorption curves very considerably . Some preliminary investigations are included in Shopov, Frolkovich and de Wilde(1993).

The above presented investigations permits to conclude that L model is able to reproduce a number of experimentally observed phenomena:

• preferential sorption of "bond" water -- for large γ

• advancing front type behaviour -- for very large γ

• sigmoidal sorption curve -- for small a .

• two-stage sorption effect -- for intermediate or small γ'

• Convex-linear-concave "swelling" -- for intermediate γ'

• Langmuir-type behaviour for large time -- for $\beta_1 \neq 0$ and it gives full agreement between the equilibrium sorption theory and the kinetic theory

5. NUMERICAL HANDLING OF MODELS

All considered models can be integrated using relatively standard numerical methods for systems of unsteady non-linear diffusion equations - e.g. Fletcher(1989). Now many mathematical tools are available, which offers also integrated graphic capabilities and we have done the presented calculations with *Mathematica* - Wolfgang (1989). The discretization in space have been done independently and the resulting system of ordinary differential equations has been solved using Mathematica solver for stiff ODF in order to insure the accuracy of the time integration.
This heavy procedure was applied in order to study the all considered models with the same code.
But very often the proposed models reduces to the solution of one non-linear evolution problem for FW.

$$(5.1) \quad \frac{\partial C'_f(z',t')}{\partial t'} = \frac{\partial^2 C'_f(z',t')}{(\partial z')^2} - F(C'_f, z', t')$$

and BW concentration is explicitly expressed as a function of FW.

For example, such formulation is possible for L-model with $\alpha_1 = \alpha_2 = \alpha_3 = 1$, for C-model for $\alpha_2 = 1$ and in some other cases - see Shopov, Frolkovich and de Wilde(1993). We shall present only the expression for L model

$$(5.2) \quad C'_b(z,t) = \mu(z,t) \; \gamma'/a \, r \left(\int_0^t \frac{C'_f(z,s)}{\mu(z,s)} \; ds \right)$$

where $\mu(z,t) = \exp[-\int_0^t \gamma \, C'_f(z,s) \; ds - \beta t]$; .

$\gamma = \gamma_1 \, l^2 / D$

$$(5.3) \quad F(C'_f, z', t') = \gamma'/a \, [C'_f (r - C'_b) - (r - 1) \, C'_b]$$

The quasi linear equation (5.1) can be integrated in a relatively standard way see e.g. Fletcher(1989). For the calculation of the history dependent integrals in (5.2) it is convenient to use a recursive procedure to link BW concentration in two consecutive moments of the discrete time - see Shopov, Frolkovich and de Wilde(1993).
In this formulation the numerical solution of the proposed models is considerably less computationally expensive.

6. CONCLUSIONS AND DISCUSSIONS

The present paper focuses on "two water" models of water sorption in epoxies. First the Carter&Kibler model is analysed because it is the most important representative of this class. It is clear that it could be applicable only if the relative amplitude of the two-stage sorption effect with respect to the pseudo-equilibrium level is constant. Even more, this model reduces to a single sorption curve in relative water uptake with respect to the equilibrium level. Hence, if this model is applicable, the sorption data in has to cluster at a single curve in relative uptake. The preferential sorption of "bond" water and Langmuir type behaviour can not be covered by this model.
If the sorption data do not cluster at a single Carter-Kibler curve, then the constants in C&KM are actually functions and this leads to the non-linear generalisation of Carter&Kibler model - C-models. They possess more flexibility but still are unable to cover Langmuir type behaviour. An important feature of these class of models is that although the behaviour of BW and FW could be rather different than with C&KM, the total water sorption curve is not qualitatively changed.
The impossibility to reproduce the Langmuir behaviour for equilibrium water uptake is primarily connected with the assumption of infinite bonding capacity of the epoxy. This is clearly unrealistic for

many epoxies and it represents a theoretical handicap of C-models.

So the class of non-linear L-models is suggested which takes into account the existing of "bonding" capacity of the epoxy. It exhibits a number of interesting properties
- terns to Langmuir models for large times
- preferential sorption of "bond" water
- double change of curvature of "swelling" graph
- advancing front phenomenon is included
- sigmoidal effect for small time could be covered

The inclusion of the equilibrium sorption theory as a limit case permits to obtain part of the governing parameters from the equilibrium sorption data and than to address the kinetic data. What is typical for this theory is that it addresses the sorption data for all RH at ones as all involved constants does not depend on RH. This permits to get direct predictions for all RH without using additional model function of the dependency of the values governing parameters on RH.

On the other side, it is clear that even "two water" models are not capable to cover fully all possible sorption phenomena. For example, all proposed non-linear models can not reproduce data with sigmoidal behaviour for small times and two-stage sorption for large time, which is possible with the simpler model of De Wilde & Shopov(1994).

The explanation for this demerit is in the simple fact that in the general case the water exists within the epoxy in many forms - often more of two of them plays a significant role:
- *free water(FW)* , which diffuses freely within the epoxy network
- *bond water* (BW)- immobilised by hydrogen bonding and other polar interactions. The swelling and plastification of the epoxy is primarily connected with the concentration of bond water in the epoxy.
- *water in "holes" (WH)* - water clustering in the regions of lower crosslinking density of the epoxy network; this water is also partially immobilised as its diffusion coefficient is much less than of FW
- *trapped water (TW)*- partially immobilised in mechanical voids of the material - microcracks and microvoids. Their relative volume will increase with the degradation of the epoxy.
- *chemically reacted water (CRW)* - normally not significant, could be included in bond water category or to be considered separately.

The assumption is that only FW can transform to some other form (BW, WH, TW, CRW) forth and back; therefore, we need "two waters" model to describe every one of the subprocesses. Clearly, water exist in many forms within the epoxies and different water forms influence the degradation unequally. On the other side, the behaviour of different water forms have to be extracted from the total water sorption curves using also some additional knowledge about the properties of the epoxies and the mathematical models.

Hence, to construct an universal model for water penetration in the epoxy and the related reduction of the mechanical properties, we need to construct "many waters" models. The present work furnishes a set of "two water" models in order to build "many water" models and C and L-models offers some choice.

The future works includes testing of C and L-model on existing data and construction of "three water" models as its seems necessary to cover at least free, bond and in "holes" waters to get more or less universal models.

REFERENCES

Adamson M. J. 1980. Thermal expansion and swelling of cured epoxy resin used in graphite/epoxy composite materials, *J. Material Science,* **15**: 1736 - 1745.

Bonniau, P. & Bunsell, A. R. 1984. A Comparative Study of Water Absorption Theories Applied to Glass Epoxy Composites, *Environmental Effects of Composite Materials,* ed. G. S. Springer, Technomic Publishing Company Inc.

Carter, H. G., and Kibler, K. G. 1978. Langmuir-type model for anomalous moisture diffusion in composite resins, J. Composite Materials, **12**: 118-131

Crank, J. 1975. *The mathematics of diffusion,* Clarendon Press, Oxford, 1975

De Wilde, W. P. & Shopov P. J. 1994. A simple model for moisture sorption in epoxies with sigmoidal and two - stage effects, *J. Composite Structures,* **27**: 12-23 .

Fletcher, C.A.J. 1989 *Computational techniques for Fluid Dynamics 1* , Springer - Verlag.

Mensitieri G., Del Nobile M.A., Apicella A. and Nicolais L. 1995. Moisture - Matrix interactions in polymer based composite materials , *Revue de l'Institute Francais du petrole*, **50**: 1-21.

Shopov P.J., Frolkowich P. and de Wilde P. 1993. Theoretical study of free and bond water models of penetrant sorption in epoxies, *Int. Report of Free University of Brussels.*

Wolfgang W. 1989. *Mathematica*, Springer Verlag, Berlin.

Accelerated evaluation of mechanical degradation behavior of GFRP in hot water

Masato Kasamori, Yoshinori Funada, Kaoru Awazu & Yoshio Watanabe
Industrial Research Institute of Ishikawa, Kanazawa, Japan

Masayuki Nakada & Yasushi Miyano
Materials System Research Laboratory, Kanazawa Institute of Technology, Japan

ABSTRACT: To evaluate the degradation behavior of GFRP accelerated in hot water, the degradation tests were carried out at various constant temperatures in hot water. It was found that the mechanical degradation behavior of the GFRP can be predicted based on the reciprocation law of time and temperature.

1 INTRODUCTION

Glass fiber reinforced plastics, GFRP has been widely used in various environments. Accordingly, reliability over a long time in various environments is required in regards to the strength of GFRP. In many cases, however, it is difficult to execute long term tests. Therefore, the predictive procedures for the degradation of GFRP strength in various environments have become a necessity.

It is well known that the predictive procedures of long term deformation and strength of resins is based on the reciprocation law of time and temperature.

The purpose of this study is to evaluate the degradation behavior of GFRP accelerated in hot water, and then to demonstrate the applicability of the reciprocation law of time and temperature to the prediction of the degradation behavior.

2 EXPERIMENTAL PROCEDURE

The materials used in this study are two kinds of glass fiber reinforced plastics, cloth and mat GFRP. The cloth GFRP consists of woven roving glass fabrics (Nippon sheet glass Co.Ltd., REW580) and an unsaturated polyester resin (Nippon shokubai Co.Ltd., Epolac G752−PTX). The mat GFRP consists of textile glass chopped strand mats (Asahi fiber Co.Ltd., #380) and an unsaturated polyester resin (Takeda chemical industries, Ltd., 8276AP).

The cloth GFRP was formed by lay up, two sheets of woven roving glass fabrics were placed in a mold and the resin was cast into about 1mm thick plate. After hardening, this flat plate was cut so that the

length and width were under 30mm and 10mm, respectively, as shown in Fig.1(a). The mat GFRP was also formed by lay up, a gel coat (Takeda chemical industries, Ltd., #8423) was first applied to the mold, then two sheets of the mats were placed. The resin was then cast into about 3mm thickness. After hardening, this flat plate of mat GFRP was cut into 1000mm long by 15mm wide specimens for three points flexural test, or 800mm long by 10mm wide specimens for Charpy impact test as shown in Fig.1(b) and (c), respectively.

Accelerated degradation tests are carried out at various constant temperatures in hot water by using a decomposition bottle for the cloth GFRP from 100 to 200°C or an autoclave for the mat GFRP from 100 to 250°C as shown in Table 1 and 2.

Table 1. Degradation test conditions of the cloth GFRP.

Temperature (°C)	100, 125, 150, 175, 200
Time (hr.)	1, 2, 4, 6, 10[*1], 24[*1]

*1 for 200°C

Table 2. Degradation test conditions of the mat GFRP.

Temperature (°C)	100, 150, 175, 200, 225, 250
Time (hr.)	0.5[*1], 1, 2, 4, 8, 24[*2]

*1 for 250°C, *2 for 100°C

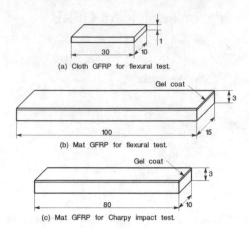

(a) Cloth GFRP for flexural test.

Gel coat

(b) Mat GFRP for flexural test.

Gel coat

(c) Mat GFRP for Charpy impact test.

Fig.1 Specimen configurations.

The degradation of the GFRP is evaluated at room temperature by gravimetry, and the three point flexural tests with a span of 20mm and a deflection rate of 0.5mm/min for the cloth GFRP, and a span of 50mm and a deflection rate of 2mm/min for the mat GFRP. Furthermore, the Charpy impact test was carried out with a span of 60mm for the mat GFRP.

3 RESULTS AND DISCUSSION

The transparency of the cloth GFRP decreased, and the crack occured and grew on the surface with progressing time and increasing temperature in hot water. Regarding the mat GFRP with gel coat, the transparency of the mat GFRP also decreased with progressing time and temperature and the crack of

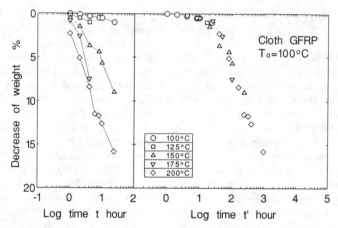

Fig.2 Master curve for decrease of weight of cloth GFRP in H_2O.

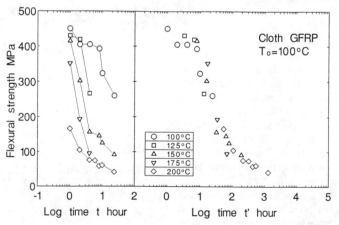

Fig.3 Master curve for flexural strength of cloth GFRP in H_2O.

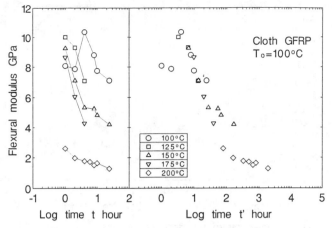

Fig.4 Master curve for flexural modulus of cloth GFRP in H₂O.

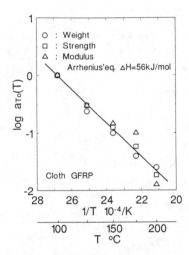

Fig.5 Time-tmperature shift factors $a_{To}(T)$ for weight, strength and modulus of cloth GFRP in H₂O.

the gel coat occured at longer time or higher temperature than the cloth GFRP.

The weights of the cloth GFRP at various temperatures in hot water are shown in the left side of Fig.2. The weight decreases markedly with progressing time and with increasing temperature. The smooth master curve of weight at a reference temperature To=100°C shown in the right side of this figure can be obtained by horizontally shifting the weights at various temperatures.

In regard to the strength and modulus, these also decrease remarkably with increasing time and temperature in the same manner as the weight.

These master curves can be obtained as shown in Figs.3 and 4, respectively.

Figure 5 shows the time−temperature shift factors $a_{To}(T)$ required for composing the master curve of weight, strength and modulus. These shift factors $a_{To}(T)$ can be described by a line which agree with each other, and the activation energies are calculated to be 56kJ/mol.

Then, phthalic acid which was separated from the hot water after the accelerated test was comfirmed by infrared spectroscope. Therefore, this result shows that the loss in weight of the GFRP in hot water is occured by hydrolysis of the unsaturated polyester resin used as the matrix in the GFRP.

Concerning the mat GFRP, the weights at various temperatures are shown in Fig.6. The weight decreases markedly with progressing time and with increasing temperature as dose the cloth GFRP. The master curve of weight at a reference temperature To=100°C can be also obtained. The strength, modulus and impact value of the mat GFRP decreases remarkably with increasing time and temperature in the same manner as the weight, and the respective master curves are shown in Figs.7 to 9. Figure 10 shows the time−temperature shift factors $a_{To}(T)$ of weight, strength, modulus and impact value of the mat GFRP. These sift factors $a_{To}(T)$ can be also described by a line, and the activation energies are calculated to be 57kJ/mol. This value is almost equal to the value of the cloth GFRP.

Therefore, these facts show that the reciprocation law of time and temperature holds for the degradation of GFRP in hot water.

275

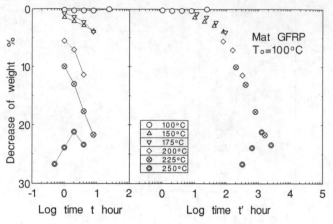

Fig.6 Master curve for decrease of weight of mat GFRP in H_2O.

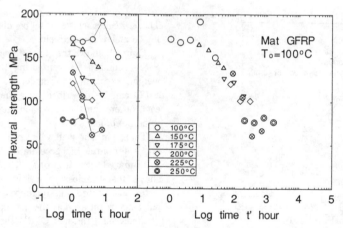

Fig.7 Master curve for flexural strength of mat GFRP in H_2O.

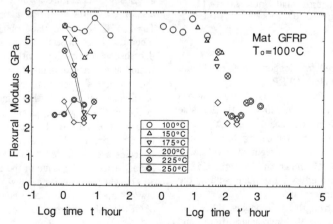

Fig.8 Master curve for flexural modulus of mat GFRP in H_2O.

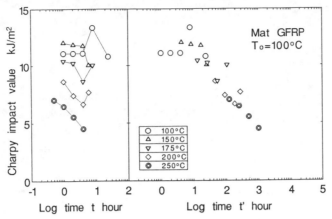

Fig.9 Master curve for charpy impact value of mat GFRP in H_2O.

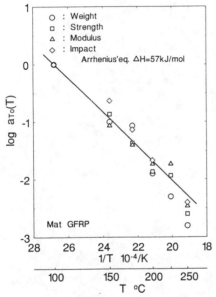

Fig.10 Time-temperature shift factors $a_{T_0}(T)$ for weight, strength, modulus and impact of mat GFRP in H_2O.

REFERENCES

Miyano,Y., M.Kanemitsu & T.kunio, 1983. Time and Temperature Dependence of Flexural Strength in Transversal Direction of Fibers in CFRP, *Fibre Sci. and Tech.* 18:65−79.

Miyano,Y., M.Kanemitsu, T.Kunio & H.A.Kuhn, 1985. A Study on Fracture Mechanisms of Unidirectional CFRP. *SAMPE J. Sep./Oct.*:33−40.

Miyano,Y., M.Kanemitsu, T.Kunio & H.A.Kuhn, 1986. Role of Matrix Resin on Fracture Strengths of Unidirectional CFRP. *J.Comp.Mat.* 20:520−538.

Miyano,Y. & M.Kasamori, 1994. A Study on Long Term Creep Behavior for Matrix Epoxy Resin and CFRP, *Proc.1994 SEM Spring Conference on Experimental Mechanics*:268−273.

Hojo,T., K.Tsuda, K.Ogasawara & K.Mishima, 1982. The Effects of Stress and Flow Velocity on Corrosion Behavior of Vinyl Ester Resin and Glass Composites, *Proc.ICCM−IV*, 2:1017−1024.

Hojo,T., K.Tsuda, K.Ogasawara & T.Takizawa, 1986. Corrosion Behavior of Epoxy and Unsaturated Polyester Resins in Alkaline Solution, *ACS Symp. Ser.* 322:314−326.

4 CONCLUSION

Accelerated degradation tests are carried out at various constant temperatures above $100\,^{\circ}C$ in hot water. The reciprocation law of time and temperature holds for the degradation behavior of the GFRP. Therefore, it is found that the degradation behavior of the GFRP can be predicted based on the reciprocation law of time and temperature.

6 Specific systems and structures

Progress in Durability Analysis of Composite Systems, Cardon, Fukuda & Reifsnider (eds)
© *1996 Balkema, Rotterdam. ISBN 90 5410 809 6*

Transient hygrothermal stresses in laminated cylinders

D. Paul & A. Vautrin
Centre Sciences des Matériaux et des Structures, Département Mécanique et Matériaux, Ecole des Mines de Saint-Etienne, France

ABSTRACT: In this paper, a simple but accurate analytical method is developed for the determination of transient hygrothermal stresses in thick laminated cylinders. Temperature and moisture distributions are approached through a finite difference method, and the compatibility of deformations are explicitly verified. Three examples are introduced at the end of the paper to illustrate some effects of curvature and transient hygrothermal states on internal stresses.

1 INTRODUCTION

Advanced fibre composites are widely used in aerospace applications and high performance structures. The severe environment to which they are submitted has lead more and more attention being paid to their thermomechanical behaviours. High hygrothermal gradients can occur, and their influence on the long-term behaviour of composite materials is still unknown. A better understanding of hygrothermal loads in multilayered structures could be of interest to the designers and becomes necessary to improve our knowledge of life-time durability.

1.1 *Multilayered or orthotropic composite plates*

The determination of curing stresses in multilayered materials has been well described by Tsai (1987) for steady states and is well understood today.

Many methods can be used for the determination of hygrothermal transient fields (Fourier cosine transform and Laplace transform, method of separation of variables...) and the results are often expressed as real series.

Ootao *et al.* (1990) have used this type of solution with the classical laminated plates theory to assess the transient stress fields in multilayered, anisotropic, laminated slabs. The compatibility of strain fields, as will be discussed in the following pages, is not expressly verified, so no certainty at all can be paid to the accuracy of their solution.

The paper of Wang and Chou (1989) studied the three-dimensional transient interlaminar thermal

stresses in elastic, angle-ply laminated composites due to sudden changes in the thermal boundary conditions, but the same kind of remarks could be made.

An interesting approach has been recently presented by Benkeddad *et al.* (1995) for the transient absorption of moisture in laminated composite plates. Each ply is divided into 'subplies' and the exact moisture concentration distribution is approached by linear segments, thanks to a finite difference method. As a result, strain and stress components are linear by parts and equations of compatibility are verified within each division.

Based upon the solution of transient temperature field, an elastic displacement-potential approach was used by Wang *et al.* (1986) in their thermal stress analysis of orthotropic medium with rectangular boundary. This approach appears to be rigorous; however, its extension to multilayered plates will lead to great mathematical difficulties.

Making use of Navier-like approach, exact three-dimensional solutions were recently obtained for cross-ply and antisymmetric angle-ply laminated rectangular plates subjected to thermomechanical loads by Savoia and Reddy (1995). To the authors knowledge, it is one of the strictest papers concerning transient internal stresses in composite structures, but unfortunately, the governing equations remain very complex.

1.2 *Composite tubes*

The effect of a uniform temperature change on the stresses for several angle-ply tube designs was considered by Hyer and Rousseau (1987) through an analytical method with an explicit verification of the compatibility equations.

An analysis of a thick cylinder consisting of homogeneous layers of different materials and subjected to a uniform thermal loading was performed by Wang and Lin (1993). Each layer was considered as an individual thin elastic shell with interfacial stresses on inner and outer surfaces. The compatibility conditions were satisfied on average through each layer, but this was basically a two-dimensional analysis.

The influence of stacking sequences on the level of burst pressures of thick laminated vessels was studied through an analytical method by Tsai (1987). Optimum angle combinations were established, but no evaluation of curing stresses was considered.

Byon and Vinson (1990) have answered this question with a simple and accurate finite cylindrical element method. High values of tensile radial strains were calculated in a thermal steady state showing that a transient study could be very useful.

The response of composite tubes to a circumferential temperature gradient was presented by Hyer and Cooper (1986). An analytical method was performed and results showed quite high values of stresses. However, the form of the stress field is strictly limited to circumferential gradients.

Chen (1983) has solved many problems of cylinders subjected to transient heat conduction by a direct power series approximation of the temperature. This method could be utilised in more complex cases such as time-dependent problems, but the relevancy of this approximation could be disputed.

At last, potential functions of displacement were utilised by Misra and Achari (1980) on the one hand, and Ootao et al. (1991) on the other hand to establish exact solutions of the transient stress fields in thick hollow cylinders (respectively anisotropic and isotropic cylinders). Although those methods appear to be efficient for simple structures, the difficulty remains very high in tensorial approaches like those required for multilayered fibre reinforced composites.

On account of this non-exhaustive description, one can realise that a growing attention is paid to hygrothermal effects on laminated structures. The purpose of this study is to present an accurate method, based on simple assumptions, to compute the transient and steady hygrothermal stresses in thick laminated cylinders. The analytical solution developed here enables us to evaluate curvature effects and transient hygrothermal states on internal stress fields.

2 FORMULATION OF THE PROBLEM

2.1 Calculation of the hygrothermal fields

Both thermal and hygroscopic transient problems can be approached with this method since the heat conduction and the Fick equations are of the same form. In this paper, only a moisture absorption under constant temperature change is assumed.

Consider an infinite length annular laminated cylinder subjected to a moist environment on its inner and outer surfaces (Fig.1).

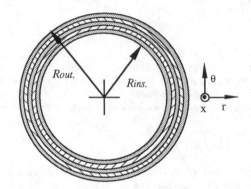

Fig.1 Laminated cylinder geometry

The diffusion phenomena is considered as axisymmetric, unidimensional and the initial moisture concentration C_{ini} is uniform at t = 0. Both surfaces of the cylinder are suddenly exposed to a moist environment. The diffusivity D is considered as a constant through the laminate thickness. The moisture concentration C inside the cylinder is then described by the Fick equation (Springer 1981).

$$\frac{\partial C}{\partial t} = D \left(\frac{1}{r}\frac{\partial C}{\partial r} + \frac{\partial^2 C}{\partial r^2} \right) \qquad (1)$$

with $C = C_{ini}$ at $t = 0$ for $R_{ins.} \leq r \leq R_{out}$. (2)
$\quad$ $C = C_{ins}$ at $t > 0$ for $r = R_{ins}$.
$\quad$ $C = C_{out}$ at $t > 0$ for $r = R_{out}$.

An explicit finite difference method, similar to the one utilized by Benkeddad et al. (1995) for laminated plates, is used to solve eqn (1) with initial and constant boundary conditions (2). Each ply is divided into 'subplies', within which the moisture concentration is assumed to vary in a logarithmic way. This last assumption is justified by the distribution of moisture content in steady state through a homogeneous cylinder subjected to different boundary conditions on its inner and outer surfaces.

The use of the explicit finite difference scheme is very simple and very easy to program but is conditionally stable. The stability condition of the calculation is drawn from the different rules of stability expressed by Suhas and Patankar :

$$\Delta t \le \frac{Rins.\Delta r^2}{D\,(\Delta r + 2Rins.)} \quad \text{where} \quad \begin{cases} \Delta r : \text{space step} \\ \Delta t : \text{time step} \end{cases} \quad (3)$$

In this study, Δr is equal to one fifth of the ply thickness and Δt is adjusted from condition (3). The iteration is automatically stopped when the moisture content in the laminate mid-plane is equal to 95% of $C_{ins.}$ or $C_{out.}$ (t 95%). Finally, the distributions of moisture content through the thickness of the cylinder at each time step are stored in a numerical file which can be utilized by the internal stresses computation program.

2.2 Compatibility of strain fields

Since the hygrothermal expansion coefficients of a fibre-reinforced composite are usually anisotropic, the laminated structures are often subjected to internal stress fields. The residual strains ε^r of a ply are defined as the difference between its non-mechanical strains ε^{nm}, that is to say its real strains in the laminated structure, and its free strains :

$$\varepsilon^r = \varepsilon^{nm} - \varepsilon^f \qquad (4)$$

Usually, the free strains of a ply from a hygrothermal state to another are assumed to be proportional to the temperature and moisture variations :

$$\varepsilon^f = \alpha\Delta T + \beta C \qquad (5)$$

with $\begin{cases} \alpha,\beta : \text{hygrothermal expansion coefficients} \\ C,\Delta T : \text{moisture and temperature variations} \end{cases}$

In most cases of transient states, $\alpha\Delta T + \beta C$ do not verify the compatibility equations of mechanics, implying that a complementary field has to be added so as to ensure that the compatibility of strains is verified (Surrel and Vautrin 1992). This is one of the origin of internal stresses in laminated structure during transient states of moisture or temperature.

2.3 Assumptions of the analytical method

The following analytical method for the calculation of transient internal stresses in thick laminated cylinders is based on several assumptions :

1. no mechanical loads are applied to the cylinder;

2. hygrothermal fields are axisymmetric and independent of the coordinate axis x;

3. the cylinder is supposed to be infinitely long, so that the determination of stress fields is performed away from the ends;

4. all the plies are made of the same material;

5. the material properties are supposed to be linear elastic and independent of moisture and temperature;

2.4 Analytical method

Consider an angle-ply laminated cylinder made of n orthotropic layers subjected to a transient moist environment C and a constant temperature change ΔT.

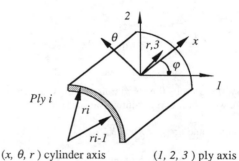

$(x,\,\theta,\,r\,)$ cylinder axis $\qquad$ $(1,\,2,\,3\,)$ ply axis

Fig.2 Reference axis

* Residual stress-strain relations of each ply in the $(x,\,\theta,\,r\,)$ axis :

$$\begin{Bmatrix} \sigma_x \\ \sigma_\theta \\ \sigma_r \\ \tau_{x\theta} \end{Bmatrix} = \begin{bmatrix} C_{xx} & C_{x\theta} & C_{xr} & C_{xs} \\ C_{x\theta} & C_{\theta\theta} & C_{r\theta} & C_{s\theta} \\ C_{xr} & C_{r\theta} & C_{rr} & C_{rs} \\ C_{xs} & C_{s\theta} & C_{rs} & C_{ss} \end{bmatrix} \begin{Bmatrix} \varepsilon_x - \alpha_x\Delta T - \beta_x C \\ \varepsilon_\theta - \alpha_\theta\Delta T - \beta_\theta C \\ \varepsilon_r - \alpha_r\Delta T - \beta_r C \\ \gamma_{x\theta} - \alpha_{x\theta}\Delta T - \beta_{x\theta}C \end{Bmatrix}$$
$$(6)$$

$$\begin{Bmatrix} \tau_{r\theta} \\ \tau_{xr} \end{Bmatrix} = \begin{bmatrix} C_{r\theta\,r\theta} & C_{r\theta\,xr} \\ C_{xr\,r\theta} & C_{xr\,xr} \end{bmatrix} \begin{Bmatrix} \gamma_{r\theta} \\ \gamma_{xr} \end{Bmatrix} \qquad (7)$$

The C_{ij} represent the rigidity matrix components of each ply in the cylinder axis. They depend on the material properties (E_i, G, ν_{ij}) and the ply orientation φ. ΔT and C represent the temperature and moisture variations.

* Strain-displacement relations :

$$\varepsilon_r = \frac{\partial w}{\partial r} \qquad \varepsilon_\theta = \frac{w}{r} \qquad \varepsilon_x = \frac{\partial u}{\partial x}$$

$$\gamma_{\theta x} = \frac{\partial v}{\partial x} \qquad \gamma_{xr} = \frac{\partial u}{\partial r} \qquad \gamma_{r\theta} = \frac{\partial v}{\partial r} - \frac{v}{r} \tag{8}$$

* Equilibrium equations :

$$\begin{cases} \dfrac{\partial \sigma_r}{\partial r} + \dfrac{\sigma_r - \sigma_\theta}{r} = 0 \\[2mm] \dfrac{\partial \tau_{r\theta}}{\partial r} + 2\dfrac{\tau_{r\theta}}{r} = 0 \\[2mm] \dfrac{\partial \tau_{xr}}{\partial r} + \dfrac{\tau_{xr}}{r} = 0 \end{cases} \tag{9}$$

* Compatibility equations :

$$\begin{cases} \dfrac{\partial^2 \varepsilon_x}{\partial r^2} = 0 \\[2mm] \dfrac{1}{r}\dfrac{\partial \varepsilon_x}{\partial r} = 0 \\[2mm] \dfrac{\partial^2 \gamma_{x\theta}}{\partial r^2} + \dfrac{1}{r}\dfrac{\partial \gamma_{x\theta}}{\partial r} - \dfrac{1}{r^2}\gamma_{x\theta} = 0 \end{cases} \tag{10}$$

By integrating the three compatibility equations (10), we obtain two conditions on strains :

$$\varepsilon_x = R_1 \quad \text{and} \quad \gamma_{x\theta} = R_2 r + K_0 \tag{11}$$

where R_1, R_2, K_0 are constants.

The two last equilibrium equations (9) give the expression of transverse shear stresses within each ply :

$$\tau_{r\theta} = \frac{R_3}{r^2} \quad \text{and} \quad \tau_{xr} = \frac{R_4}{r} \quad R_3, R_4 \text{ are constants.} \tag{12}$$

The expressions of the displacements $u(x,r)$ and $v(x,r)$ can then be drawn from the strain-stress relations (7) and the strain-displacement equations (8) :

$$\begin{cases} u(x,r) = -S_{r\theta\,xr}\dfrac{R_3}{r} + S_{x\,r\,x\,r}\,R_4 Lnr + R_1 x + R_5 \\[2mm] v(x,r) = R_2 x\,r - \dfrac{S_{r\theta\,r\theta}}{2}\dfrac{R_3}{r} - S_{r\theta\,xr}\,R_4 + R_6 r \\[2mm] K_0 = 0 \end{cases} \tag{13}$$

where R_5, R_6 are constants and $[S_{ijkl}] = [C_{ijkl}]^{-1}$.

At this stage of the analysis, one is reminded that the transient hygroscopic profil within a ply is assumed to be logarithmic : $C(r) = A_i \ln r + B_i \quad r_{i-1} \le r \le r_i$ (14)
The reason of this choice has been previously given. The coefficients A_i and B_i are determined at each time, for each ply thanks to the interface hygroscopic values calculated with the finite difference method.

Finally, the remaining equilibrium equation leads to the following differential equation :

$$r^2 \frac{\partial^2 w}{\partial r^2} + r\frac{\partial w}{\partial r} - \frac{C_{\theta\theta}}{C_{rr}}w = \frac{(A_1 + A_2)r + Br^2 + Cr\ln r}{C_{rr}} \tag{15}$$

$$\begin{cases} K_1^T = C_{xr}\alpha_x + C_{r\theta}\alpha_\theta + C_{rr}\alpha_r + C_{rs}\alpha_{x\theta} \\ K_1^m = C_{xr}\beta_x + C_{r\theta}\beta_\theta + C_{rr}\beta_r + C_{rs}\beta_{x\theta} \\ K_2^T = C_{x\theta}\alpha_x + C_{\theta\theta}\alpha_\theta + C_{r\theta}\alpha_r + C_{s\theta}\alpha_{x\theta} \\ K_2^m = C_{x\theta}\beta_x + C_{\theta\theta}\beta_\theta + C_{r\theta}\beta_r + C_{s\theta}\beta_{x\theta} \\ A_1 = (C_{x\theta} - C_{xr})R_1 + K_1^m A_i + (K_1^m - K_2^m)B_i \\ A_2 = (K_1^T - K_2^T)\Delta T \\ B = (C_{s\theta} - 2C_{rs})R_2 \\ C = (K_1^m - K_2^m)A_i \end{cases}$$

This second order differential equation can be solved by means of a simple variable change :

$$r = e^z \quad \text{and} \quad D = \frac{d}{dz} \tag{16}$$

The resolution of the constant coefficients differential equation obtained gives the expression of the radial displacement :

$$w = R_7 r^{\sqrt{K_s}} + R_8 r^{-\sqrt{K_s}} + Er + Fr^2 + Gr\ln r \tag{17}$$

$$\begin{cases} K_3 = C_{\theta\theta}/C_{rr} \\ K_4 = (C_{x\theta} - C_{xr})/(C_{rr} - C_{\theta\theta}) \\ K_5 = K_1^m(A_i + B_i) + K_1^T\Delta T - K_2^m B_i - K_2^T\Delta T \\ K_6 = 2(K_1^m - K_2^m)A_i C_{rr} \\ K_7 = (K_5(C_{rr} - C_{\theta\theta}) - K_6)/(C_{rr} - C_{\theta\theta})^2 \\ K_8 = (C_{s\theta} - 2C_{rs})/(4C_{rr} - C_{\theta\theta}) \\ E = K_4 R_1 + K_7 \quad \text{and} \quad F = K_8 R_2 \\ G = (K_1^m - K_2^m)A_i/(C_{rr} - C_{\theta\theta}) \\ R_7, R_8 : \text{constants} \end{cases}$$

Finally, at time t, eight unknown constants R_1, R_2, R_3, R_4, R_5, R_6, R_7, R_8 are to be determined for each ply. This resolution can be performed by satisfying several mechanical conditions described in the following lines.

* Rigid body motion

It can be suppressed by setting
$$R_5 = R_6 = 0 \text{ in the innermost layer.} \qquad (18)$$

* Continuity of u and v displacements

$$\begin{cases} u^{(i)}(r_i) = u^{(i+1)}(r_i) \\ v^{(i)}(r_i) = v^{(i+1)}(r_i) \end{cases} \Rightarrow \begin{cases} R_1^{(i)} = R_1^{(i+1)} \\ R_2^{(i)} = R_2^{(i+1)} \\ R_5^{(i)} = R_5^{(i+1)} \\ R_6^{(i)} = R_6^{(i+1)} \end{cases} i \in [1, n-1]$$

where i refers to the ith ply and r_i represents its outer radius.
With conditions (18) we have :

$$\begin{cases} R_1^{(i)} = \text{constant} \\ R_2^{(i)} = \text{constant} \\ R_5^{(i)} = 0 \\ R_6^{(i)} = 0 \end{cases} i \in [1, n] \qquad (19)$$

* Continuity of displacement w

$$w^{(i)}(ri) = w^{(i+1)}(ri) \quad i \in [1, n-1] \qquad (20)$$

* Continuity of $\tau_{\theta r}$ and τ_{xr}

$$\begin{cases} \tau_{\theta r}^{(1)}(R_{ins.}) = \tau_{xr}^{(1)}(R_{ins.}) = 0 \\ \tau_{\theta r}^{(n)}(R_{out.}) = \tau_{xr}^{(n)}(R_{out.}) = 0 \end{cases}$$

and $\qquad (21)$

$$\begin{cases} \tau_{\theta r}^{(i)}(r_i) = \tau_{\theta r}^{(i+1)}(r_i) \\ \tau_{xr}^{(i)}(r_i) = \tau_{xr}^{(i+1)}(r_i) \end{cases} i \in [1, n-1]$$

From equation (12), we deduce :

$$R_3^{(i)} = R_4^{(i)} = 0 \quad i \in [1, n] \qquad (22)$$

Then, it is not possible to assess the transverse shear stress with the assumptions made in this method.

* Boundary conditions and continuity of normal stress σ_r

$$\begin{cases} \sigma_r^{(1)}(R_{ins.}) = 0 \\ \sigma_r^{(n)}(R_{out.}) = 0 \end{cases}$$

and $\qquad (23)$

$$\sigma_r^{(i)}(r_i) = \sigma_r^{(i+1)}(r_i) \quad i \in [1, n-1]$$

* Integral boundary conditions

$$\begin{cases} 2\pi \sum_{i=1}^{n} \int_{ri-1}^{ri} \sigma_x (r) r \, dr = 0 \\ 2\pi \sum_{i=1}^{n} \int_{ri-1}^{ri} \tau_{x\theta} (r) r^2 \, dr = 0 \end{cases} \qquad (24)$$

Finally, the determination of the transient hygrothermal displacement field u, v, w at time t for an n-layer tube imposes the resolution of a 2n+2 equations-2n+2 unknowns. The transient stress field σ_1, σ_2, σ_3, τ_{12} is then directly deduced from the transient residual strains :

$$\begin{Bmatrix} u \\ v \\ w \end{Bmatrix} \Rightarrow \begin{Bmatrix} \varepsilon_x \\ \varepsilon_\theta \\ \varepsilon_r \\ \gamma_{x\theta} \end{Bmatrix} \Rightarrow \begin{Bmatrix} \varepsilon_1 - \alpha_1 \Delta T - \beta_1 C \\ \varepsilon_2 - \alpha_2 \Delta T - \beta_2 C \\ \varepsilon_3 - \alpha_3 \Delta T - \beta_3 C \\ \gamma_{12} \end{Bmatrix} \Rightarrow \begin{Bmatrix} \sigma_1 \\ \sigma_2 \\ \sigma_3 \\ \tau_{12} \end{Bmatrix}$$

3 RESULTS

In this section, short validations of the program developed above are presented. Then, numerical examples are discussed to introduce the effects of curvature and transient states on internal stress fields in composite laminated structures.

3.1 Validation of the method

The accuracy of the finite difference method for the calculation of the transient hygroscopic profile inside the laminate was already checked by Benkeddad et al. (1995). The difference between this method and an analytical solution given by Crank (1975) is less than 0.2%.

The level and the distribution of stresses inside a thin laminated cylinder calculated by Hyer and Rousseau (1987) for a uniform change in temperature were exactly verified by the present method which is based on similar assumptions.

The thickness effects on thermal stress fields were compared with Timoshenko's analytical solution (1961) for an isotropic cylinder. The maximum discrepancy between the two methods was 1.7%, which can be considered as a correct result.

Finally, the distribution of internal stresses during moisture absorption or desorption was computed for a thin laminated cylinder ($R_{out.} / R_{ins.}$ <1.05). Good agreements were shown with results obtained in thin laminated plates by Benkeddad et al. (1995).

3.2 Material properties

The thermo-mechanical properties of the material used in this study come from Tsai (1987) and are listed in Table 2. Only angle-ply laminated cylinders [$\pm\varphi$] are studied here. Note that by lack of data on radial properties, E_3, v_{23}, α_3 and β_3 were reasonably chosen by the authors. It should also be remembered that hygroscopic and thermal independent material properties were assumed.

3.3 Simulations

Fig. 3 represents the evolution of thermal stresses in fibre direction σ_1 for different angle-ply laminates. The material is assumed to be completely dry and the residual stresses are computed for a ΔT equal to 100°C below the "free stress temperature". Only stresses at the inner radius $R_{ins.}$ of the cylinders are represented and three different cases listed in Table 1 are considered.

Note that a cylinder is considered as a thick cylinder when its outer-inner radius ratio $R_{out.} / R_{ins.}$ is greater than 1.1 (Tsai 1987). Fig. 3 clearly indicates that over a maximum angle of about 30° the levels of thermal stresses can reach very high values for thick cylinders. This phenomena can be justified by the growing difference existing between the radial and the circumferential expansion coefficients from φ =30° to φ =90°. In the case $R_{out.} / R_{ins.}$ = 3, the maximum value of 160 MPa is obtained for an angle not far from 55°. It is interesting to note that the "best angle" determined by Tsai (1987) for thick pressure vessels in those type of angle-ply laminates is near from 60°, without taking into account thermal stresses.

Figs. 4 and 5 represent respectively σ_1 (along the fibres) at the inner radius $R_{ins.}$ and σ_2 (transverse to the fibres) in the laminate mid-plane. The laminate considered is a [$\pm\varphi$]$_{2s}$. Its inner radius is equal to 2 cm and its $R_{out.} / R_{ins.}$ ratio has a value of 1.4. At time t = 0, the whole dry cylinder is subjected on its inner and outer surfaces to symmetric moisture conditions listed in Table 3. One should point out from fig. 4 that for a 90° angle-ply laminate, the presence of moisture has made σ_1 drop from a positive to a negative value. Therefore, this study confirms that there probably exists a content of moisture which minimizes the level of thermal stresses.

Fig. 5 shows that for angles smaller than 20°, much higher values of transverse stresses σ_2 can be found in the transient state than in the dry state or in the moisture saturation state. Similar results have been found by Benkeddad et al. (1995) on laminated plates.

Table 1. Laminated structures

	laminate	$R_{out.}/R_{ins.}$	$R_{ins.}$ m
Case 1	[$+\varphi,-\varphi,+\varphi,-\varphi,+\varphi$]$_s$	1.05	0.1
Case 2	[$+\varphi,-\varphi,+\varphi,-\varphi,+\varphi$]$_s$	1.5	0.1
Case 3	[$+\varphi,-\varphi,+\varphi,-\varphi,+\varphi$]$_s$	3	0.1

Table 2. Mechanical properties of the material

Material	E_1 GPa	E_2 GPa	E_3 GPa	v_{12}	v_{13}	v_{23}	G_{12} GPa	α_1 10^{-6} K^{-1}	α_2, α_3 10^{-6} K^{-1}	β_1	β_2, β_3
Graphite/epoxy	181	10.3	10	0.28	0.28	0.43	7.17	0.02	22.5	0	0.6

Table 3. Environmental conditions and characteristics of the moisture absorption

Material	Temperature °C	Temperature change °C	Relative humidity %	D_r mm^2/s	C_{ini} %	C_{max} %
Graphite/epoxy	40	-100	100	3.9x10^{-8}	0	1.5

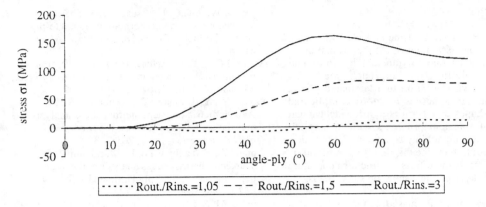

Fig. 3 $\sigma 1$(Rins.) function of ply angle for $\mathsf{c} = 0$

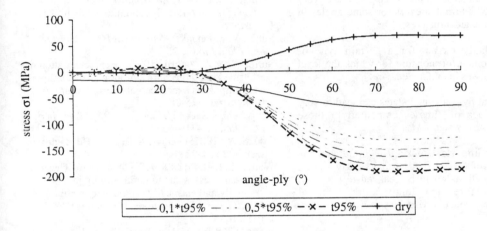

Fig. 4 $\sigma 1$(Rins.) function of ply angle during absorption - Rout./Rins.=1,4

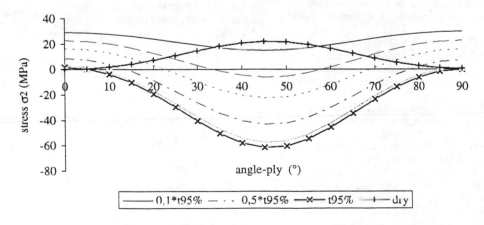

Fig. 5 $\sigma 2$((Rins.+Rout.)/2) function of ply angle during absorption - Rout./Rins.=1,4

CONCLUSION

In this paper an analytical method was presented, based on simple assumptions, for the evaluation of transient stresses due to transient hygrothermal states. Completing the approach made by Benkeddad *et al.* (1995) on thin laminated plates, this method gives access to more complicated structures shapes and permits a strict mechanical approach of transient problems with an explicit verification of compatibility equations of mechanics. The validation has shown that this method is efficient to assess the effects of structures curvature and transient hygrothermal states on stress fields.

The three examples discussed at the end of this paper have illustrated three main points :

1. The curvature of cylinders can give rise to very high values of internal stresses for some angles in angle-ply laminated structures.

2. It probably exists, for a certain type of laminate, a content of moisture for which the level of internal stresses is minimized.

3. Transient hygrothermal states can lead to very different values of stresses compared to those calculated in steady states.

Finally, it is important to keep in mind that the development of this method is at its early stages. Some modifications could be realised to take into account moisture and temperature material properties dependence or viscoelastic effects.

REFERENCES

Benkeddad, A., M. Grédiac & A. Vautrin 1995. On the transient hygroscopic stresses in laminated composite plates. *Composite Structures.* 30 : 201-205.

Byon, O & J.R. Vinson 1990. Stress analyses of thick walled cross-ply composite cylindrical shells taking account of curing stresses. *Achievements in Composites in Japan and the United States* : 257-264. Tokyo.

Chen, P.Y.P. 1983. Axisymmetric thermal stresses in an anisotropic finite hollow cylinder. *Journal of Thermal Stresses . 6* : 197-205.

Cranck, J. 1975. *The Mathematics of Diffusion.* 2nd edition. Oxford University Press. London.

Hyer, M.W. & D.E. Cooper 1986. Stresses and deformations in composite tubes due to a circumferential temperature gradient. *Journal of Applied Mechanics.* 53 : 757-764.

Hyer, M.W. & C.Q. Rousseau 1987. Thermally induced stresses and deformations in angle-ply composite tubes. *Journal of Composite Materials.* 21 : 454-480.

Misra, J.C. & R.M. Achari 1980. On axisymmetric thermal stresses in an anisotropic hollow cylinder. *Journal of Thermal Stresses.* 3 : 509-520.

Ootao, Y., Y. Tanigawa & T. Fukuda 1991. Axisymmetric transient thermal stress analysis of a multilayered composite hollow cylinder. *Journal of Thermal Stresses.* 14 : 201-213.

Ootao, Y., Y. Tanigawa & H. Murakami 1990, Transient thermal stress analysis and bending behavior of an angle-ply laminated slab. *Journal of Thermal Stresses.* 13 : 177-192.

Savoia, M. & J.N. Reddy 1995. Three-dimensional thermal analysis of laminated composites plates. *Int . J. Solid Structures.* 32 (5) : 593-608.

Springer, G.S. 1981. *Environmental effects on composite materials.* Technomics Publishing. Westport.

Suhas, V. & Patankar. *Numerical Heat Transfert and Fluid Flow* : 25-73.

Surrel, Y. & A. Vautrin 1992. La caractérisation des effets de l'humidité sur les propriétés mécaniques des composites à matrice polymère. *Annale des Composites* : 15-28.

Timoshenko, S. & J.N. Goodier 1961. *Théorie de l'Elacticteé* : 431-446.

Tsai, S.W. 1987. *Composite Design.* Third edition. section 23 : 1-22

Wang, H., R.B. Pipes & T.S. Chou 1986. Thermal transient stresses due to rapid cooling in a thermally and elastically orthotropic medium. *Metallurgical Transactions A.* Volume 17A : 1051-1055.

Wang, J.T.S & C.C. Lin 1993. Thermal stresses in layered cylinders. *Modelling and Processing Science..* ICCM 9. 3 : 104-111.

Wang, Y.R. & T.W. Chou 1989. Three-dimensional transient interlaminar stresses in angle-ply composites. *Journal of Applied Mechanics.* 56 : 601-608

Progress in Durability Analysis of Composite Systems, Cardon, Fukuda & Reifsnider (eds)
© 1996 Balkema, Rotterdam. ISBN 90 5410 809 6

Influence of delamination on geometrically nonlinear composite shells

O. M. M. Teyeb & D. Weichert
Laboratoire de Mécanique de Lille, CNRS-URA, Villeneuve d'Ascq, France

R. Schmidt
Bergische Universität GH Wuppertal, Germany

ABSTRACT : A geometrically non-linear formulation for axisymmetric shells using a total Lagrangian approach is presented. A nine node finite element model has been developed that takes into account the rotation of normals to the midsurface. A model of delamination using a representative local volume is used. Numerical examples are presented to illustrate the element behaviour and the accuracy of the approach.

1 INTRODUCTION

Plate and shell structures are found frequently in aerospace applications. Quite often these structures are made of composite materials. In many practical cases, the analysis of plate and shell structures requires computational methods like the finite element method and many papers have been written on shell element formulations and their extensions to geometric nonlinearity. These formulations assume in general nodal rotations to be small. With this assumption, the element displacement field becomes a linear function of the nodal rotations. The geometrically nonlinear shell formulation based on this linearized displacement field restricts the magnitude of nodal rotations and requires small load increments, so that between two successive load increments the nodal rotations remain small. The accuracy of analysis with such elements begins to deteriorate when the elements start experiencing finite rotations.

In this paper a total Lagrangian formulation is presented for the axisymmetric shell elements for which the displacements are quadratic functions of nodal rotations. The effect of interply delamination enters the formulation through modifications of strains in the constitutive law. The corresponding internal state variables are defined utilizing the kinematics of the interply delamination region. These internal state variables depend on the components of the displacements created by the delamination.

2 KINEMATICS OF DEFORMATION

Fig. 1 shows a portion of the shell in the undeformed configuration. Two co-ordinate systems are used. The global Cartesian co-ordinate is represented by (x,y,z) axes fixed in space. On the other hand a local co-ordinate system $(\vartheta^1, \vartheta^2, \vartheta^3)$ is defined by vectors a_1, a_2, a_3, with a_1 and a_2 tangent to the shell midsurface and a_3 normal to the shell midsurface.

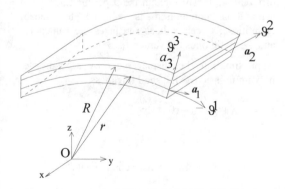

Fig. 1. : Kinematics of a shell element.

The position of any point in space can the be given by

$$R(\theta^1, \theta^2, \theta^3) = r(\theta^1, \theta^2) + \theta^3 a_3(\theta^1, \theta^2) \qquad (2.1)$$

where R is the spatial position vector and r is the position vector on the reference surface.

The spatial components of Green-Lagrange's strains can be written as

$$E_{ij} = \frac{1}{2}\left(V_i \big\|_j + V_j \big\|_i + V^k \big\|_i V_k \big\|_j \right) \qquad (2.2)$$

where V_i are the components of displacement spatial vector. Concerning the relationships between covariant derivatives of space and surface quantities, we refer to Librescu (1975).

The covariant derivatives of the space tensors may be expressed in terms of surface quantities, as follows

$$V_{\alpha\|\beta} = \mu_\alpha^\beta \left(v_{\alpha\|\beta} - b_{\gamma\beta} v_3 \right)$$

$$V^\alpha \big\|_\beta = \left(\mu^{-1}\right)^\alpha_\gamma \left(v^\gamma \big\|_\beta - b^\alpha_\beta v_3 \right)$$

$$V_{\alpha\|3} = \mu_\alpha^\gamma v_{\gamma,3}$$

$$V^\alpha \big\|_3 = \left(\mu^{-1}\right)^\alpha_\gamma v^\gamma,_3$$

$$V^3 \big\|_\alpha = V_{3\|\alpha} = v^3,_\alpha + b^\lambda_\alpha v_\lambda = v_{3,\alpha} + b_{\alpha\lambda} v^\lambda$$

$$V^3 \big\|_3 = V_{3\|3} = v^3,_3 = v_{3,3}$$

where $\mu_\alpha^\beta = \delta_\alpha^\beta - \theta^3 b_\alpha^\beta$ is the shifter tensor. (2.3)

The double and single vertical bars are used to identify the covariant differentiation with respect to the space and surface undeformed metrics, respectively, while a comma denotes partial differentiation. Throughout the paper the Einsteinian summation convention is adopted for tensor quantities : Greek indices run takes the values 1,2 and the Latin ones the values 1,2,3.

Approximating the displacement by second-order polynomials across the shell thickness

$$v_\alpha = \overset{(0)}{v}_\alpha + \theta^3 \overset{(1)}{v}_\alpha + (\theta^3)^2 \overset{(2)}{v}_\alpha$$

$$v_3 = \overset{(0)}{v}_3 + \theta^3 \overset{(1)}{v}_3 + (\theta^3)^2 \overset{(2)}{v}_3 \qquad (2.4)$$

we obtain as the strain-tensor components

$$E_{\alpha\beta} = \overset{(0)}{E}_{\alpha\beta} + \theta^3 \overset{(1)}{E}_{\alpha\beta} + (\theta^3)^2 \overset{(2)}{E}_{\alpha\beta}$$
$$\quad + (\theta^3)^3 \overset{(3)}{E}_{\alpha\beta} + (\theta^3)^4 \overset{(4)}{E}_{\alpha\beta}$$

$$2E_{\alpha3} = 2\overset{(0)}{E}_{\alpha3} + 2\theta^3 \overset{(1)}{E}_{\alpha3}$$
$$\quad + 2(\theta^3)^2 \overset{(2)}{E}_{\alpha3} + 2(\theta^3)^3 \overset{(3)}{E}_{\alpha3}$$

$$E_{33} = \overset{(0)}{E}_{33} + \theta^3 \overset{(1)}{E}_{33} + (\theta^3)^2 \overset{(2)}{E}_{33} \qquad (2.5)$$

where for example $\overset{(0)}{E}_{\alpha\beta}$ is given by

$$\overset{(0)}{E}_{\alpha\beta} = \frac{1}{2}\left\{ \overset{(0)}{v}_{\alpha|\beta} + \overset{(0)}{v}_{\beta|\alpha} - 2b_{\alpha\beta} \overset{(0)}{v}_3 \right.$$
$$\left. + \left(\overset{(0)}{v}_{3,\alpha} + b_{\alpha\lambda} \overset{(0)}{v}^\lambda \right)\left(\overset{(0)}{v}_{3,\beta} + b_{\beta\delta} \overset{(0)}{v}^\delta \right) \right.$$
$$\left. + \left(\overset{(0)}{v}^\lambda \big|_\alpha - b^\lambda_\alpha \overset{(0)}{v}_3 \right)\left(\overset{(0)}{v}_{\lambda|\beta} - b_{\lambda\beta} \overset{(0)}{v}_3 \right) \right\}$$

3 MODELING OF DAMAGE

It is assumed that material inelasticity is contained within small zones surrounding the delamination. The effect of delamination is accounted for via the local volume average of the scalar product of the delamination opening displacement vector v^D and the delamination face normal n^D (ALLEN 1987)

$$\alpha^D_{ij} = \frac{1}{V_L} \int_{S_2} v^D_i \, n^D_j \, dS \qquad (2.6)$$

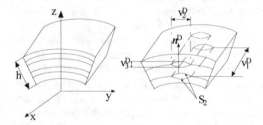

Fig. 2 : Local volume before and after Interply Delamination.

where V_L is the local volume for which delamination can be considered statically homogeneous and S_D is the surface area of delamination in V_L. The local ply stress-strain relations are then given by

$$\sigma_{ij} = C^{ijkl}\left(E_{kl} - \alpha^D_{kl} \right) \qquad (2.7)$$

The kinematics of the delamination are defined in Fig. 2 by

$$v^D_i = \begin{Bmatrix} v^D_1 \\ v^D_2 \\ v^D_3 \end{Bmatrix} \qquad \text{and} \qquad n^D_j = \begin{Bmatrix} 0 \\ 0 \\ 1 \end{Bmatrix} \qquad (2.8)$$

Then it follows

$$\alpha^D_{i\beta} = 0 \qquad (2.9)$$

and only the components α_{i3}^D are not equal to zero

$$\alpha_{i3}^D = \frac{1}{V_L}\int_{S_2} v_i^D\, dS \qquad (2.10)$$

Using the O'Brien (ALLEN 1987) delamination strain energy release rate model as a first approximation, the delamination is given by

$$\frac{\partial \alpha_{i3}^D}{\partial E_{i3}} = -\frac{n}{2}\frac{\left(E_X - E^*\right)}{\left[\left(C^{i3}\right)^T + \left(C^{i3}\right)^B\right]}\left(\frac{S_D}{S}\right) \qquad (2.11)$$

where n is the number of plies in the laminate, S_D is the total delamination surface area experimentally determined by X-ray radiographs, S is the total interface surface area of the local volume, E^* is the modulus of sublaminates formed by the delamination, given by $E^* = \frac{1}{h}\sum_{i=1}^{d}E_i h_i$. Here, d is the number of sublaminates and h is the laminate thickness, E_i is equal to E_X which is Young's modulus of the lamina and $\left(C^{ij}\right)$ collects the standard elastic material properties of the undamaged plies.

4 FINITE ELEMENT FORMULATION

For a solid, the equation of static equilibrium can be written as

$$\int_{\Omega}\delta\{E\}^T\{\sigma\}d\,\Omega - \delta\{v\}^T\{\bar{R}\}$$
$$= \delta U - \delta\{v\}^T\{\bar{R}\} = 0 \qquad (2.12)$$

where $\{\sigma\}$ is the second Piola-Kirchhoff stress vector, $\{\bar{R}\}$ is the applied load term, $\{\delta E\}$ is the virtual strain vector expressed in terms of virtual displacement $\delta\{v\}$ and U is the strain energy.
Here both the U and $\{\bar{R}\}$ in general depend upon nodal displacement $\{v\}$. Particularly U is a complex function of $\{v\}$.

$$\delta U = \int_{\Omega}\delta\{E\}^T\{\sigma\}d\Omega \qquad (2.13)$$

Using the previous decomposition of components of deformation we can write under matrical form

$$\{E\} = \{E_L\} + \{E_{NL}\}$$
$$= [B_0]\{v\} + \frac{1}{2}[A(\{v\})][G]\{v\} \qquad (2.14)$$

$$\delta\{E\} = [B_0]\delta\{v\} + \frac{1}{2}\delta[A(\{v\})]$$
$$[G]\{v\}\frac{1}{2}[A(\{v\})][G]\delta\{v\} \qquad (2.15)$$

where $[B_0]$, $[A]$ and $[G]$ are differential operator matrices (see appendix 1). $[A]$ and $[G]$ are such that

$$\delta[A(\{v\})][G]\{v\} = [A(\{v\})][G]\delta\{v\} \qquad (2.16)$$

$$\{\sigma\} = [H]\{E\} \qquad (2.17)$$

$$\{v\} = [N]\{\underline{v}\} \qquad (2.18)$$

with $\{v\}$ as the displacement function and $\{\underline{v}\}$ as nodal displacement. $[N]$ are the shape functions. The matrix $[H]$ is defined in Appendix 3.
Using the previous results we can write

$$\delta U = \delta\{\underline{v}\}^T[K_D]\{\underline{v}\} \qquad (2.19)$$

$[K_D]$ is the Direct Stiffness matrix defined by

$$[K_D] = \int_{\Omega}[N]^T\left[[B_0] + [A(\{v\})][G]\right]^T$$
$$[H]\left[[B_0] + \frac{1}{2}[A(\{v\})][G]\right][N]d\Omega \qquad (2.20)$$

We can rearranged the term $\left[\delta A(\{v\})\right]^T\{\sigma\}$ (see Appendix 2) and obtain

$$\left[\delta A(\{v\})\right]^T\{\sigma\} = [S][G]\delta\{v\} = [S][G][N]\delta\{\underline{v}\} \qquad (2.21)$$

Taking the second variation of the strain energy, we find

$$\delta^2 U = \delta\{\underline{v}\}^T[K_T]\,\delta\{\underline{v}\}$$
$$= \delta\{\underline{v}\}^T\left[[K_\sigma] + [\bar{K}]\right]\delta\{\underline{v}\} \qquad (2.22)$$

with

$$[K_\sigma] = \int_{\Omega}\left\{[N]^T[G]^T[S][G][N]\right\}d\Omega \qquad (2.23)$$

$$[\bar{K}] = \int_{\Omega}\left\{[N]^T\left[[B_0] + [A(\{v\})][G]^T\right][H]\right.$$
$$\left.[A(\{v\})][G][N]\right\}d\Omega \qquad (2.24)$$

K_T is the tangent stiffness matrix.

5 NUMERICAL EXAMPLES

The performance of the presented nine node shell element was tested by solving example problems of delaminated composite plates and shells with linear and nonlinear geometry taking account of rotations and normal deformations (nine parameter theory). In all examples only the details of the deflections are provided. Either the Newton-Raphson method or the modified Riks-Wemper method was used for iterative solution. Results were compared to known solutions available.

5.1 Cylindrical panel under point load

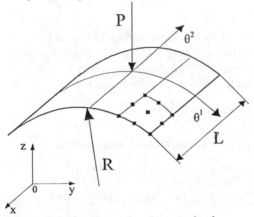

Fig. 3 : Cylindrical panel under point load

A cylindrical panel is under point load applied at the centre. The boundary conditions are such that curved edges are free while the straight edges are clamped. Owing to the symmetry of the geometry and loading, only a quarter of the panel was modelled.

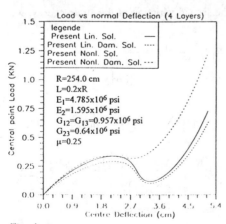

Fig. 4 :

Fig. 4 shows the normal deflections at the central point for 4 layers : (90/0/90/0). The discrepancy between this present linear solution and nonlinear one appears to be due to the neglected terms in the linear deformation of shell. We remark also that the damaged model follows the undamaged one with a little difference.

5.2 Clamped plate with a concentred load at the center (8 layers).

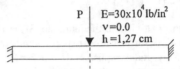

Fig 5 : A clamped plate with a concentred load

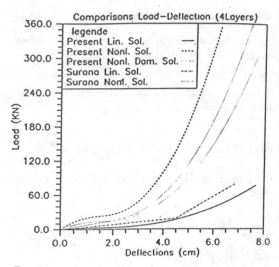

Fig. 6 :

Fig. 6 shows the deflection at the central point. We remark that damage has a considerable influence on the nonlinear behaviour. Compared to the nonlinear solution, the linear one is rather simple to obtain and demands a third of calculation time.

5.3 A spherical shell under loads at two point opposite poles.

A pinched spherical shell is a special problem in that there regions of large bending stresses, regions where membranes actions predominates and the regions of high stress concentration.

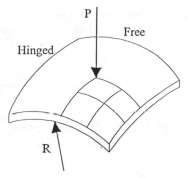

P

Free

Hinged

R

Fig. 7 : A radial displacement at the centre of pinched spherical shell under loads at two opposite points.

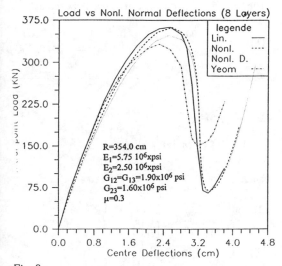

Load vs Nonl. Normal Deflections (8 Layers)

legende
Lin. —
Nonl. ----
Nonl. D. —·—
Yeom —·—

R=354.0 cm
E_1=5.75 10^6xpsi
E_2=2.50 10^6xpsi
G_{12}=G_{13}=1.90x10^6 psi
G_{23}=1.60x10^6 psi
μ=0.3

Fig. 8 :

Fig. 8 shows the results of this present linear, nonlinear and nonlinear damaged solutions compared with the linear and nonlinear solutions obtained by Surana (1989).

6 CONCLUSIONS

The result of numerical tests indicates that the present formulation can provide interesting solutions for geometrically nonlinear plates and shells with delamination model. For the example problems tested in the paper, only a small number of elements were needed to obtain convergent solutions.

REFERENCES

Allen, D.H., C.E. Harris & S.E. Groves 1987. A Thermomechanical Constitutive Theory for Elastic Composites with Distributed Damage - Part I : Theoretical Development; Part II : Application to Matrix Cracking in Laminated Composites. *Int. J. Solids & Struct.* 23:9:1301-1338.

Allen, D.H., C.E. Harris & S.E. Groves 1990. Damage Modelling in Laminated Composites, Yielding Damage and Fatigue of Anisotropic Solids. In J.P. Boehler (eds), *Mechanical Engineering Publications Limited*: 535-550. London.

Allix, O. & P. Ladevèze 1989. A Damage Prediction Method for Composite Structures. *Int. J. for Num. Meth. in Engng.* 27:271-283.

Bathe, K.J. & E.L. Wilson 1976. *Numerical Methods in Finite Element Analysis*. New-Jersey:Prentice-Hall.

Green, A.E. & W. Zerna 1968. *Theoretical Elasticity*, second edition, Oxford:Clarendon press.

Harris, C.E. & D.H. Allen 1988. A Continuum Damage Model of Fatigue Induced Damage in Laminated Composites, *SAMPE Journal* July/August. 43-51.

Harris, C.E., D.H. Allen & E.W. Nottorf 1987. Damage-Induced Changes in the Poisson's Ratio of Cross-Ply Laminates an Application of Continuum Damage Mechanics Model for Laminated Composites, Damage Mechanics in Composites. In A.S.D. Wang & G.K. Haritos (eds), 12:17-23, New York:ASME.

Ju, J.W. 1989. On Energy-based Coupled Elastoplastic Damage Theories Constitutive Modeling and Computational Aspects, *Int. J. Solids Struct.*, 25:7:803-833.

Kachanov, L.M. 1981. Crack Growth under Conditions of Creep and Damage. In Creep in Structures. 520-524. Berlin:Springer-Verlag.

Ladevèze, P. 1990. A Damage Approach for Composite Structures Theory and Identification. Proceedings of EUROMECH N°269 Experimental Identification of the Mechanical Characteristics of Composite Materials and Structures. Saint-Etienne.

Lee, J.W., D.H. Allen & C.E. Harris 1991. The Upper Bounds of Reduced Axial and Shear Moduli in Cross-Ply Laminates with Matrix Cracks, Composite Materials Fatigue and

Fracture , Vol. 3, ASMT STP 1110, T.K. O'Brien ed., American Society for Testing and Materials, Philadelphia.

Lemaître, J. & J.L. Chaboche 1990. Mechanics of Solids Materials, Cambridge:University Press.

Librescu, L. 1975. Elastostatics and Kinetics of Anisotropic and Heterogeneous Shell-type Structures, Leyden:Noordhoff Int. Publ..

Lo, D.C., D.H. Allen & C.E. Harris Modeling the Progressive Failure of Laminated Composites with Continuum Damage Mechanics, 23rd National Symposium on Fracture Mechanics.

Neale, K.W. 1972. A General Variational Theorem for the Rate in Elasto-Plasticity, *Int. J. Solids Struct.*, Vol. 8, pp. 865-876.

Palmerio, A.F. 1988. On a Moderate Rotation Theory for Anisotropic Shells. Engineering Science & Mechanics Department, Virginia Tech., Blacksburg University Thesis.

Reddy, J.N. 1982. Bending of Laminate Anisotropic Shells by a Shear Deformable Finite Element, Fibre Science and Technology. 17:19-24.

Reissner, E. 1953. On a Variational Theorem for Finite Elastic Deformations, *J. Math. Phys.* 32:129:865-876.

Schmidt, R. & J.N. Reddy. 1988. A Refined Small Strain and Moderate Rotation Theory of Elastic Anisotropic Shells. *J. of Appl. Mech. (ASME).* 55:611-616.

Schmidt, R. & D. Weichert 1989. Refined Theory Elastic-Plastic Shells at Moderate Rotations. *ZAMM.* 69:11-21.

Simo, J.C. & J.W. Ju 1987. Strain and Stress-based Continuum Damage Models - I Formulation, *Int. J. Solids Struct.* 23:7:821-840.

Talreja, R. 1985. A Continuum Mechanics Characterization of Damage in Composite Materials. Proc. R. Soc., 399A:195-216, London.

Teyeb, O.M. 1993. Analyse Géométriquement Non Linéaire des Coques et Plaques : Application aux Structures Laminées, Thèse de Doctorat, Université des Sciences et Technologies de Lille.

Valid, R. 1977. *La Mécanique des Milieux Continus et le Calcul des Structures*, In Collection des Etudes et Recherche D'Électricité de France, Eyrolles (eds).

Washizu, K. 1968. *Variational Methods in Elasticity and Plasticity*. Oxford:Pergamon Press.

Wempner, G. 1981. *Mechanics of Solids with Applications to Thin Bodies*, Sijthoff & Noordhoff, Alphen aan den Rijn, The Netherlands.

Yang H.T.Y. & Y.C. Wu. 1989. A Geometrically Non-linear Tensorial Formulation of a Skewed Quadrilateral Thin Shell Finite Element. *Int. J. Num. Meth. in Engng.* 28:2855-2895.

Yeom C.H. & S.W. Lee. 1989. An Assumed Strain Finite Element Model for Large Deflection Composite Shells. *Int. J. Num. Meth. in Engng.* 28:1749-1768.

Zienkiewicz, O.C. 1971. *The Finite Element Method in Engineering Science*, London:Mc. Graw-Hill.

APPENDIX 1

We give here the spherical components which can be simplified and give the cylindrical or Cartesian ones. The bending components can be decomposed as

$$\{E\}^T = \left\{ \overset{(0)}{E}_{11}; \ \overset{(0)}{E}_{22}; \ 2\overset{(0)}{E}_{12}; \ \overset{(1)}{E}_{11}; \ \overset{(1)}{E}_{22}; \ \overset{(1)}{E}_{12}; \ \overset{(2)}{E}_{11}; \right.$$
$$\left. \overset{(2)}{E}_{22} \ \overset{(2)}{E}_{12}; \ \overset{(3)}{E}_{11}; \ \overset{(3)}{E}_{22}; \ 2\overset{(3)}{E}_{12}; \ \overset{(4)}{E}_{11}; \ \overset{(4)}{E}_{22}; \ 2\overset{(4)}{E}_{12} \right\}$$

$$\{E\} = \{E_L\} + \{E_{NL}\}$$

$$= [B_0]\{v\} + \frac{1}{2}[A(\{v\})][G]\{v\}$$

where

$$[B_0] = \begin{bmatrix} [B_{01}] & [0_1] & [0_1] \\ & [B_{01}] & [0_1] \\ sym. & & [B_{01}] \end{bmatrix}$$

$$[B_{01}] = \begin{bmatrix} h_{,1} & 0 & \frac{h}{R} \\ \frac{\cot g\phi}{R}h & \frac{h_{,2}}{R\sin\phi} & \frac{h}{R} \\ \frac{h_{,2}}{R\sin\phi} & \left(h_{,1} - \frac{\cot g\phi}{R}h\right) & 0 \\ \frac{h_{,1}}{R} & 0 & \frac{h}{R^2} \\ \frac{\cot g\phi}{R^2}h & \frac{h_{,2}}{R^2\sin\phi} & \frac{h}{R^2} \\ \frac{h_{,2}}{R^2\sin\phi} & \left(\frac{h_{,1}}{R} - \frac{\cot g\phi}{R^2}h\right) & 0 \\ 0 & 0 & 0 \\ 0 & 0 & 0 \\ 0 & 0 & 0 \end{bmatrix}$$

and $\left[0_1\right]$ is a zero matrix [3x6].

$$[A] = \begin{bmatrix} [A_1] & [0_2] & [0_2] \\ & [A_1] & [0_2] \\ sym. & & [A_1] \end{bmatrix}$$

$$[A_1] = \begin{bmatrix} A_1 & A_7 & A_{13} & 0 & 0 & 0 \\ 0 & 0 & 0 & A_{19} & A_{25} & A_{31} \\ A_2 & A_8 & A_{14} & A_{20} & A_{26} & A_{32} \\ A_3 & A_9 & A_{15} & 0 & 0 & 0 \\ 0 & 0 & 0 & A_{21} & A_{27} & A_{33} \\ A_4 & A_{10} & A_{16} & A_{22} & A_{28} & A_{34} \\ A_5 & A_{11} & A_{17} & 0 & 0 & 0 \\ 0 & 0 & 0 & A_{23} & A_{29} & A_{35} \\ A_6 & A_{12} & A_{18} & A_{24} & A_{30} & A_{36} \end{bmatrix}$$

$A_1 = v_{3,1}^{(0)} - \dfrac{v_1^{(0)}}{R}$; $A_2 = \dfrac{v_{3,2}^{(0)} - \sin\phi\, v_2^{(0)}}{R\sin\phi}$; $A_3 = v_{3,1}^{(1)} - \dfrac{v_1^{(1)}}{R}$

$A_4 = \dfrac{v_{3,2}^{(1)} - \sin\phi\, v_2^{(1)}}{R\sin\phi}$; $A_5 = v_{3,1}^{(2)} - \dfrac{v_1^{(2)}}{R}$; $A_6 = \dfrac{v_{3,2}^{(2)} - \sin\phi\, v_2^{(2)}}{R\sin\phi}$

$A_7 = v_{1,1}^{(0)} + \dfrac{v_3^{(0)}}{R}$; $A_8 = \dfrac{v_{1,2}^{(0)} - \cos\phi\, v_2^{(0)}}{R\sin\phi}$; $A_9 = v_{1,1}^{(1)} + \dfrac{v_3^{(1)}}{R}$

$A_{10} = \dfrac{v_{1,2}^{(1)} - \cos\phi\, v_2^{(1)}}{R\sin\phi}$; $A_{11} = v_{1,1}^{(2)} + \dfrac{v_3^{(2)}}{R}$; $A_{12} = \dfrac{v_{1,2}^{(2)} - \cos\phi\, v_2^{(2)}}{R\sin\phi}$

$A_{13} = \dfrac{v_{2,1}^{(0)}}{R\sin\phi}$; $A_{14} = \dfrac{R\sin\phi\, v_{2,2}^{(0)} + R\sin\phi\cos\phi\, v_1^{(0)} + R\sin^2\phi\, v_3^{(0)}}{(R\sin\phi)^3}$

$A_{15} = \dfrac{v_{2,1}^{(1)}}{R\sin\phi}$; $A_{16} = \dfrac{R\sin\phi\, v_{2,2}^{(1)} + R\sin\phi\cos\phi\, v_1^{(1)} + R\sin^2\phi\, v_3^{(1)}}{(R\sin\phi)^3}$

$A_{17} = \dfrac{v_{2,1}^{(1)}}{R\sin\phi}$; $A_{18} = \dfrac{R\sin\phi\, v_{2,2}^{(2)} + R\sin\phi\cos\phi\, v_1^{(2)} + R\sin^2\phi\, v_3^{(2)}}{(R\sin\phi)^3}$

$A_{19} = \dfrac{v_{3,2}^{(0)} - \sin\phi\, v_2^{(0)}}{(R\sin\phi)^2}$; $A_{20} = \dfrac{v_{3,1}^{(0)} - \frac{v_1^{(0)}}{R}}{R\sin\phi}$; $A_{21} = \dfrac{v_{3,2}^{(1)} - \sin\phi\, v_2^{(1)}}{(R\sin\phi)^2}$

$A_{22} = \dfrac{v_{3,1}^{(1)} - \frac{v_1^{(1)}}{R}}{R\sin\phi}$; $A_{23} = \dfrac{v_{3,2}^{(2)} - \sin\phi\, v_2^{(2)}}{(R\sin\phi)^2}$; $A_{24} = \dfrac{v_{3,1}^{(2)} - \frac{v_1^{(2)}}{R}}{R\sin\phi}$

$A_{25} = \dfrac{v_{1,2}^{(0)} - \sin\phi\, v_2^{(0)}}{(R\sin\phi)^2}$; $A_{26} = \dfrac{v_{1,1}^{(0)} + \frac{v_3^{(0)}}{R}}{R\sin\phi}$; $A_{27} = \dfrac{v_{1,2}^{(1)} - \cos\phi\, v_2^{(1)}}{(R\sin\phi)^2}$

$A_{28} = \dfrac{v_{1,1}^{(1)} + \frac{v_3^{(1)}}{R}}{R\sin\phi}$; $A_{29} = \dfrac{v_{1,2}^{(2)} - \cos\phi\, v_2^{(2)}}{(R\sin\phi)^2}$; $A_{30} = \dfrac{v_{1,1}^{(2)} + \frac{v_3^{(2)}}{R}}{R\sin\phi}$

$A_{31} = \dfrac{R\sin\phi\, v_{2,2}^{(0)} + R\sin\phi\cos\phi\, v_1^{(0)} + R\sin^2\phi\, v_3^{(0)}}{(R\sin\phi)^4}$; $A_{32} = \dfrac{v_{2,1}^{(0)}}{(R\sin\phi)^2}$

$A_{33} = \dfrac{R\sin\phi\, v_{2,2}^{(1)} + R\sin\phi\cos\phi\, v_1^{(1)} + R\sin^2\phi\, v_3^{(1)}}{(R\sin\phi)^4}$

and $\left[O_2\right]$ is a zero matrix [3x6].

$$[G] = \begin{bmatrix} [G_1] & [0_3] & [0_3] \\ & [G_1] & [0_3] \\ sym. & & [G_1] \end{bmatrix}$$

where

$$[G_1] = \begin{bmatrix} -\dfrac{h}{R} & 0 & h_{,1} \\ h_{,1} & 0 & \dfrac{h}{R} \\ 0 & h_{,1} R\sin\phi & 0 \\ 0 & -h\sin\phi & h_{,2} \\ h_{,2} & h\cos\phi & 0 \\ hR\sin\phi\cos\phi & h_{,2} R\sin\phi & hR\sin^2\phi \end{bmatrix}$$

$$\{v\}^T = \left\{ v_1^{(0)};\ v_2^{(0)};\ v_3^{(0)};\ v_1^{(1)};\ v_2^{(1)};\ v_3^{(1)};\ v_1^{(2)};\ v_2^{(2)};\ v_3^{(2)} \right\}$$

et $\left[0_3\right]$ is a zero matrix 6x3.

APPENDIX 2

$$[S] = \begin{bmatrix}
s_1 & 0 & 0 & s_2 & 0 & 0 & s_3 & 0 & 0 & s_4 & 0 & 0 & s_5 & 0 & 0 & s_6 & 0 & 0 \\
 & s_1 & 0 & 0 & s_2 & 0 & 0 & s_3 & 0 & 0 & s_4 & 0 & 0 & s_5 & 0 & 0 & s_6 & 0 \\
 & & s_7 & 0 & 0 & s_8 & 0 & 0 & s_9 & 0 & 0 & s_{10} & 0 & 0 & s_{11} & 0 & 0 & s_{12} \\
 & & & s_{13} & 0 & 0 & s_4 & 0 & 0 & s_{14} & 0 & 0 & s_6 & 0 & 0 & s_{15} & 0 & 0 \\
 & & & & s_{13} & 0 & 0 & s_4 & 0 & 0 & s_{14} & 0 & 0 & s_6 & 0 & 0 & s_{15} & 0 \\
 & & & & & s_{16} & 0 & 0 & s_{10} & 0 & 0 & s_{17} & 0 & 0 & s_{18} & 0 & 0 & s_{19} \\
 & & & & & & s_5 & 0 & 0 & s_6 & 0 & 0 & s_{20} & 0 & 0 & s_4 & 0 & 0 \\
 & & & & & & & s_5 & 0 & 0 & s_6 & 0 & 0 & s_{20} & 0 & 0 & s_4 & 0 \\
 & & & & & & & & s_{11} & 0 & 0 & s_{18} & 0 & 0 & s_{22} & 0 & 0 & s_{12} \\
 & & & & & & & & & s_{15} & 0 & 0 & s_6 & 0 & 0 & s_{23} & 0 & 0 \\
 & & & & & & & & & & s_{15} & 0 & 0 & s_6 & 0 & 0 & s_{23} & 0 \\
 & & & & & & & & & & & s_{24} & 0 & 0 & s_{12} & 0 & 0 & s_{25} \\
 & & & & & & & & & & & & s_{26} & 0 & 0 & s_{27} & 0 & 0 \\
 & & & & & & & & & & & & & s_{26} & 0 & 0 & s_{27} & 0 \\
 & & & & & & & & & & & & & & s_{28} & 0 & 0 & s_{29} \\
 & & & & & & & & & & & & & & & s_{30} & 0 & 0 \\
 \text{SYM.} & & & & & & & & & & & & & & & & s_{30} & 0 \\
 & & & & & & & & & & & & & & & & & s_{31}
\end{bmatrix}$$

where

$$S_1 = R^{11}_{(0)}; \; S_2 = \frac{R^{12}_{(0)}}{R\sin\phi}; \; S_3 = R^{11}_{(1)}; \; S_4 = \frac{R^{12}_{(1)}}{R\sin\phi}$$

$$S_5 = R^{11}_{(2)}; \; S_6 = \frac{R^{12}_{(2)}}{R\sin\phi}; \; S_7 = \frac{R^{11}_{(0)}}{(R\sin\phi)^2}; \; S_8 = \frac{R^{12}_{(0)}}{(R\sin\phi)^3}$$

$$S_9 = \frac{R^{11}_{(1)}}{(R\sin\phi)^2}; \; S_{10} = \frac{R^{12}_{(1)}}{(R\sin\phi)^3}; \; S_{11} = \frac{R^{11}_{(2)}}{(R\sin\phi)^2}; \; S_{12} = \frac{R^{12}_{(3)}}{(R\sin\phi)^3}$$

$$S_{13} = \frac{R^{22}_{(0)}}{(R\sin\phi)^2}; \; S_{14} = \frac{R^{22}_{(1)}}{(R\sin\phi)^2}; \; S_{15} = \frac{R^{22}_{(2)}}{(R\sin\phi)^2}; \; S_{16} = \frac{R^{22}_{(0)}}{(R\sin\phi)^4}$$

$$S_{17} = \frac{R^{22}_{(1)}}{(R\sin\phi)^4}; \; S_{18} = \frac{R^{12}_{(2)}}{(R\sin\phi)^3}; \; S_{19} = \frac{R^{22}_{(2)}}{(R\sin\phi)^4}; \; S_{20} = R^{11}_{(3)}$$

$$S_{21} = \frac{R^{12}_{(3)}}{R\sin\phi}; \; S_{22} = \frac{R^{11}_{(3)}}{(R\sin\phi)^2}; \; S_{23} = \frac{R^{22}_{(3)}}{(R\sin\phi)^2}; \; S_{24} = \frac{R^{22}_{(2)}}{(R\sin\phi)^4}$$

$$S_{25} = \frac{R^{22}_{(3)}}{(R\sin\phi)^4}; \; S_{26} = R^{11}_{(4)}; \; S_{27} = \frac{R^{12}_{(4)}}{R\sin\phi}; \; S_{28} = \frac{R^{22}_{(2)}}{(R\sin\phi)^4}$$

$$S_{29} = \frac{R^{12}_{(4)}}{(R\sin\phi)^3}; \; S_{30} = \frac{R^{22}_{(4)}}{(R\sin\phi)^2}; \; S_{31} = \frac{R^{22}_{(4)}}{(R\sin\phi)^4}$$

with $R^{ij}_{(n)} = \int_{-h/2}^{h/2} \mu S^{ij} (\theta^3)^n d\theta^3$

and $\mu = \left|\mu^{\beta}_{\alpha}\right| = \left(1 + \frac{\theta^3}{R}\right)^2$

APPENDIX 3

$$[H] = \begin{bmatrix}
[A] & [B] & [D] & [E] & [F] \\
 & [D] & [E] & [F] & [G] \\
 & & [F] & [G] & [H] \\
\text{Sym.} & & & [H] & [K] \\
 & & & & [L]
\end{bmatrix}$$

where

$$[A] = \begin{bmatrix}
a_{11} & a_{12} & a_{16} \\
 & a_{22} & a_{26} \\
\text{Sym.} & & a_{66}
\end{bmatrix}$$

$[B]; [D]; [E]; [F];$
$[G]; [H]; [K] \, and \, [L]$

are the same form that $[A]$

$$\left(a_{ij}; b_{ij}; d_{ij}; e_{ij}; f_{ij}; g_{ij}; h_{ij}; k_{ij}; l_{ij}\right)$$
$$= \int_{-h/2}^{h/2} C^{ij} \left(1 + \frac{\theta^3}{R}\right)^2 (\theta^3)^n d\theta^3$$

with n={0;1;2;3;4;5;6;7;8;9}.

Progress in Durability Analysis of Composite Systems, Cardon, Fukuda & Reifsnider (eds)
© *1996 Balkema, Rotterdam. ISBN 90 5410 809 6*

Durability analysis of cement composite system exposed to a physico-chemical environment

Tomasz Błaszczyński
Institute of Structural Engineering, The Technical University of Poznań, Poland

ABSTRACT: Durability analysis of RC system by field and laboratory tests based on data collected over many years are described. All structures are situated in environment, which can be divided into physical, chemical and physico-chemical. The problem of physico-chemical influence on the reinforced concrete is not well known. It is found that the crude oil products with very low neutralisation number are the physico-chemical active agents on the concrete. That environment is linked basically to the marine and industrial structures. Experimental analysis affecting compressive strength are described. Comparing the influence of various oil products on compressive strength of concrete and its bond, leads to the conclusion that there are large differences in effects. From vaseline oil with almost no influence to some mineral oils with the serious influence, even to 30 or 50% loss of initial values. Contamination of concrete by hydrocarbons gives a new material, which behaves differently.

1 INTRODUCTION

Thinking about structures today and beyond 2000, it has to be remember about those, who are exist. From that point the durability of cement composite constructional material in different environments is now recognised as a very important part of the design process. From all kinds of them the little known is the physico-chemical one, which is linked to the organic active polar molecules. It is found that crude oil products with a very low neutralisation number (lower then 0.25 mg KOH/g) consist some of them. These products are basically linked to marine and industrial RC structures.

In technical literature, effects of crude oil products on concrete are classified either as non-harmful or only mildly harmful, but there is evidence that serious damage can be caused. For example an investigation of damage to a large turbine (200 MW) at one of the national power generating plants in Poland concluded that the damage was due to large decrease in the dynamic stiffness of the RC foundation frame caused by the heavy contamination of the concrete by mineral oil.

A construction is always in specified environment, which is govern by its own rules and has its own variable features. It can not be taken from the environment as it can not be taken from the ground. The properties of this material are closely connected not only to its composition and inner structure but also to the environment.

In case of physico-chemical environment usually physico-chemical bonds are affected and because of that the process can be reversible sometime.

Long term investigations therefore were put in hand to determine the effect of a group of crude oil products - mineral oils - on properties of concrete and reinforced concrete. The paper reports our detailed findings.

2 DESTRUCTION MECHANISM

The influence of different crude oil products on concrete in comparison to water investigated by the author and other researchers is shown on figure 1.

The most common conclusion is that concrete behaves as a "molecular sieve" where different sized molecules penetrate differently into the concrete pore system (Mills 1968). In particular, small sized molecules as in water penetrate the smaller inner gel pores not reached by larger molecules. Cook and

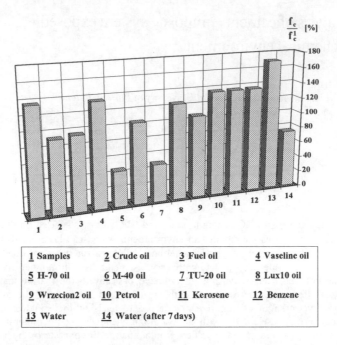

Fig. 1. Effects of different crude oil products influence on concrete (in comparison to water)

Haque (1974) also concluded that the reason why benzene and kerosene do not cause any changes to the properties of concrete similar to those caused by water was due to the molecular sieve effect witch allowed water particles to penetrate gel pores not reached by other fluids.

Feldman (1970) is less certain about the molecular sieve effect and has suggested that even fluids of similar molecular size to water such as, nitrogen and methanol cannot penetrate pores reached by water. The ability of water to penetrate the inner layer space is attributed to its special "affinity".

Short time changes in properties of concrete saturated by water is due to lowering of cement matrix strength and they are reversible (see 14 in fig. 1). In long term water acts positively on concrete and strengthening occurs (13 in fig. 1) (Sierych 1982).

It is known that every mineral oil consists of basic oil and improves. The basic oil is electrically almost neutral (non-polar), only the improves are active including surface - active polar molecules.

In spaces unfilled with water, oil penetrates and deposits the polar molecules on the surfaces. When water fills these spaces, the penetration of oils take place according to the minimum of dielectric constant e gradient as shown by figure 2 (Błaszczyński 1992).

The mechanism of failure of the concrete inner structure and its bond to reinforcement can be explained by the effect of lowering the surface energy and by the effect, which takes place at the polar fluid-solid interface. When a microcrack has occurred, the polar molecules penetrate along its

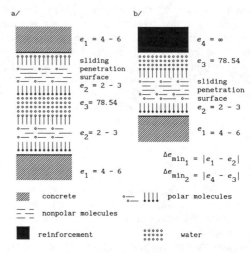

Fig. 2. Mild change of phases rule for mineral oil and water in concrete (a) and reinforced concrete (b) microcracks and microspaces

298

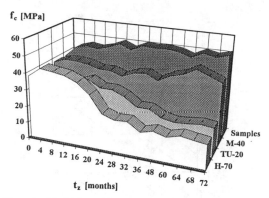

f_c [MPa]

Samples
M-40
TU-20
H-70

t_z [months]

Fig. 3. Variation of concrete C-25 compressive strength during the period of exposure to different oils

surface to the apex. Therefore, at their apexes, microcracks and other microdefects become subjected to a pressure action. The polar active molecules adsorbed on the surface give a reduction in surface energy, which can make this surface more ductile (Błaszczyński 1991).

The results of long term action of different mineral oils are presented in figure 3. They clearly show the significant decrease of concrete C-25 compressive strength in time (Błaszczyński 1994).

Contamination of concrete by hydrocarbons gives a new material, which behaves differently. The results of the stress (σ) - strain (ε) relation in dry state and in

fully saturated in function of the longitudinal strains are different. The nonlinear behaviour of strength and strain variations depends on the contents of hydrocarbon and its type. The schematic graph of the σ - ε relation for dry and for oil saturated concrete is shown in figure 4.a; it can be noticed that the strain ε_R, corresponding to the maximum stress, is lower for oil saturated concrete than for dry concrete; on the other hand, the maximum limit strain ε_m is correspondingly higher. Figure 4.b presents real graph for dry and oil TU-20 saturated concrete C-20.

Comparison of physico-chemical influences between water and oils leads to a conclusion, that polar molecules within the hydrocarbon chain are harmful. Water molecules are small dipoles geometrically and when acted positively on concrete strengthening occurs. The hydrocarbon chain is non-polar and non-harmful, but in connection with hydrophilic part gives the problem. This explains why petrol, kerosene, benzene and vaseline oil are not corrosive to concrete since they contain non-harmful short hydrocarbon chains. First of all in fuels water is the polar element or there are a very small part of other polar molecules, but with very short chain, which are not long enough for hydrophobing of the concrete inner structure.

The existence of organic polar active molecules in liquid may be detected directly by IR spectroscopy. The degree of polarization of mineral oil can be measured also indirectly by its lubricity or demulgation method.

a/

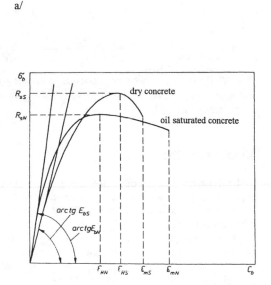

b/

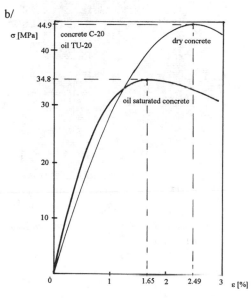

Fig. 4. σ - ε diagram for dry and oil saturated concrete a/ schematic proposal, b/ real graph for oil TU-20 and concrete C-20

3 ASSESSMENT METHODS

Using the infra-red spectroscopy with the small amount of oil one can get the whole plot of the examined fluid. Comparing the basic oil and oil with improvers it can be seen the significant peaks in the case of all the most active polar molecules. Figure 5 shows a comparison of vaseline oil (basic non-polar oil) IR spectra with the spectra of mixtures based on it, by adding 1%, 5% or 10% of oleinic acid (very polar active improver).

The degree of polarization of acting fluids can be also measured indirectly by its lubricity P_z. Lubricity depends on viscosity and for statistical multiple regression analyses the kinematic viscosity η_k and the lubricity P_z are used as dependent variables, giving a chart presented in figure 6. The chart can be used to find compressive strength of concrete grade C-25 after a long term action of any crude oil products.

The simplest method is a demulgation method, which is based on assumption, that polar molecules keeping emulsion. The last figure shows the dependence of concrete compressive strength upon the water phase exuded from oil-water emulsion after 60 min.

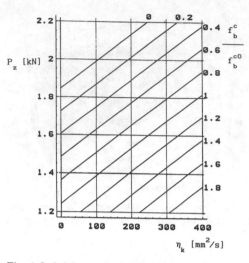

Fig. 6. Lubricity method chart for concrete C-25

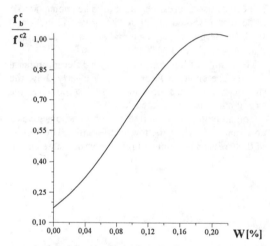

Fig. 7. Relationship between concrete compressive strength and amount of exuded water phase from demulgation method

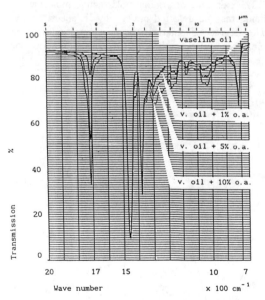

Fig. 5. Comparison of vaseline oil IR spectra with the spectra of other mixtures

4 CONCLUSIONS

The durability analysis of cement composite system exposed to a physico-chemical environment shows that significant reduction in compressive strength and bond to the reinforcement can occur.

When designing structure from cement constructional material in contact with crude oil products, apart from checking the value of neutralisation number, checks should be made on presence of organic polar active molecules by the infra-red spectroscopy, lubricity or demulgation method. Some of the oil products are safe but some are clearly very aggressive.

REFERENCES

Błaszczyński, T.Z. 1991. Durability of cement based composite systems influenced by physico-chemical agents. in A.H. Cardon & G. Verchery (eds), *Durability of polymer based composite systems for structural application:* 504-513. London: Elsevier Science Publisher Ltd.

Błaszczyński, T.Z. 1992. Influence of physico-chemical agent on the cement composite materials. *Physico-Chemical Mechanics.* 20: 19-23.

Błaszczyński, T.Z. 1994. Durability analysis of RC structures exposed to a physico-chemical environment. *Proceedings of the Third International Conference on Global Trends in Structural Engineering:* 67-70. Singapore: CI-Premier.

Cook, D.J. & M.N. Haque 1974. The effect of sorption on the tensile creep and strength reduction of desiccated concrete. *Cement and Concrete Research.* 5: 735-744.

Feldman, R.F. & P.J. Sereda 1970. A new model for hydrated portland cement and its practical implications. *Engineering Journal.* 8/9: 53-56.

Mills, R.H. 1968. Molecular sieve effect in concrete. *Proc. Symp. Chemistry of Cement:* 74-84. Tokyo.

Sierych, P.J. 1982. The influence of moisture content on concrete compressive strength. *Beton i Zelezobeton.* 8: 16-17. (in Russian).

Progress in Durability Analysis of Composite Systems, Cardon, Fukuda & Reifsnider (eds)
© *1996 Balkema, Rotterdam. ISBN 90 5410 809 6*

Determination of the optical anisotropy parameters in transparent composite systems

S.Yu. Berezhna
Lviv State I. Franko University, Ukraine

I.V. Berezhnyi, O.M. Krupych & O.G. Vlokh
Institute for Physical Optics, Lviv, Ukraine

Abstract : new optical polarization technique being developed on the basis of the computerized tomography approach is proposed to be used for the state of stress determination in transparent composite systems. Light waves propagation process has been analysed by means of geometrical optics method. Inverse problem is considered for zero-order approximation of transfer equations (equations for amplitudes). Originally invented polarimetry technique is applied for experimental data obtaining. Some peculiarities of HIIP (high informative invariant polarimetry) technique are discussed.

1 INTRODUCTION

Peculiarities of the composite systems technological processes cause an appearance of the internal residual stresses in these materials. We deal with transparent composites. The problem under consideration is a state of stress study in such systems. More correctly we research anisotropical properties of the medium. It is more general problem, because presence of stresses is the only one origin of the anisotropy appearance or change. Since the most typical case for analysis is that one when stressed state is a volume (or three-dimensional) one, commonly used integrated research techniques become unsufficient. It is necessary to draw in computerized tomography approach for the analysis of such a case.

Computerized tomography techniques are well known as very promising and powerful ones for different diagnostic applications. Advances and problems of the optical computerized tomography as a method for non-destructive testing and inspection are widely discussed in recent times (Aben 1986, Puro 1992, Andrienko and Dubovikov 1994).

In this paper we present a new optical computerized tomography technique that is adjusted for the purposes of optical anisotropy parameters determination in weakly inhomogeneous transparent media.

The application of computerized tomography approach means following principal problems solution:

1) analysis of the process of sounding radiation propagation in studied medium; deduction of the equations describing the process; formulation of the direct problem;

2) formulation and solution of the inverse problem, deduction of relations between parameters to be determined and known (experimentally obtained) ones;

3) development of experimental technique being fitted purposes of the problem;

4) choise and adjustment of reconstruction algorithm.

Now we discuss results of the above three items realization. We have worked out main principles of new optical tomography technique and tested their validity by model experiments. The forth task solution is being in progress at the present time.

2 LIGHT WAVES PROPAGATION IN WEAKLY INHOMOGENEOUS WEAKLY ANISOTROPIC MEDIA

Optical tomography technique deals with light waves as a sounding radiation. So, we have to research a problem of electromagnetic waves propagation in the media under consideration. The most general case is a one of non-stationary inhomogeneous anisotropic medium with dispersion (both time and spatial ones). This general problem is as interesting as it is difficult. But one has to consider all of above mentioned factors when the problem needs it. Studied medium is stationary and non-dispersive one in most cases a scientist deals in practice. So, the problem narrows to consideration of inhomogeneous and anisotropic medium. Last one is still sufficiently general since it includes weakly inhomogeneous and anisotropic, weakly inhomogeneous and weakly anisotropic or strongly inhomogeneous anisotropic cases. Given division is essential because each of above mentioned cases possesses of special peculiarities and light wave propagation analysis differs for them.

Here we consider weakly inhomogeneous and weakly anisotropic media. These are the media where above properties are "undesirable" ones. It is a case of technologically induced residual stresses being parasitic values. To this class also belong media where anisotropy is a functional parameter, therefore it has to be of strictly defined form and value. So we have analysed light wave propagation just in such a medium: stationary non-dispersive weakly inhomogeneous and weakly anisotropic one.

Optical anisotropy of the medium is described by dielectric permeability tensor $\hat{\varepsilon}$. Weakly anisotropic is a medium for which following inequalities are true:

$$\left|\varepsilon_{aa}-\varepsilon_{bb}\right| << \varepsilon_{aa},\varepsilon_{bb}, \ a,b = 1,2,3$$

$$\varepsilon_{ab} << \varepsilon_{aa},\varepsilon_{bb} \qquad (2.1)$$

where $\varepsilon_{aa},\varepsilon_{bb}$ - diagonal components of the tensor $\hat{\varepsilon}$, ε_{ab} - nondiagonal ones. Really $\left|\varepsilon_{aa}-\varepsilon_{bb}\right|$ and ε_{ab} is of $10^{-3}-10^{-4}$ order.

Weakly inhomogeneous is such a medium, where dielectric permeability $\hat{\varepsilon}$ weakly changes within studied volume V. That means fulfilment of next inequality:

$$|\nabla\varepsilon| = |\varepsilon(\vec{r}) - \overline{\varepsilon}| << \overline{\varepsilon} \qquad (2.2)$$

where $\overline{\varepsilon}$ is a middle value of $\hat{\varepsilon}$ in the limits of V.

If a medium being wekly inhomogeneous is also smoothly inhomogeneous, one can successfully applies geometrical optics method (Kravtsov 1980, Kline 1965) for electromagnetic waves field analysis in such a medium. Smoothly inhomogeneous is a medium for which following is true:

$$\lambda|\nabla\varepsilon| << \overline{\varepsilon} \qquad (2.3)$$

where λ is a lightwave length in a medium.

For weakly anisotropic medium certain geometrical optics method is inapplicable because of small difference between refractive indeces values Δn ($n=\sqrt{\varepsilon}$ for non-magnetic medium). Small Δn value leads to divergence of $\vec{E}_1, \vec{H}_1$

fields in the first approximation of geometrical optics technique ($\vec{E}, \vec{H}$ - respectively electric and magnetic field vectors of light wave). In order to avoid above divergence different authors have proposed some ways (Aben 1979, Kubo 1978).The most successive electromagnetic waves theory standpoint consists in application of quasiisotropic approximation of certain geometrical optics technique (Kravtsov 1968). Here we present such an approach.

Within quasiisotropic approximation of geometrical optics general dielectric tensor $\hat{\varepsilon}$ is regarded as consisting of a big isotropic $\varepsilon \cdot \delta_{ab}$ and a small anisotropic part χ_{ab}:

$$\varepsilon_{ab}= \overline{\varepsilon}\cdot\delta_{ab} + \chi_{ab} , \qquad (2.4)$$

with a tensor χ_{ab} being of $1/ik_0$ order :

$$\chi_{ab}= \frac{1}{ik_0}\cdot\xi_{ab} , \qquad (2.5)$$

where $\overline{\varepsilon}= \frac{1}{3}\cdot Sp(\varepsilon_{ab})$, δ_{ab} is a Kroneker symbol, $k_0 = \omega_0/c$ - wave number. Tensor χ_{ab} is named as anisotropy one. Above we have conditioned dielectric tensor to be real (non-gyrotropic and non-absorbing medium).

Electromagnetic waves propagation is described by following vector equation:

$$\vec{\nabla}\times\vec{\nabla}\times\vec{E} - k_0^2\cdot\hat{\varepsilon}(\vec{r})\cdot\vec{E}= 0 , \qquad (2.6)$$

or in an index form:

$$e_{ikm}\cdot e_{mnj}\cdot\frac{\partial^2 E_j}{\partial x_k \partial x_n} - k_0^2\cdot\varepsilon_{ij}\cdot E_j = 0 ,$$

where e_{ikm} are Levi-Chivit symbols, $i,j,k,m,n = 1,2,3$.

Application of geometrical optics technique means that solution of the above equation has to be finded in the form:

$$\vec{E}(\vec{r})= \vec{A}(\vec{r})\cdot\exp\left(i\cdot k_0\cdot\varphi(\vec{r})\right) , \qquad (2.7)$$

where $\varphi(\vec{r})$ and $\vec{A}(\vec{r})$ are slowly variable functions of coordinate $\vec{r}$:

$$\vec{A}(\vec{r}) = \vec{A^0}(\vec{r}) + \frac{1}{i\cdot k_0}\cdot\vec{A^1}(\vec{r}) + \dots \qquad (2.8)$$

Above procedure is known as Debay-expansion (Debay 1911). After substituting of (2.7) and (2.8) expressions to equation (2.6), where $\hat{\varepsilon}$ is described by (2.4)-(2.5) relations, and equating of coefficients with the same degree of k_0 one obtains :

$$\left(e_{ikm}\cdot e_{mnj}\cdot\frac{\partial\varphi}{\partial x_n}\cdot\frac{\partial\varphi}{\partial x_k}+\overline{\varepsilon}\cdot\delta_{ij}\right)\cdot A_j^0=0 \tag{2.9}$$

and

$$e_{ikm}\cdot e_{mnj}\cdot[\frac{\partial A_j^0}{\partial x_n}\cdot\frac{\partial\varphi}{\partial x_k}+\frac{\partial A_j^0}{\partial x_k}\cdot\frac{\partial\varphi}{\partial x_n}+$$

$$+A_j^0\cdot\frac{\partial^2\varphi}{\partial x_k\partial x_n}]+\xi_{ij}\cdot A_j^0=0 \tag{2.10}$$

Linear simultaneous equations (2.9) will have a non-zero solution when determinant of the simultaneous equation is equal to zero :

$$\det\left|q_{ij}\right|=\det\left|e_{ikm}e_{mnj}\frac{\partial\varphi}{\partial x_n}\frac{\partial\varphi}{\partial x_k}+\overline{\varepsilon}\cdot\delta_{ij}\right|=0 \tag{2.11}$$

Equations (2.10) and (2.11) describe light waves field in the zero approximation of the quasiisotropic geometrical optics technique. Equation (2.11) is an eikonal equation and (2.10) are transfer equations of zero approximation for waves in inhomogeneous weakly anisotropic medium. Next order approximations are described by the similar equations :

$$e_{ikm}e_{mnj}[\frac{\partial A_j^{(s+1)}}{\partial x_n}\cdot\frac{\partial\varphi}{\partial x_k}+\frac{\partial A_j^{(s+1)}}{\partial x_k}\cdot\frac{\partial\varphi}{\partial x_n}+$$

$$+A_j^{(s+1)}\frac{\partial^2\varphi}{\partial x_k\partial x_n}+\frac{\partial^2 A_j^{(s)}}{\partial x_k\partial x_n}]+\xi_{ij}A_j^{(s+1)}=0, \tag{2.12}$$

where s is a number of approximation.

Equation (2.11) leads to :

$$\det\left|q_{ij}\right|=-\overline{\varepsilon}\left(\frac{\partial\varphi}{\partial x_i}\cdot\frac{\partial\varphi}{\partial x_j}\cdot\delta_{ij}-\overline{\varepsilon}\right)=0, \tag{2.13}$$

Equation (2.13) is true when :

$$\varepsilon=0 \quad \text{or} \quad \frac{\partial\varphi}{\partial x_i}\cdot\frac{\partial\varphi}{\partial x_j}\cdot\delta_{ij}-\overline{\varepsilon}=0, \tag{2.14}$$

Condition $\varepsilon=0$ indicates possibility of a longitudinal waves existence (essential for plasma) and $\frac{\partial\varphi}{\partial x_i}\cdot\frac{\partial\varphi}{\partial x_j}\cdot\delta_{ij}-\overline{\varepsilon}=0$ or $(\vec{\nabla}\varphi)^2=\overline{\varepsilon}$ correspond to well known eikonal equation for isotropic medium (Born & Wolf 1964).

Have analysed ray equation (Born & Wolf 1964, Kravtsov 1980) in a case under consideration

(weakly inhomogeneous medium) one can show that rays trajectory are straight lines. Then we have

$$\vec{\nabla}\varphi=\sqrt{\varepsilon}\cdot\vec{l},$$

where $\vec{l}$ is an ort, coinciding with ray direction. Respectively

$$\varphi=\varphi^0+\int_{\vec{r}^0}^{\vec{r}}\overline{n}\,d\vec{r}$$

$$\varphi(\vec{r})=\varphi^0+|\vec{r}-\vec{r}^0|\cdot\overline{n},$$

where φ^0 is an initial value of eikonal at $\vec{r}^0$, $r=|\vec{r}-\vec{r}^0|$ is a distance along the ray and $\overline{n}=\sqrt{\varepsilon}$.

Transfer equations can be written in more useful form :

$$\begin{cases}\dfrac{dA_i^0}{dr}=\dfrac{ik_0}{2\sqrt{\varepsilon}}\left(\chi_{ii}A_i^0+\chi_{ij}A_j^0\right)\\[2mm]\dfrac{dA_j^0}{dr}=\dfrac{ik_0}{2\sqrt{\varepsilon}}\left(\chi_{ji}A_i^0+\chi_{jj}A_j^0\right)\end{cases} \tag{2.15}$$

If the initial ray propagates along OZ axis of rectangular coordinate system XOZ (respectively $\vec{l_0}=(0,0,1)$) equations (2.15) coincide with Aben equations (Aben 1979) of integrated photoelasticity :

$$\begin{cases}\dfrac{dA_x}{dz}=\dfrac{1}{2}\dfrac{ik_0}{2\sqrt{\varepsilon}}\left(\dfrac{\varepsilon_{11}-\varepsilon_{22}}{2}A_x+\varepsilon_{12}A_y\right)\\[3mm]\dfrac{dA_y}{dz}=\dfrac{1}{2}\dfrac{ik_0}{2\sqrt{\varepsilon}}\left(\varepsilon_{21}A_x-\dfrac{\varepsilon_{11}-\varepsilon_{22}}{2}A_y\right)\end{cases} \tag{2.16}$$

Problem of quasiisotropic approximation error needs special research. Here we only note that main source of the error is the second degree of $\Delta n\approx\chi_{ab}$ members being neglected in quasiisotropic approximation (Kravtsov 1980). These members influence becomes most of all perceptible for phase of wave :

$$\psi_1\approx k_0\int_0^r(\Delta n)^2 dr .$$

As to parameters of polarization ellipse (being determined in most applications) they are less sensitive to the square members of Δn than phase. Experimental technique, we are going to use, deals just with the parameters of the polarization ellipse.

So, in this section we have analysed light wave propagation in weakly inhomogeneous weakly anisotropic media and obtained equations describing the process.

3 THE INVERSE PROBLEM

The inverse problem solution is a central procedure of any technique being apllied to internal structure testing. Principal equation of the approach is following :

$$Ax = y, \tag{3.1}$$

where A is an operator described processes taken places during data obtaining, y vector is experimentally determined data set, x vector is unknown parameters. Often operator A is a superposition of some operators. Mathematical aspects of computerized tomography are widely discussed in monographs of Tikhonov (1987) and Natterer (1986). Principal equation (3.1) in computerized tomography is written in form :

$$\int_{\Gamma(\vec{p})} f(\vec{r}) \, d\sigma = p(\vec{p}), \tag{3.2}$$

where $\Gamma(\vec{p})$ are some curves in R^n (n>1), $d\sigma$ is respective differential form for $\Gamma(\vec{p})$, $f(\vec{r})$ is the function to be determined. Equation (3.2) decribes relation between unknown function $f(\vec{r})$ and its integral $p(\vec{p})$ (or $f(\vec{r})$ projection in another words). For R^n (n=2,3) $\Gamma(\vec{p})$ are recpectively straight lines or planes and problem of $f(\vec{r})$ determination was solved by Radon (1917). Radon integral transform is a fundamental of computerized tomography deals with image reconstruction from their integral projections (Hermen 1980).

The inverse problem of $f(\vec{r})$ function determination is ill-conditioned one. A number of approaches has been elaborated for such a problem solution (Tikhonov 1987). Procedure of $f(\vec{r})$ determination in computerized tomography is treated as a reconstruction algorithm creation. Large number of known reconstruction algorithmes are divided on two groups : finite series-expansion reconstruction methods (Censor 1983) and integral transform ones (Lewitt 1983). We are going to apply the first-type one. The application of such an algorithm means initial discretisation of the inverse problem. According to above technique inhomogeneous medium is regarded as a set of identical cubic cell within each one the medium can be considered as homogeneous one (Censor 1983, Kubo 1979, Berezhna 1994). In our case each cell is described by its own $\hat{\varepsilon}$ tensor components being

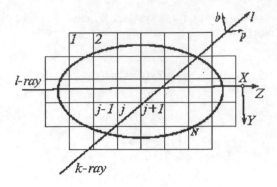

Fig. 1.
Plane illustration of cell model.

constant values within the cell (Fig.1). Constant components of $\hat{\varepsilon}$ change from cell to cell discretely. Number of cell (N) depends on the inhomogenity smoothness. So, dielectric tensor computerized tomography problem for non-gyrotropic medium got a form of 6N unknown tensor $\hat{\varepsilon}$ components determination.

The inverse problem in optical tomography can be considered for any of the equation : eikonal, ray or transfer one. We deal with transfer equations. As it was noted in section 2 of the paper quasiisotropic approximation error is the smallest for polarization ellipses parameters. Experimental technique we have worked is based on the polarimetry measurements. Last ones draw in polarization ellipses parameters operation.

The inverse problem for zero-order approximation of transfer equations (2.15) we have investigated early (Berezhna 1994). Equations (2.15) reduce to simultaneous differential equations with constant coefficients since $\hat{\varepsilon}$ (or $\hat{\chi}$) tensor components were supposed to be invariable within each j-th cell (j - number of cell, j=1,2,..,N). We have considered initial value problem for equation (2.15) in the form :

$$\begin{bmatrix} A_1(r) \\ A_2(r) \end{bmatrix} = \begin{bmatrix} t_{pp}^{jk} & t_{pb}^{jk} \\ t_{bp}^{jk} & t_{bb}^{jk} \end{bmatrix} \cdot \begin{bmatrix} A_1(0) \\ A_2(0) \end{bmatrix}, \tag{3.3}$$

where unknown coefficients t_{pb}^{jk} are constant values, subscripts j and k corresponds to the cell number and sounding ray number respectively (more detailed see section 4 of the paper), $A_1(0) \equiv A_1(r=0)$, $A_2(0) \equiv A_2(r=0)$. We have obtained (Berezhna 1994) expressions that relate t^{jk} matrix components with χ_{pb}^{jk} components of anisotropy tensor $\hat{\chi}^j$. Recently it was proved (Berezhna 1995) that t^{jk} matrices such a way calculated can be treated as Jones matrices (JM). Since t^{jk} matrix describes optical properties of j-th cell we have named it as "micro" JM.

The inverse problem solution has been reduced to determination of 6N unknown components χ_{pb}^{jk} of anisotropy tensor $\hat{\chi}$ from the expressions related them with micro JM-es components. Set of the last ones can be obtained by means of experimental technique.

4 EXPERIMENTAL TECHNIQUE AND SOME PECULIARITIES OF ITS APPLICATION.

In general any computerized tomography technique is two-level one. At the first stage experimental data are obtained and accumulated. At the second one these data are used for calculation. Experimental technique we have worked out is based on the originally invented kind of polarimetry method. This method we have named as hight informative invariant polarimetry one (Berezhna 1991), HIIP-technique. By means of this technique we can determine normalized JM (NJM) for any ray passing the object. Thanks to this ability the technique is suitable just for the tomography purposes . NJM-es being determined as results of the measuring procedures we have called as "macro"-NJM-es. After determination of necessary number of "macro"-NJM-es we use them for calculation of "micro"-NJM-es that describe elementary cubic cells. Final expressions contains "micro"-NJM-es components as coefficients and anisotropy tensor $\hat{\chi}$ components as unknown values.

Now we describe some details of measuring procedure. Directly in experiment we measure characteristic angles and some other angles being used for calculation of "macro"-NJM components. In order to increase number of independent values could be measured for one ray passing (that makes possible NJM determination after only one ray passing) we use elliptically polarized incident light. For given ray direction and known ellipticity of incident light we find emergent light intensity minimum. Procedure of last one identification consist in coordinated rotation of P,C and A, P is a polariser, A is an analyser, C is a compensator. At the moment of minimum intensity registration we measure characteristic angle. After that for the same ray direction we change incident light ellipticity and repeat above procedure that results in determination of one more characteristic angle. Cycle is ended by the third repetition of the procedure. So, for three different ellipticities and one the same ray direction we measure set of characteristic angles. On the basis of these data we solve six simultaneous non-linear equations and determine

three complex-valued components of NJM.

It is principle restriction of any polarimetry technique (Azzam 1977) that only NJM can be determined by means of its application. So, optical tomography technique we are developing provides determination of only 5N from 6N unknown components of $\hat{\chi}$ tensor. For determination of rest N unknown values it is necessary to draw in phase measuring techniques or other additional equations that relates unknown parameters. In the case of stress induced anisotropy additional data can be obtained from mechanical equations.

As it was noted in section 3 we use discrete model for inhomogenity representation. Each cell (Fig.1.) is described by χ_{mn} of its centre. Principal item of such an approach is a problem of cells number N choice. N-value directly affects an accuracy of the anisotropy tensor components reconstruction. Bigger number of cells that inhomogeneous medium is divided on provides more accurate restoration of inhomogeneous distribution of studied parameter. At the same time N-value increasing causes enlargement of χ_{mn}^{j} components number to be determined. So, quantity of simultaneous equation to be solved increases also. The problem of big number simultaneous equation solution is known as extremely complicated one. Beside that N-value is also limited by real accuracy of measurement procedure. We have tried to make some criterions for N-value choice. In particular we have suggested to take some preliminary measurements for the determination of sufficient N-value. Such measurements are not directly related with the reconstruction process. A priori information about the dielectric (or stress) tensor distribution in the studied object is intended to be used. Usually researcher has such information since residual stresses are technologically caused ones.

Main idea of research we have taken consists in comparison of defined characteristic values being obtained by measuring and calculating ways. Calculations have been fulfilled by using of the algorithm based on division of light ray path in medium on N segments (linear analog of mentioned cubic cells). We changed number of N and analysed responsive variations of the calculated characteristics about measured one. Simulation experiments have been carried out by using diametrally stressed glass disk. Measurement set is illustrated on Fig.2.

NJM suggested to be measured by application of HIIP-technique has diagonal form (in its own coordinate system) and its T_{22} component can be written as :

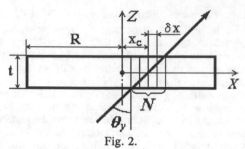

Fig. 2.

Measurement set of diametrally stressed glass disc. N- number of cells, θ_y - angle between incident ray and OZ axis.

$$T^e_{22} = \cos\Delta + i\sin\Delta \qquad (4.1)$$

where $\Delta = 2\cdot\pi\cdot\Delta n\cdot d /\lambda$ is an optical phase retardation for given ray direction. Measurement accuracy for Δ is $d\Delta^e = \pm 0.01^o$, superscript "e" means "experimental". At the same time for given ray direction optical phase retardation Δ^c can be calculated on the basis of known stress tensor components distribution and C_0 (C_0 is photo-elastic constant, "c" near Δ means "calculated"). Calculation algorithm was described in (Berezhna 1994). Respective calculations result in obtaining of "theoretical" "macro" NJM, which T_{22} component is equal to :

$$T^e_{22} = \exp\left(i\left\{\left[\sum_{j=1}^{N}(\chi_1-\chi_2)_j\right]\cdot C\cdot\Delta d\right\}\right), \qquad (4.2)$$

where $C = k_0 /2\sqrt{\varepsilon_0}$, k_0 - wave number, ε_0 - dielectric constant of unstressed medium, Δd is segment length.

Since optical anisotropy of stressed disk is caused by stress distribution expression (4.2) can be written (Ginzburg 1944) in the form :

$$\mathrm{Re}(T_{22}) = \cos\left\{c\cdot\left[\sum_{j=1}^{N}(\chi_1-\chi_2)_j\right]\cdot\Delta d\right\} =$$

$$= \cos\left\{C_0\cdot\left[\sum_{j=1}^{N}(\sigma_1-\sigma_2)_j\right]\cdot\Delta d\right\} = \cos\Delta^c \qquad (4.3)$$

In the set being used $(\sigma_1-\sigma_2)$ difference is equal to:

$$(\sigma_1-\sigma_2) = \sigma_x\cdot\cos^2(\theta_y) - \sigma_y,$$

where σ_x and σ_y - main stresses.

We have simulated weakly anisotropic medium , so phase retardation in the centre of stressed disk

was equal to $\pi/2$ ($C_0 = 2\cdot C\cdot C_1\cdot\sqrt{\varepsilon}$, C_1 is photo-elastic constant).

As it was assumed we have establish that divergence between calculated Δ^c and measured Δ^e depends on the value of N. This dependence is illustrated in Fig.3.

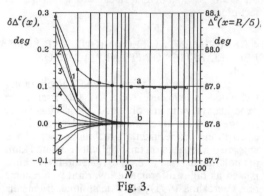

Fig. 3.

a) Dependence of computed Δ^c from the number of cells N for $x=R/5$;

b) Dependence of Δ^c calculation error, $\delta\Delta^c$, from the number of cells N for different x : 1-0, 2-0.1R, 3-0.2R, 4-0.3R, 5-0.4R, 6-0.5R, 7-0.6R, 8-0.8R

We attempted to explanate obtained behaviour of "restoration" accuracy by involving Δ^c gradient. Coordinate dependences of Δ^c and $grad\Delta^c(x)$ are represented in the Fig.4. By comparing $\delta\Delta^c(x)$ (Fig.3,b) and $grad\Delta^c(x)$ (Fig.4,b) behaviours one can note that the biggest value of Δ^c calculation error corresponds to more rapid $grad\Delta^c$ change (x coordinate changes within 0 - 3R/10), minimum error is for the section, where $grad\Delta^c$ change is the slowest (x within 4R/10 - 6R/10) and $\delta\Delta^c$ is "saturating" for x values (7R/10 - R) corresponded to $grad\Delta^c(x)$ uniform change.

We have finded interesting to draw in $grad(grad\Delta^c)$ coordinate dependence for further researching of the "restoration" accuracy question. Second derivative $d^2\Delta^c /dx^2(x /R)$ and error of Δ calculation $\delta\Delta^c(x /R)$ coordinate dependences are shown in Fig.5 (a,b). Set of curves in the Fig.5. correspond to defined N values. By comparing of these characteristic behaviours $(\delta\Delta^c$ and $d^2\Delta^c /dx^2$) we have stated that irrespectively of N value the largest errors of Δ^c calculation take places for the coordinates (or ray passing) that

308

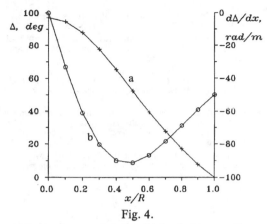

Fig. 4.

a) Dependence of Δ from the relative coordinate x/R: by "+" for measured Δ^e; by unbroken line for calculated Δ^c, N=100; measurement error $\delta\Delta^e = 0.01°$, XOZ section, angle between OZ and incident ray $\theta = 30°$.

b) Dependence of $\partial\Delta^c/\partial x$ from the relative coordinate x/R for Δ^c being calculated with N=100; $\theta = 30°$; XOZ section.

corresponds to maximum value of the Δ^c second derivative.

We have concluded some prescriptions for simplification of N cell choise procedure. By comparising of Δ^e and Δ^c data it was finded that correlation between that ones depends on the number of cells N and light ray direction. Also we have stated that additional information for solution of N number choice question can be obtained from analysis of $grad\Delta^c(x/R)$ dependences. We suggest to use $grad\ grad\ \hat{\sigma}(\vec{r})$ rough distribution for evaluation of N, since $\Delta(r)$ is directly related with $\sigma(\vec{r})$. Researcher is able to analyse rough distribution of $\sigma(\vec{r})$ (has to be determined) in most number of cases on the basis of *a priori* information.

Our idea consists in utilization of so named "analysing" ray. Such a way we have titled the ray for which unknown N value is going to be determined. Since this ray corresponds to the direction for which necessary number of cells is of a maximum value, one can state that N value being calculated for this direction will satisfy all other ones. Last statement means that measured Δ^e and calculated Δ^c values correlate within measurement accuracy, i.e.

$$\delta\Delta^c\left(N\right) = \delta\Delta^e,$$

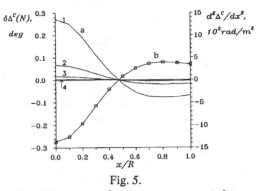

Fig. 5.

a) Dependence of Δ^c calculation error $\delta\Delta^c$ from the relative coordinate x/R for different N: 1 - N=1; 2 - N=2; 3 - N=4; 4 - N=8;

b) Dependence of the second derivative of Δ^c, $d^2\Delta^c/dx^2$, from the relative coordinate being calculated for Δ^c with N=100, $\theta = 30°$, XOZ section

where $\delta\Delta^c(N)$ is an error of Δ calculation with N cells number, $\delta\Delta^e$ is an error of Δ measurement. It is natural to use $grad\ grad\ \hat{\sigma}(\vec{r})$ rough dependence for determination of "analysing" ray coordinates since maximum N value is necessary for the section where $\sigma(\vec{r})$ change is the most steep. So, "analysing" ray coordinates will coincide with $grad\ grad\ \hat{\sigma}(\vec{r})$ maximum.

Thus, we suggest to pass only one analysing ray in the direction needs the biggest number of cells to be divided on. How to choose this direction we have discussed above.

Suggested way for simplification of cells number N choice is not definitive one and rather give information for further research and discussion.

One more peculiarity of presented optical tomography technique application was discussed in (Berezhna, Berezhnyi, Krupych, Vlokh 1995). It concerns problem of coordinate system choise in which measured data for arbitrary ray have to be presented. We have proposed to determine own coordinate system for each ray. Such system axes coincide with asimuthes of polarization ellipses of eigenpolarization states. Set of NJM is supposed to be represented in set of each ray eigenpolarization coordinate systems. Establishment of mutual orientation of own coordinate systems provides representation of all measured data in one of them. Often such a way chosen coordinate system is appeared to be related with external shape of an object.

309

5 CONCLUSIONS

We have described new optical technique being applicable to anisotropy parameters determination in transparent inhomogeneous media. Suggested technique is based on application of tomography approach. Main principles of the technique operation and some particular problems of its application have been discussed.

REFERENCES

Aben, H. K. 1979. *Integrated photoelasticity.* London : McGrow-Hill.

Aben H. K. 1986. Integrated photoelasticity as tensor field tomography. *Proc of the Internat. Sympos. on Photoelasticity.* Tokyo: 243-250.

Andrienko, Y. A. &M. S. Dubovikov 1994. Optical tomography of tensor fields: the general case. *J. Opt. Soc Am.* 11, No 5: 1628-1631.

Azzam, R. M. A. & N. M. Bashara 1977. *Ellipsometry and polarized light.* Amsterdam, North-Holland Publishing Company.

Berezhna, S. Yu., I. V. Berezhnyi 1991. Reconstruction of Jones-matrix of an object by means of PCSA-polarimeter. *Optica i Spectroscopia.* 70, No 5 : 1107-1111.

Berezhna, S. Yu., I. V. Berezhnyi & O. G. Vlokh 1994. Optical tomography of anisotropic inhomogeneous medium. *Proc of 10th Int. Conf. on Experim Mechan.* Lisbon 18-22 July 1994. Rotterdam, A. A. Balkema : 431-435.

Berezhna, S. Yu., I. V. Berezhnyi, O. M. Krupych & O. G. Vlokh 1995. Some peculiarities of dielectric tensor field optical tomography technique (to be published in *Proc. of SPIE*).

Berezhna, S. Yu., I. V. Berezhnyi & O. G. Vlokh 1995. Use of Jones-matrix formalism in an optical tomography problem. *Optika i Spectroscopia.* (To be published).

Born, M and E. Wolf 1964. *Principles of optics.* Oxford - London : Pergamon Press.

Censor, Ya, 1983. Finite series-expansion reconstruction methods. *Proc. of IEEE* 71 : 409-419.

Debay, P 1911. Comments to the article : Sommerfeld A., Runge J. Answendung der Vektorrechnung auf die grundlagen der geometrischen optik.. *Ann. Phys.* 35 : 277-279.

Ginzburg, V. L. 1944. About stress investigation by optical method. *J. of Tech. Physics.* 14, No 3 : 181-192.

Herman, G. T. 1980. *Image Reconstruction from Projections: The Fuundamentals of Computerized Tomography.* New York, NY: Academic Press.

Kline, M. & I. W. Kay 1965. *Electromagnetic theory and geometrical optics.* NY: John Wiley & Sons.

Kravtsov, Yu 1968. "Quasiisotropic" approximation of geometrical optics. *Dokl AN SSSR.* 183, No 1 : 74-76.

Kravtsov, Yu. A. & Yu. I. Orlov 1980. *Geometrical optics of inhomogeneous media.* Moscow: Nauka.

Kubo, H. and R. Nagata 1978. Equations of light propagation in an inhomogeneous crystal. *Opt. Commun.* 27, No 2 : 201-206.

Kubo, H. and R. Nagata 1979. Determination of dielectric tensor fields in weakly inhomogeneous anisotropic media. *J. Opt. Soc. Am.* 69, No 4: 604-610.

Lewitt, R.M. Reconstruction algorithms: transform methods. 1983. *Proceed of IEEE* 71, No 3: 390-408.

Natterer, F 1986. *The mathematics of computerized tomography. New York* : B. C. Teubner, Stuttgart and John Wiltey & Sons Ltd.

Puro, A. E. 1992. Investigation of the stressed state of elastic bodies by optical tomography method. *Prikladnaya Mechanika (Applied Mechanics).* 28, No 3: 46-51.

Radon, J. 1917. Uber die bestimmung von funktionen durch ihre integralwerte langs gewisser mannigfaltigkliten. *Ber. Verb. Sachs. Akad. Wiss. Leipzig.* 69 : 262-277.

Tikhonov, A. N., V. Ya. Arsenin & A. Timonov 1987. *Mathematical problems of computerized tomography.* Moscow : Nauka.

Progress in Durability Analysis of Composite Systems, Cardon, Fukuda & Reifsnider (eds)
© 1996 Balkema, Rotterdam. ISBN 90 5410 809 6

Influence of environmental aging on mechanical properties of single lap joints

J. Bonhomme, F. Manrique & F. J. Belzunce
Instituto Tecnológico de Materiales, Llanera (Asturias), Spain

ABSTRACT: water absorption in the epoxy adhesives studied in this work follow the Fick Law. Their mechanical properties decrease in a linear mode over the studied range, and after dried, they recover the initial properties. In the laminates, the loss of material makes water absorption does not follow Fick Law so closely. The effect of other aging phenomena as marine water and UV radiation and their effect on mechanical properties are disscused too.

1. INTRODUCTION

The materials studied were two cold curing epoxy adhesives: one is 3M Scotch-weld 3520 B/A and the other is IIM 3/3 fom Paniker. The adhesives are refered as 3M and Paniker in the present work. On the other hand, we have two symmetric glass fiber-vinylester laminates named as lam6 and lam7 and the single lap joints by combination of these materials (ASTM D 1002).

We present first the materials interaction with distiled and marine water, then, we'll discuss the action of UV radiation in short-term laboratory tests.

2. WATER ABSORPTION

2.1 The kinetic of water absorption

Rectangular shape specimens of 1.5 mm thickness of each material have been introduced in distiled water at 50, 60 and 75 °C for 90 days and in a solution of water and 2.5 % Wt of NaCl at 60 °C to simulate marine water (ASTM 1183). The samples were weighted periodically in order to measure water absorption. Together with these specimens, we have immersed samples of all the materials to perform mechanical tests.

Fick Law can be modelized by:

$$M\% = M_{max}\%\left\{1 - \frac{8}{\pi^2}\sum_{n=0}^{\infty}\left(\frac{1}{2n+1}\right)^2 \cdot \right.$$

$$\left. \cdot \exp\left(-\frac{Dt}{h^2}\pi^2(2n+1)^2\right)\right\}$$

(Ec. 2.1)

where:

$M\%$ = water percentage in the sample
$M_{max}\%$ = water percentage in the equilibrium
D = diffusion coefficient
h = sample thickness
t = time

We can divide the curve in two parts: for short times $Dt/h^2 \ll 0.05$ and then ec. 2.1 can be expresed as:

$$M\% = M_{max}\%\frac{4}{\sqrt{\pi}}\left(\frac{Dt}{h^2}\right)^{1/2} \text{ (Ec. 2.2)}$$

and for long times is given by:

$$M\% = M_{max}\%\left\{1 - \frac{8}{\pi^2}\exp\left(-\frac{Dt}{h^2}\pi^2\right)\right\} \text{ (Ec. 2.3)}$$

With the results obtained in the absorption tests we can have drawn the curves as shown in figures 1 to 3.

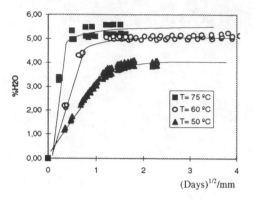

Figure 1. Water absorption of 3M adhesive at different temperatures

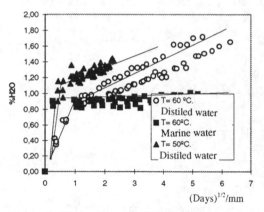

Figure 2. Water absorption in laminate 7

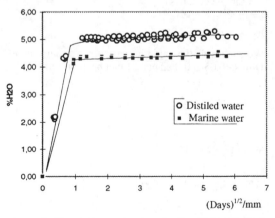

Figure 3. Water absorption in adhesive 3M in distiled and marine water

Some conclusions are derived from these curves. As temperature of water is increased, the rate of water intake in adhesives and laminates is higher and the water amount in the equilibrium is higher too. On the other hand, we can observe that the presence of NaCl in water decrease the water in the equilibrium. Finally when comparing the behaviour of the adhesives we have found that Paniker adhesive absorve less amount of water but his behaviour is not so close to the Fickian Law as in 3M adhesive.

An analysis in FTIR and TGA showed that the Paniker adhesive contained 46 % Wt of $BaSO_4$ probably to increase its tixotropic properties to improve the applicability on joints in vertical walls.

On the other hand, we can see in figure 2 typical laminates absorption curves. During the tests, loss of material was observed so the curves should be corrected to fit closer the Fickian Law. Is important to consider too, that composites are not a continuum media because are composed of two phases: fibers and matrix. So, to increase the accuracy of the job, it should be necessary to apply another model as the rule of mixtures. Another explication to the anomalus behaviour is the presence of voids in the laminates.

Now, from the tests results we can calculate the constants of diffusion in function of the temperature of the adhesives by means of the next expressions:

For the water content in the equilibrium we have:

$$M_{max} \% = Ae^{BT} \qquad \text{(Ec. 2.4)}$$

And for the diffusion coefficient, it is known to follow the Arhenius Law by means of:

$$D = D_0 \exp(-E_a/RT) \qquad \text{(Ec. 2.5)}$$

Tables 1 to 4 shows the calculated constants for both adhesives.

Table1. Constants of equation 2.4 for the adhesives in distiled water.

	$M_{max} \% = Ae^{BT}$	
	A	B
Adhesive 3M	2.4747	0.0107
Adhesive Paniker	0.9908	0.0220

Table 2. Constants of equation 2.5 for the adhesives in distiled water.

	$D = D_0 \exp(-E_a/RT)$	
	D_0 (cm^2/s)	E_a/R (°K)
Adhesive 3M	3.50×10^5	9.8629×10^3
Adhesive Paniker	1.53×10^{-2}	4.8392×10^3

Table 3. Constants of equation 2.4 for the adhesives in marine water.

	$M_{max} \% = Ae^{BT}$	
	A	B
Adhesive 3M	2.3557	0.0103
Adhesive Paniker	0.3501	0.0282

Table 4. Constants of equation 2.5 for the adhesives in marine water.

	$D = D_0 \exp(-E_a/RT)$	
	D_0 (cm^2/s)	E_a/R (°K)
Adhesive 3M	7.19×10^5	1.0070×10^4
Adhesive Paniker	2.29×10^6	1.0808×10^4

Now, by means of the above equations and calculated constants, it is possible to predict water absorption in function of water temperature for long periods of time in distiled and marine water.

In the next step we'll try to relate water content and mechanical properties in order to predict mechanical properties over long periods of time.

2.2 The loss in mechanical properties

As said before, together with the absorption samples we have immersed specimens to perform tensile tests in the adhesives (ASTM D 638), laminates (ASTM D 3039) and joints (ASTM D 1002). It is important to remember that water temperature influence the water rate intake but all mechanical tests were carried out at room temperature to eliminate the influence of this variable on the mechanical properties. The results are shown in tables 5 to 7.

Table 5. Tensile strength in MPa in adhesives and laminate in function of water temperature and time.

	3M		Paniker		Laminate 7	
Days	50 °C	60 °C	50 °C	60 °C	50 °C	60 °C
0	25.5	25.5	27.5	27.5	142.1	142.1
17	17.8	-----	24.0	-----	-----	-----
18	-----	14.1	-----	22.45	-----	108.8
21	-----	-----	-----	-----	113.4	----
60	13.2	-----	21.2	-----	101.3	----
90	-----	11.7	-----	14.86	-----	78.97

As we can see from the results of the tests, initially, the Paniker adhesive shows better mechanical properties than 3M adhesive, but, joints made of 3M behave some better than Paniker adhesive due to the fact that former adhesive has a quite higher ultimate strain and so the stress field in the joint is more favorable.

Table 6. Elastic modulus in GPa in adhesives and laminate in function of water temperature and time.

	3M		Paniker		Laminate 7	
Days	50 °C	60 °C	50 °C	60 °C	50 °C	60 °C
0	1.27	1.27	3.87	3.87	9.54	9.54
17	0.68	-----	2.99	-----	-----	-----
18	-----	0.33	-----	2.88	-----	10.83
21	-----	-----	-----	-----	9.35	-----
60	0.43	-----	3.20	-----	11.57	-----
90	-----	0.25	-----	2.06	-----	8.36

Table 7. Shear stregth in MPa in sinle lap joints in function of water temperature and time.

	3M-laminate 7		Paniker-laminate7	
Days	50 °C	60 °C	50 °C	60 °C
0	8.29	8.29	6.14	6.14
18	-----	7.92 (1)	-----	7.07 (2)
21	5.60	-----	4.75	-----
60	5.35	-----	4.91	-----
90	-----	6.97 (3)	-----	4.59

(1) One of the samples broke in the sustrate in tensile mode at 73 MPa
(2) One sample broke 80 % cohesive at a shear stress of 4 MPa. Another one broke in tensile mode in the sustrate.
(3) All samples broke in the sustrate in tensile mode at 64 MPa. The table shows the shear stress at maximum load to compare with the other results.

We can observe too that mechanical properties decrease as the immersion time or water temperature is increased (the water content is higher in the first case and the rate of water absorption is higher in the second one).

This effect is reversible in the adhesives but not in the laminates as we can see in table 8 were the tensile test results of dried samples after 90 days immersed at 60 °C are shown.

Table 8. Mechanical properties of dried adhesives and laminates after 90 days immersed in water at 60 °C.

	3M	Paniker	Lam 7
Tensile stress (MPa)	29.5	29.0	73.7
Elastic modulus (GPa)	1.75	4.91	11.44
Ultimate strain (%)	3.1	0.9	0.9

As we said above, the adhesives recover the initial properties after dried and even the values are higher due to a postcured effect. So we can conclude that the water acts as a plastizier and no structural or chemical changes have happened.

On the other hand, the dried laminate shows the initial rigidity but it does not recover the initial stregth. Since the interfacial bond strength between fibers and matrix affects mainly the transverse strength but not the transverse modulus, we can conclude that the reduction in the tensile stregth may be a result of the weakening of the interfacial bond by absorbed moisture (Tsai, Hahn, 1980) (we should remember that our laminate is composed of 0/90° cloth and mats).

With the different amount of absorbed water reached at each temperature we have constructed the curves shown in figures 4, 5, 6 and 7.

As we observe, the resuts show that there are no discontinuity in the curves. This means that mechanical properties are independent of immersion temperature, being a function exclusively of absorbed water (let's remember that all mechanical tests were carried out at room temperature).

On the other hand, the evolution of mechanical properties are quite linear until saturation in the range of time and temperature studied, so

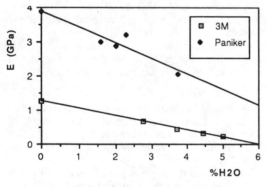

Figure 4. Elastic modulus of adhesives in function of absorbed water.

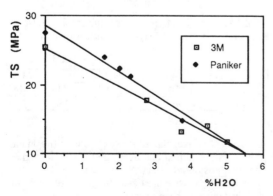

Figure 5. Tensile stregth of adhesives in function of absorbed water

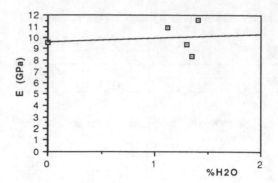

Figure 6. Elastic modulus of laminate 7 in function of absorbed water

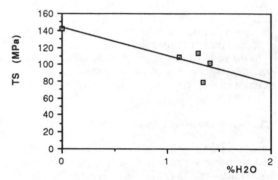

Figure 7. Tensile stregth of laminate 7 in function of absorbed water.

constructing the regresion line we are able to predict the mechanical properties at other quantities of absorbed water inside the studied range.

In this way it will be possibly to predict the loss of mechanical properties at lower temperatures and longer periods of time using the model and coefficients shown in tables 1 and 2 and the above curves.

Table 9 shows the linear equations corresponding to the above curves.

Table 9. Mechanical properties in function of absorbed water (%).

	Elastic modulus (GPa)	Tensile strength (MPa)
3M	$E = 1.256-0.208 \cdot M$	$\sigma_y = 25.23-2.75 \cdot M$
Paniker	$E = 3.878-0.455 \cdot M$	$\sigma_y = 28.49-3.37 \cdot M$
Lam 7	$E = 9.597+0.322 \cdot M$	$\sigma_y=142.83-32.74 \cdot M$

In the case of laminates other research groups (G. Pitchard and S. D. Spaeake, 1987) have found that for higher absorption ranges the curves loss their linearity and are closed modeled by:

$$p = a \cdot (1 - \exp(-b \cdot \exp(-M))) + d \qquad \text{(EC. 2.6)}$$

2.3. Marine water

As we have seen above, the water saturation in marine water is lower than that in distiled water. In order to compare the degradation suffered by the materials immersed in marine water, table 10 shows the mechanical properties of laminate and adhesives after 60 days of immersion in saline water at 60 °C.

Table 10. Mechanical properties of laminate and adhesives after 60 days immersed in marine water at 60 °C.

	3M	Paniker	Lam 7
Water absorption (%)	4.37	1.75	0.92
Tensile stregth (MPa)	6.98	12.43	80.61
Elastic modulus (GPa)	0.108	1.083	9.132

This is illustrated in figure 8 in the case of the Paniker adhesive. We can conclude that in the studied case, the adhesives and laminates show a significant higher degradation at the same level of absorbed water compared with distiled water.

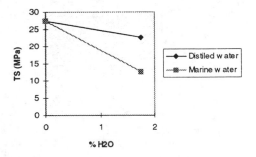

Figure 8. Tensile stregth of Paniker adhesive in distiled and marine water

3. UV RADIATION

In order to quantify the effect of UV radiation, samples of adhesives, laminates and joints were introduced in an environmental chamber to be exposed to UV radiation during a period of 60 days (ASTM G 26). Samples were tested every 10 days in order to monitorize properties changes.

Figure 9 shows the results in the adhesives. As we can see, in general the materials present a slightly increase in their properties due to a post-cured effect until a limit is reached.

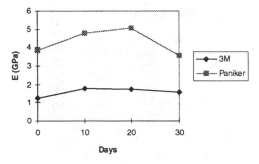

Figure 9. Effect of UV radiation in elastic modulus of adhesives.

By means of ATR technique it was verified that the degradation induced by UV radiation afected only to a thin external layer of the materials, so its mechanical properties have not been significant modified.

4. CONCLUSSIONS

1. A model was developed to predict mechanical property changes due to absorbed moisture over long periods of time outside the experimental range.

2. Empirical equations to determine the mechanical properties in function of absorbed moisture were developed.

3. Changes experimented in adhesives due to absorbed water were recovered when dried. On the other hand, changes in laminate strength are not reversible due to a possible degradation of the bond interface between fibres and matrix.

4. Failures experimented by the joints were by delamination of laminates in most of the cases so its behaviour is governed by laminate property degradation. As joints absorbe water, some laminate failures in tensile mode were observed.

5. UV radiation only affects the external surface so the mechanical properties do not present significant changes.

5. FURTHER WORK

In order to verify the use of water absorption kinetic to predict laminate properties changes, we are performing absorption and mechanical tests at 20 °C for long periods of time.

The next step is to study the combined effect of water, UV radiation and temperature.

On the other hand, samples of all materials are also exposed to natural outdoor aging in order to compare the results with that obtained in accelerated laboratory tests.

6. ACKNOWLEDGEMENT

We would like to thank the Spanish company MEFASA for providing the materials and financial aid for the present work.

7. REFERENCES:

1. Tsai, S. W.; Hahn, H. T. (1980). *Introduction to composite materials.* Connectica: Technomic
2. Pritchard, G.; Speake, S. D. (1987). *The use of water absorption kinetic data to predict property changes.* Composites, vol.18, nº 3.

Progress in Durability Analysis of Composite Systems, Cardon, Fukuda & Reifsnider (eds)
© *1996 Balkema, Rotterdam. ISBN 90 5410 809 6*

Approach to the compatibility of polymer composite-cement concrete (PC-CC) system

L. Czarnecki
Warsaw University of Technology, Poland

W. Głodkowska
Technical University, Koszalin, Poland

ABSTRACT: Polymer concretes could be a valuable means to enhance the durability of contemporary cement concretes under the action of chemical aggressive environmental factors. They can be considered as a high performance repair materials for portland cement concrete structures, usually used under severe service conditions. Particularly, surface repair, partial depth repair and structure repair are typical kinds of application. A two component system: portland cement concrete (substrate) in contact with repair polymer composite is produced. The lack of understanding of the polymer composite - cement concrete interaction is frequently the of main cause many failures in practice. The compatibility of PC - CC system is the topic of the project.

Several requirements of positive CO - operation between polymer and portland cement concrete can be formulated in the form of mathematical inequalities. The material control features of Polymer Concrete - Portland Cement Concrete system will be selected and the range of representative values of the main features of PC and CC will be defined. On this base, the response surface of the satisfactory compatibility between those both types of materials can be computed and characterised. This will provide a technical basis for selection of polymer repair materials which will give good results in durable repairs.

1. INTRODUCTION

Constant development in civil engineering and growing industrial activities create constant demand for building materials that satisfy increasingly stringent requirements. More and more chemical pollution in the atmosphere has been added to other aggressive environmental factors. This has significantly increased deterioration of materials in existing building structures. The contemporary concretes are ready-to-use and high performance building materials, but the vanished durability of concrete under service conditions seems to be the price for its universality. In this situation the polymer composites appear as useful protective and repair materials. The repairing and anti - corrosion protection of the building structure is one of the most important application of the polymer composites. They can be considered as high performance repair materials for portland cement concrete structures, usually used under severe service conditions. Particularly, surface repair, partial depth repair, structure repair and strengthening are typical examples of application. A two component system:

portland cement concrete (substrate) in contact with polymer composite is produced. The lack of understanding of the polymer composite - cement concrete interaction is frequently the source of many failures in practice. The compatibility of PC - CC systems is here the main problem.

2. BASIC APPROACH

The scientific aim of the study is to improve the understanding of the nature of the polymer composite - portland cement concrete system. From the engineering point of view results should be the useful tool for prediction of the behavior of the two - component PC - CC system and their durability, as well as for selection of the proper repair/protective material for the specific application.

The concrete substrates are different one from the other in age, quality and exposure: from the relatively new concrete to the almost deteriorated ones exposed to various temperatures, relative humidity, chemical aggressive environment and mechanical loads. There are two basic problems involved with usability of polymer composites and durability of the system as a whole:

- proper formulating of the polymer concrete with the aim to obtain a material with the properties which meet requirements; it is a material design and optimization problem,
- the compatibility of the system; it is the problem of materials selection and evaluation of the system behavior viz. "getting along well together".

Constant development in civil engineering and growing industrial activities create a continual demand for building materials that satisfy increasingly stringent requirements. Polymer composites seem to be a good way to meet these requirements. A two - component system: portland cement concrete (substrate) in contact with polymer composite results from the application of the polymer composite to the ordinary concrete. The lack of understanding of the nature of this system is frequently the source of many failures in practice. The compatibility of PC - CC systems is still a difficult problem. There is a great need to develop a tool, which can be used to predict the behavior of the system and to select the proper components, particularly polymer composites, for the given application. The significance of compatibility problem have been stressed also by some other authors in recently published papers.

On the base of the suitable literature data and some own experiences the material features that influence the PC - CC system have been selected and their probable value ranges have been defined. The compatibility of the polymer composite - portland cement concrete system has been characterized by several mathematical inequalities. On this base, the model based on response surfaces of the satisfactory compatibility between both types of materials and its output has been analysed in terms of its predictability.

Conversion of the conceptual model into a mathematical one could be done and verified by the experimental program. That will enable to obtain the model, which in comparison with some empirical approach, will be more predictive and more universal (applicable to many materials).

3. SERVICEABILITY

The material serviceability in the construction could be considered according to various requirements defined by environmental service conditions. In the space: chemical-, mechanical-, and physical (mainly thermal) loads could be recognized. Due to the balance between materials and service loads the subspace of the material usability could be defined.

If the value of loads increased then the failure and fracture space could be recognized. In the case of the polymer repaired portland cement concrete substrate the situation is even more complicated due to the compatibility problem. Various repair jobs resulted into the two components: polymer composites - portland cement concrete system. In engineering practice significant variability of behavior of such a system is observed. Sometimes it is reliable and durable; in other cases failures occurs after different periods of service time. There is the need for defining of the set of requirements towards materials which treated simultaneously will determine "compatibility space" (Fig.1).

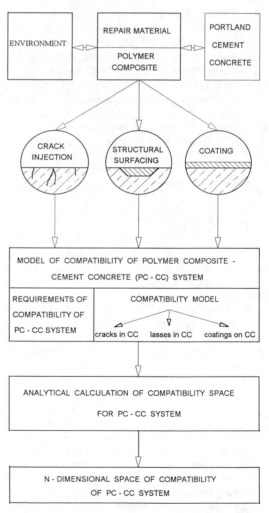

Fig.1 Basis of general research approach of the compatibility of polymer composite - portland cement substrate system

Suitable material control parameters (Tabl.1) and requirements are considered to include:
- mechanical parameters:
  - bond strength,
  - cracking resistance,
  - interlayer adhesion,
  - ultimate strength in glue joints,
- thermal parameters:
  - setting shrinkage,
  - thermal compatibility,
  - thermal resistance including shock resistance,
- chemical resistance controlled by the mass transportation process. This set of requirements is still open and could be completed, e.g. by thaw/freeze resistance. The thaw/freeze resistance -in turn - could be characterized by the balance between water permeability and tensile of given material. The benefit of the study should occur in a proper equilibrium between reliability, durability and economy of the system. Various assumption should be done in order to create the compatibility model at the sufficient accuracy level and simplicity such as, e.g. Hook' law valid, lack of synergic effects of various loads, service temperature below glass transition, etc.

The "compatibility model" will consist of above ten equations and/or inequations given in Table 2. More than ten parameters on polymer concrete and a close to number of parameters on portland cement concrete (see Tabl.1) will control the model. It means that the N (>10) -dimensional "compatibility space" of polymer composite - portland cement concrete system will be calculated. Searching for compatibility space consist in subsequent substitution of value of suitable material features. The value of material features will be chosen from the range of values which occur in practice.

Table 1. Material control parameters for calculation of compatibility space of polymer composite - cement concrete (PC - CC) system

MATERIAL CONTROL PARAMETERS

PC		CC	
R_{pr}	- tensile strength, MPa	R_b	- compressive strength, MPa
$R_\tau^{p/p}$	- PC - CC interface shear strength, MPa	R_{br}	- tensile strength, MPa
E_{pr}	- tensile elasticity modulus, MPa	E_{br}	- tensile elasticity modulus, MPa
E_p	- compressive elasticity modulus, MPa	α_{tb}	- linear expansion coefficient, K^{-1}
α_{tp}	- linear expansion coefficient, K^{-1}	λ_b	- thermal conductivity, $W/m \cdot K$
λ_p	- thermal conductivity, $W/m \cdot K$	a_r	- crack width, mm
ε_{pr}	- ultimate tension strain, %	l_f	-length of substrate crack covering layer, mm
ε_s	- linear shrinkage, $^o/_{oo}$	l_u	- length of concrete decrement, mm
ν_p	- Poisson's coefficient		
D	- diffusion coefficient, $cm^2 s^{-1}$		
t	- time of penetration, s		
h_p	- thickness, mm		

$R_\tau^{p/b}$ - interface shear strength, MPa
$R_{pr}^{p/b}$ * - interface tensile strength, MPa
ΔT - temperature rise limit, K

* For injection and glue joints

Table 2. Model of compatibility of polymer composite - portland cement substrate (PC - CC) system

Requirements of compatibility	Cracking injection	Structural surfacing	Coatings
$\varepsilon_{pr} \cdot l_f \geq \Delta a$	+	+1) for $l_u \geq l_f$	+
$\dfrac{R_{pr} \cdot h_p}{R_{br}} \cdot \varepsilon_{pr}\left(1+\varepsilon_{pr}\right) \geq \Delta a_r$	-	+1)	+1)
$\dfrac{R_{pr} \cdot h_p}{R_{br}} \cdot \varepsilon_{pr}\left(1+\varepsilon_{pr}\right) \geq a_r + \Delta a_r$	-	-	+2)
$R_\tau^{p/b} \geq R_{br}$	-	+	+
$R_{pr}^{p/b} \geq R_{br}$	+	+	-
$R_\tau^{pi/pi+1} \geq R_{br}$	-	+	+
$R_\tau^{p/b3)} > \dfrac{\left(\alpha_{tp} - \alpha_{tb}\right) E_{pr} \cdot E_{br}}{E_{pr} + E_{br}} \cdot \Delta T$	+	+	+
$R_\tau^{p/b3)} \geq \dfrac{\left(\varepsilon_{pr-} \dfrac{R_{br}}{E_{br}}\right)}{E_{pr} + E_{br}} E_{pr} \cdot E_{br}$	+	+	+
$\dfrac{\lambda_b}{\lambda_p} < \dfrac{E_{br} \cdot \alpha_{tb}}{B}$ $B = \dfrac{\left(\alpha_{tp} - \alpha_{tb}\right)}{E_{pr} + E_{br}} E_{pr} \cdot E_{br}$	-	-	+
$R_{pr} \geq \dfrac{0,3 \cdot E_p \cdot \varepsilon_s}{\left(1 - v_p\right)}$	+	+	+
$R_\tau^{p/b3)} \geq \dfrac{0,3 \cdot E_p \cdot \varepsilon_s}{\left(1 - v_p\right)}$	+	+	+
$R_\tau^{pi/pi+1} \geq \dfrac{0,3 \cdot E_p \cdot \varepsilon_s}{\left(1 - v_p\right)}$	-	+	+
$h_p \geq \pi\sqrt{D \cdot t}$	-	-	+

+ - condition that determines the compatibility of polymer composite with cement concrete,
- - the condition does not occur,
1) - in regard to the cracked concrete surface,
2) - in regard to the non - cracked concrete surface,
3) - in case of cracks injection $R_{pr}^{p/b}$ is employed.

4. EXAMPLES OF CALCULATION OF COMPATIBILITY SUBSPACE

Conditions describing the polymer composite - cement concrete co - operation space in its mathematical sense create three separate systems of non-linear and linear inequalities (compare Tabl. 2), in which the polymer composite features are their unknowns. Three calculation and two graphic programs have been worked out in order to solve the inequalities system (in order to determine the compatibility space of PC - CC system).

The characteristic examples of the compatibility subspace for given PC - CC system and given service conditions of system are presented in Fig. 2 - 5.

320

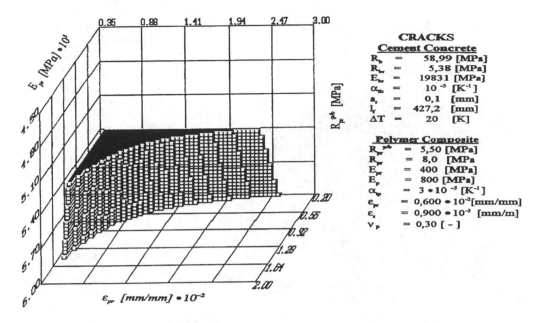

CRACKS
Cement Concrete
R_b	=	58,99 [MPa]
R_{bv}	=	5,38 [MPa]
E_{bv}	=	19831 [MPa]
α_{tb}	=	10^{-5} [K^{-1}]
a_r	=	0,1 [mm]
l_r	=	427,2 [mm]
ΔT	=	20 [K]

Polymer Composite
$R_r^{p/b}$	=	5,50 [MPa]
R_{pv}	=	8,0 [MPa]
E_{pr}	=	400 [MPa]
E_p	=	800 [MPa]
α_{vp}	=	$3 \bullet 10^{-5}$ [K^{-1}]
e_{pr}	=	$0,600 \bullet 10^{-2}$[mm/mm]
ε_s	=	$0,900 \bullet 10^{-3}$ [mm/m]
v_P	=	0,30 [-]

Fig. 2. The example of the compatibility space of polymer composite and cement concrete in the case of cracks injections in concrete, determined for the usability temperature changes at $\Delta T = 20$ K

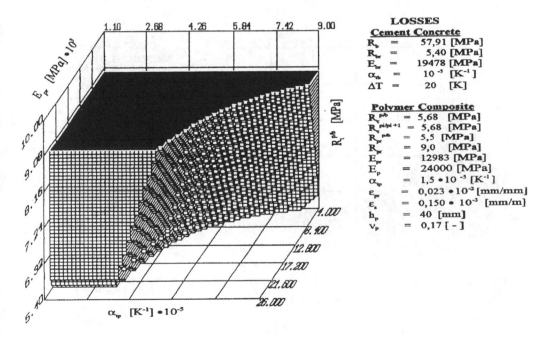

LOSSES
Cement Concrete
R_b	=	57,91 [MPa]
R_{bv}	=	5,40 [MPa]
E_{bv}	=	19478 [MPa]
α_{tb}	=	10^{-5} [K^{-1}]
ΔT	=	20 [K]

Polymer Composite
$R_r^{p/b}$	=	5,68 [MPa]
$R_r^{pi/pi+1}$	=	5,68 [MPa]
$R_r^{p/b}$	=	5,5 [MPa]
R_{pr}^v	=	9,0 [MPa]
E_{pr}	=	12983 [MPa]
E_p	=	24000 [MPa]
α_{vp}	=	$1,5 \bullet 10^{-5}$ [K^{-1}]
ε_{pr}	=	$0,023 \bullet 10^{-2}$ [mm/mm]
ε_s	=	$0,150 \bullet 10^{-3}$ [mm/m]
h_p	=	40 [mm]
v_P	=	0,17 [-]

Fig. 3. The example of the compatibility space of polymer composite and cement concrete in the case of the non - cracked surface concrete decrement repairs, determined for the usability temperature changes at $\Delta T = 20$ K

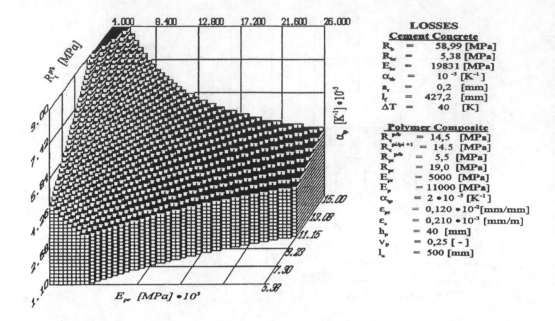

LOSSES
Cement Concrete

R_b	=	58,99 [MPa]
R_{br}	=	5,38 [MPa]
E_{br}	=	19831 [MPa]
α_b	=	10^{-5} [K^{-1}]
a_r	=	0,2 [mm]
l_r	=	427,2 [mm]
ΔT	=	40 [K]

Polymer Composite

$R_r^{p/b}$	=	14,5 [MPa]
R_r^{p/p_i+1}	=	14.5 [MPa]
$R_t^{p/b}$	=	5,5 [MPa]
R_{pr}	=	19,0 [MPa]
E_{pr}	=	5000 [MPa]
E_p	=	11000 [MPa]
α_{vp}	=	$2 \cdot 10^{-5}$ [K^{-1}]
ε_{pr}	=	$0,120 \cdot 10^{-2}$ [mm/mm]
ε_s	=	$0,210 \cdot 10^{-3}$ [mm/m]
h_p	=	40 [mm]
ν_p	=	0,25 [-]
l_v	=	500 [mm]

Fig. 4. The example of the compatibility space of polymer composite and cement concrete in the case of cracked surface (a_r = 0,2 mm) concrete decrement repairs, determined for the usability temperature changes at $\Delta T = 40$ K

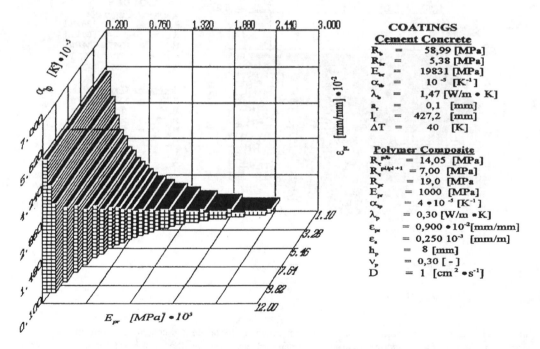

COATINGS
Cement Concrete

R_b	=	58,99 [MPa]
R_{br}	=	5,38 [MPa]
E_{br}	=	19831 [MPa]
α_b	=	10^{-5} [K^{-1}]
λ_b	=	1,47 [W/m • K]
a_r	=	0,1 [mm]
l_r	=	427,2 [mm]
ΔT	=	40 [K]

Polymer Composite

$R_r^{p/b}$	=	14,05 [MPa]
R_r^{p/p_i+1}	=	7,00 [MPa]
R_{γ}	=	19,0 [MPa]
E_{pr}	=	1000 [MPa]
α_{vp}	=	$4 \cdot 10^{-5}$ [K^{-1}]
λ_p	=	0,30 [W/m •K]
ε_{pr}	=	$0,900 \cdot 10^{-2}$ [mm/mm]
ε_s	=	$0,250 \ 10^{-3}$ [mm/m]
h_p	=	8 [mm]
ν_p	=	0,30 [-]
D	=	1 [cm^2 •s^{-1}]

Fig. 5. The example of the compatibility space of polymer composite and cement concrete in the case of non - cracked concrete surface protection where cracks of a_r = 0,1 mm width are being accepted, determined for the usability temperature changes at $\Delta T = 40$ K

5. CONCLUDING REMARKS

Following conclusions can formulated on the basis of the obtained results of computer simulation of compatibility space for polymer composite and cement concrete:

1. There is possibility for such selection of technical properties of polymer composites, that makes the determination of compatibility space for given polymer composite and cement concrete possible.

2. The obtaining of polymer composite fulfilled requirements of compatibility with cement concrete, is difficult but practically possible.

3. The determined compatibility space of polymer composite and cement concrete should be treated rather as a necessary condition than sufficient one. The compatibility space approach can be developed and improved for example, by additional introduction of material factors including their changes in time, effect of adhesion decreasing from moisture concrete substrate, etc.

4. The analytical selection of polymer composite for repair and anti - corrosion protection can be used for introductory evaluation of polymer composite usability in given application.

REFERENCES

Allen R. & Edwards E. 1987. The repair of concrete structures. Blackie, Glasgow, London.

Czarnecki L. 1985. The status of polymer concrete. Concrete Int. Design and Construction. 7: 47 - 53.

Czarnecki L. 1987. Large area of polymer concrete using in transportation, hydrotechnics and industry. TIZ - Fachberichte. 6: 410 - 416.

Czarnecki L., Głodkowska W. & Wiąckowska A. 1991. Model of compatibility of polymer composite - cement concrete (PC - CC) system, p. 484 - 493 Int. Colloquium "Durability of Polymer Based Composite Systems for Structural Applications". Elsevier Appl. Sci. (ed. Cardon A. H., Verchery G).

Czarnecki L. & Grabowski J. 1986. Criterion of Cracking resistance of glass fiber reinforced resin, a comparative study. ISAP'86, Int. Symposium Adhesion Between Polymers and Concrete. Aix - en Provance.

Czarnecki L. & Garbacz A. 1995. Evaluation of polymer coating - crack - bridging ability, p. 703 - 709. 3rd Int. colloquium "Industrial floors '95". Esslingen.

Czarnecki L & Clifton J. 1991. Polymer concretes; material design and optimization problems, p. 63 - 71. Int. Symposium on Concrete Polymer Composites. Bochum.

Czarnecki L., Clifton J. & Głodkowska W. 1992. Problem of compatibility polymer mortars and cement concrete system, p. 964 - 971. Int. Colloquium Material Science and Restoration" . Esslingen.

Czarnecki L. & Głodkowska W. 1992. Polymer mortar - cement concrete system; problem of compatibility, p. 420 - 429. VII Int. Congress on Polymers in Concrete. Moscow.

Emmons P. H., McDonald J. E. & Vaysburd A. M. 1992. Some compatibility problems in repair of concrete structures - a fresh look, p. 836 - 848 Int. Colloquium "Material Science and Restoration". Esslingen:

Głodkowska W. 1994. Ph.D. Thesis, Warsaw.

Ignatiev N. & Chatereji S. 1992. On the mutual compatibility of mortar and concrete in composite members. Cement & Concrete Composites. 14: 179 - 183.

Van Gemert D., Czarnecki L. & Bares R. 1988. Basis for selection of PC and PCC for concrete repair. Int. Journal of Cement Composites and Lightweight Concrete. 10: 121 - 123.

ACKNOWLEDGEMENTS

Authors would like to thank Dr J. Clifton (NIST, Gaithersburg USA) and Prof. Piątek Z. (WSI, Koszalin, PL.) for Their valuable remarks.

This work has been granted by Maria Skłodowska - Curie US - PL. Joint Fund II: MEN/NIST - 95 - 234.

Author index